HIGH-T_c SUPERCONDUCTORS

HIGH-T_C
SUPERCONDUCTORS

Edited by
Harald W. Weber
Atomic Institute of the Austrian Universities
Vienna, Austria

PLENUM PRESS • NEW YORK AND LONDON

Library of Congress Cataloging in Publication Data

High-T[subscript c] superconductors / edited by Harald W. Weber.
   p.   cm.
   Proceedings of an International Discussion Meeting on High-T[subscript c] Super-
conductors, held Feb. 7–11, 1988, at the Castle of Mauterndorf, in Salzburg, Austria.
   Bibliography: p.
   Includes index.
   ISBN 0-306-43091-6
   1. High temperature superconductors—Congresses.I. Weber, Harald W. II. Interna-
tional Discussion Meeting on High-T[subscript c] Superconductors (1988: Salzburg,
Austria)
QC611.98.H54H54 1988   88-30730
537.6'23—dc19   CIP

Proceedings of an International Discussion Meeting on
High-T_c Superconductors, held February 7–11, 1988,
at the Castle of Mauterndorf, in Salzburg, Austria.

© 1988 Plenum Press, New York
A Division of Plenum Publishing Corporation
233 Spring Street, New York, N.Y. 10013

PREFACE

In the past two years conferences on superconductivity have been characterized by the attendance of hundreds of scientists. Consequently, the organizers were forced to schedule numerous parallel sessions and poster presentations with an almost unsurveyable amount of information. It was, therefore, felt that a more informal get-together, providing ample time for a thourough discussion of some topics of current interest in high-temperature superconductivity, was timely and benefitial for leading scientists as well as for newcomers in the field.

The present volume contains the majority of papers presented at the International Discussion Meeting on High-T$_c$ Superconductors held at the Mauterndorf Castle in the Austrian Alps from February 7 to 11, 1988. Each subject was introduced in review form by a few invited speakers and then discussed together with the contributed poster presentations. These discussion sessions chaired by selected scientists turned out to be the highlights of the meeting, not only because all the participants truly appreciated the possibility of an information exchange, but mainly because of the magnificent job done by the discussion chairmen, John A. Mydosh (Leiden), Martin Peter (Geneva) and Ken E. Gray (Argonne). First results on the just discovered Bi-superconductors and the clarification of electron resonance experiments on (123)-compounds should be mentioned in particular.

The relaxed atomosphere favoring free discussions was certainly promoted by the surroundings offered in the Mauterndorf Castle, which dates back to 1253. Poster presentations and a conference banquet in historic knight's halls are certainly not found everyday in conference routines. Special thanks are due to the staff of the castle and the State of Salzburg, which offered the conference location and provided financial support. Furthermore, I wish to acknowledge the encouragement of Prof.Dr.Hans Tuppy, Minister of Science and Research, who also attended to open the meeting, the support of the Federal Ministry of Science and Research, the Austrian Physical Society and the Atomic Institute of the Austrian Universities. The invaluable help of the staff members Dr.Erwin Seidl, Dr.Franz M. Sauerzopf, Mr.Hans Niedermaier, Mrs.Ilse Futterer and, in particular, Mrs.Brigitta Buchberger as well as of my graduate students Trixi Vlcek, Marcus Frischherz, Peter Gregshammer, Michael Wacenovsky and Heinrich Wiesinger are gratefully acknowledged.

Finally, I wish to thank Plenum for the smooth cooperation during the production of this book, Brigitta Buchberger for her patience with the typing of numerous papers and my wife Ann for her help with the editing of this book.

Harald W. Weber

CONTENTS

SECTION 5: THIN FILMS

SECTION 1: OVERVIEW

THE SOLID STATE CHEMISTRY OF MIXED VALENCE

COPPER OXIDE HIGH T_C SUPERCONDUCTORS

B. Raveau, F. Deslandes, C. Michel, M. Hervieu

Laboratoire de Cristallographie et Sciences des Matériaux
Institut des Sciences de la Matière et du Rayonnement
Université de Caen, Bd du Maréchal Juin, 14032 Caen Cedex, France

G. Heger, G. Roth

Kernforshungszentrum Karlsruhe, INFP, P.O.B. 3640
D-7500 Karlsruhe, F.R.G.

INTRODUCTION

The recent investigations of superconductivity in ternary or pseudo ternary copper oxides recently performed by numerous staffs all over the world, show that up to the present, only two structural families belonging to the La_2CuO_4 and $YBa_2Cu_3O_7$ types, exhibit high critical temperatures, i.e. greater than 30K. The main factors which govern the superconducting properties are discussed here in a first point. $YBa_2Cu_3O_{7-\delta}$, appears as most fascinating owing to its high T_C (92K), but also very complex owing to its large possible oxygen deviation from stoichiometry. Its structural evolution versus oxygen non stoichiometry involving extended defects is presented in connection with its superconducting properties. The substitution of copper by a magnetic element, iron, very recently performed is finally presented here, taking into account the oxygen stoichiometry in the oxides $YBa_2Cu_{3-x}Fe_xO_7$.

La_2CuO_4-TYPE AND $YBa_2Cu_3O_{7-\delta}$-TYPE, TWO LOW DIMENSIONAL SUPERCONDUCTORS

The formal mixed valence of copper Cu(II)-Cu(III) necessary for a hole delocalization, i.e. for the existence of metallic or semi-metallic properties of copper oxides, is not sufficient for the appearance of superconductivity. It is now well established that mixed valence copper oxides such as $La_4BaCu_5O_{12+\delta}$[1a] or $La_{8-x}Sr_xCu_8O_{20}$[1b] do not exhibit any superconductivity in spite of their metallic conductivity contrary to the oxides $La_{2-x}A_xCuO_{4-y}$ (A = Ba, Sr, Ca)[1a,2] and $YBa_2Cu_3O_{7-\delta}$[1a,3]. These latter oxides are characterized by a layered structure. The "40K"-La_2CuO_4-type superconductor exhibits indeed a K_2NiF_4 structure (**Fig. 1a**) built up from SrO-type innsulating layers intergrown with single oxygen deficient superconducting perovskite layers. In the same way the "92K" $YBa_2Cu_3O_7$ superconductor has its structure (**Fig. 1b**) formed of triple infinite $[Cu_3O_8]_\infty$ layers built up of corner-sharing CuO_5 pyramids and CuO_4 square planar groups, whose cohesion is ensured by yttrium planes. This low dimensionality of the structure appears as a second necessary factor for superconductivity as very early pointed out by Labbë and Bok[4].

The question which can be set out deals with the role of Cu(III) content in the superconducting properties of these low-dimensional oxides. The evolution of the critical temperature versus oxygen non-stoichiometry δ in $YBa_2Cu_3O_{7-\delta}$ established by several authors[5a,b-7] shows that T_C decreases as Cu(III) content decreases (Fig. 2). However a contradition appears for $\delta \geq 0.50$ if one takes only into consideration the classical behavior of copper : $YBa_2Cu_3^{II}O_{6.50}$ would only contain Cu(II) from the charge balance and should not be a superconductor. The explanation to this phenomenon takes its origin in the particular structure of $YBa_2Cu_3O_{7-\delta}$ which will be in fact an intergrowth between the $YBa_2Cu_3O_7$ structure (Fig. 1b) and the $YBa_2Cu_3O_6$ one (Fig. 3).

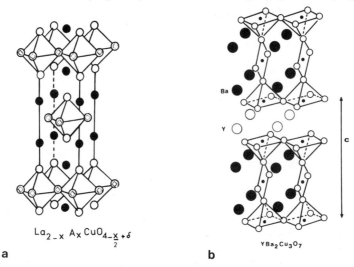

Fig. 1. Crystal structure of the oxides K_2NiF_4-type (a), $YBa_2Cu_3O_7$-type (b)

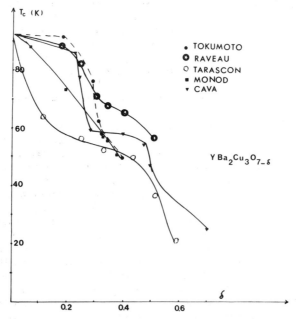

Fig. 2. Evolution of T_C vs δ in the $YBa_2Cu_3O_{7-\delta}$ compounds, from different authors.

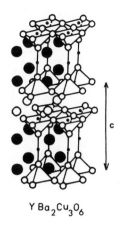

$Y Ba_2Cu_3O_6$

Fig. 3. Crystal structure of $YBa_2Cu_3O_6$.

Consequently, whatever δ may be, different from 0 and 1, one observes a disproportionation of Cu(II) into Cu(III) and Cu(I) leading to Cu(II)-Cu(III) superconducting regions and Cu(II)-Cu(I) insulating domains coexisting in the same crystal. Contrary to $YBa_2Cu_3O_{7-\delta}$, the La_2CuO_4-type oxides will not present this disproportionation: the formation of anionic vacancies, according to the formulation La_2CuO_{4-y} does not lead to a linear coordination of copper, characteristic of Cu(I), but either to a pyramidal or to a square planar coordination compatible with Cu(II) and Cu(III). The observation of noticeable amounts of Cu(I) in $YBa_2Cu_3O_{7-\delta}$ by X-ray absorption spectroscopy[8,9], whereas no Cu(I) was detected by Bianconni et al.[1b] in La_2CuO_4-type oxides, is in agreement with this point of view.

Superconductivity observed for La_2CuO_4[1a-1b] corresponds to a particular behaviour of this structure. The oxide La_2CuO_4, prepared by annealing in air or under an oxygen flow, is only a surface superconductor, due to the existence of Schottky defects in the normal semi-conductor $La_{2-2\epsilon}Cu_{1-\epsilon}O_{4-4\epsilon}$, whose surface is then more easily oxidized by long time annealing, leading to the limiting formulation $La_{2-2\epsilon}Cu_{1-\epsilon}O_4$. On the opposite the bulk superconductor obtained by annealing under high oxygen pressure the compound corresponding to an excess of CuO, can be formulated $La_{1.95}CuO_4$, and thus has a hole concentration i.e. a Cu(III) content close to that of $La_{1.85}Sr_{0.15}CuO_4$, in agreement with its similar value of T_C (37K).

STRUCTURAL EVOLUTION OF $YBa_2Cu_3O_{7-\delta}$ VERSUS OXYGEN STOICHIOMETRY

The ability of $YBa_2Cu_3O_{7-\delta}$ to adopt a large oxygen non stoichiometry, $0 \leq x \leq 1$, is now a well known feature. Several studies showed that the oxygen content and/or the symmetry (orthorhombic or tetragonal) have a direct influence on the superconducting properties of these oxides, corresponding to a decrease of the T_C vs δ[5a,b-7]. However, if such a mean effect is observed, the values of the parameters differ from one author to the other, owing to the accuracy of the oxygen content measurement and the thermal treatment. Moreover, these overall results do not take into account the real state of the matrix. In that way, the structures of the limiting compounds, $YBa_2Cu_3O_7$ and $YBa_2Cu_3O_6$, were well characterized as well by X-ray single crystal as by neutron

5

diffraction techniques [1a,1b,10]. The first one ($0 \leq \delta \leq 0.2$) was systematically investigated by means of electron diffraction and high resolution microscopy by several groups[1b,17,18] showing local variation of the oxygen stoichiometry, twins and oriented domains and various extended defects. On the opposite few studies were performed near the composition $YBa_2Cu_3O_6$[1b]. The electron diffraction investigation of numerous particles in the $YBa_2Cu_3O_6$ matrix, corresponding to samples with nominal compositions $\delta = 1$ obtained from different thermal treatments (argon, vacuum, sealed tube with zirconium) showed to us that all the phases exhibit a large part of amorphous or bad crystallized grains, besides the tetragonal crystals which can be picked up for X-ray or H.R.E.M. studies. This feature can indeed be correlated with an inhomogeneous removing of the oxygen atoms from one grain to the other, depending on the nature of the precursor $YBa_2Cu_3O_{7-\delta}$. Moreover, the E.D. patterns of some of these crystals exhibit extra spots corresponding to a tetragonal $2a\sqrt{2} \times 2a\sqrt{2}$ superstructure (Fig. 4). Such a feature can be correlated to the existence of residual oxygens in the matrix, rearranged in a short range order, corresponding to a well defined composition such as $\delta = 0.875$ or $\delta = 0.75$. Such an assumption is reinforced by the behavior of those crystals under the electron beam : the extra spots quickly disappear to give the tetragonal subcell spots ($a \times a$), after a loss of the residual oxygen. Unfortunatly, this phenomenon prevents accurate HREM images to be recorded, which would help to understand this ordering.

Fig. 4. [001] electron diffraction pattern, brighter dots correspond to the tetragonal subcell and the smaller ones to a $2a\sqrt{2} \times 2a\sqrt{2}$ superstructure.

More difficult, because more complex, is the understanding of the intermediate compositions, i.e., $0.2 < \delta < 0.7$, where the thermal treatments are, once more, of high significance (particularly under reduction or reoxidization way of preparation). For instance, "mixed" crystals, partly orthorhombic partly tetragonal, were observed for a sample of nominal composition $YBa_2Cu_3O_{6.7}$ obtained by reoxidation of $YBa_2Cu_3O_6$ (Fig. 5) ; they appear as half twinned on the image. Such crystals are assumed to be formed of the intercalation of oxygen atoms in a well crystallized $YBa_2Cu_3O_6$ matrix. For the same mixed composition other particles appear as "polycrystalline", as shown on the E.D. pattern and image in Fig.6. Such a feature is correlated to local recrystallisations in an amorphous matrix of the reduced sample $YBa_2Cu_3O_6$, reduced and reoxidized particles exhibiting exactly the same plate-like morphology.

6

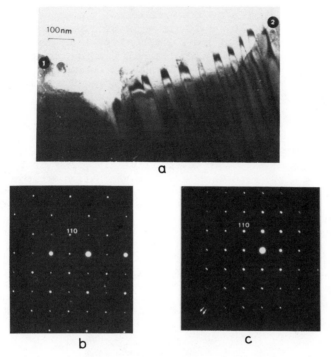

Fig. 5. Low resolution image and corresponding SAED patterns of mixed tetragonal (1-b), orthorhombic (2-c) crystals.

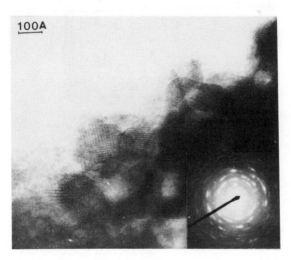

Fig. 6. Low resolution image and ED pattern of a polycrystalline particle.

$YBa_2Cu_3O_{6.5}$ is an other interesting example of the structural evolution of these oxides with their "thermal history". Two sorts of samples must be distinguished : tetragonal semiconducting samples obtained by quenching the orthorhombic superconductor $YBa_2Cu_3O_{6.9}$ in air from 950°C to room temperature and the orthorhombic form obtained by reoxidation of the oxide $YBa_2Cu_3O_6$ at low temperature. The powder neutron diffraction study showed that the crystalline part of the "tetragonal" sample corresponds to $\delta = 0.75$ and that the 0.25 oxygen atoms are statistically distributed in the plane of Cu(1), between the barium atoms (Fig. 7) ; however the important background and the abnormally high values of some anisotropic thermal parameters suggest the existence of an amorphous part in the matrix and some inhomogeneities in the crystalline part. The electron diffraction and HREM studies carried out in order to understand these features revealed, besides well crystallized grains, systematically coated with an amorphous barium rich layer, the existence of highly disturbed crystals, partly polycrystalline states and streaks along c ; moreover careful examination of the particles along [001] showed that the matrix is composed of both true tetragonal crystals and orthorhombic ones ; such phenomena which suggest indeed local variations of the oxygen content were confirmed by the HREM images[15]. Such results support the hypothesis of disproportionation of Cu(II) into Cu(III) and Cu(I) leading to the coexistence in the matrix of both superconducting (Cu(II)-Cu(III)) and "insulating" (Cu(II)-Cu(I)) regions according to the formulation $YBa_2(Cu_2^{II}Cu^{III}O_7)_{0.25}$ $(Cu_2^{II}Cu^IO_6)_{0.75}$.

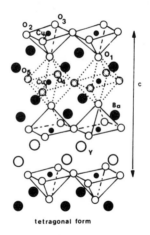

tetragonal form

Fig. 7. Crystal structure of the tetragonal oxide $YBa_2Cu_3O_{6.25}$

The HREM study of the orthorhombic form of nominal composition $YBa_2Cu_3O_{6.60}$ carried out in collaboration with Tarascon[11] is also in agreement with this disproportionation of Cu(II). Moreover systematical modulations (2a × b) appear in the contrast of the [001] HREM images (Fig. 8a) ; they can be easily explained by an ordering of the oxygen vacancies (Fig. 8b). Whereas in other samples, some crystals show a × 2b supercells (Fig. 8c, d) witness of a complete ordering all over the crystal.

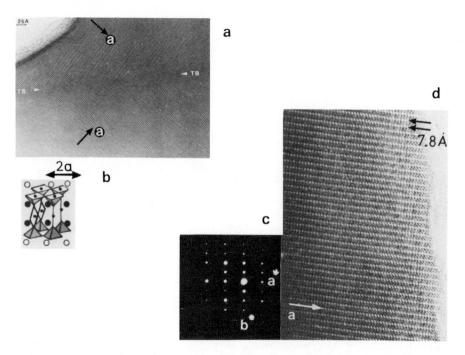

Fig. 8. Modulations (2a × b) : a) [001] HREM image, b) schematic model. Superstructure (a × 2b) : c) ED pattern, d) [001] HREM images.

SUBSTITUTION OF COPPER BY IRON in $YBa_2Cu_3O_7$

The samples of composition $YBa_2Cu_{3-x}Fe_xO_{7-\delta}$ ($0 \leq x \leq 0.45$ in steps of 0.05) were prepared by standard ceramic procedures. Starting from $BaCO_3$, Y_2O_3, CuO and Fe_2O_3, materials were first synthesized at 950°C in air for 24 hours, then ground, pressed, sintered two hours at 950°C and annealed one day at 400°C, both under pure oxygen flow. The oxygen content in these oxides was determined by microthermogravimetric analysis in a flow of reducing gas (Ar + 10 % H_2) up to 1000°C. For all the compositions except x = 0.45 (Fig. 9), δ remains close to zero, leading to approximately 7 oxygen atoms per cell. These results are in agreement with those of Kistenmacher et al.[12] and Roth et al.[13]. These latter authors studied the structure of samples of the composition $YBa_2Cu_{2.85}Fe_{0.15}O_{7+\delta}$ by various techniques, all converging to δ ≈ 0, within the experimental error. This means that the mean oxidation state of the 3d metallic atoms remains 2.33 whatever the iron content lower than 13 % (x = 0.4). Assuming that iron is mainly in +III oxidation state, the introduction of one Fe(III) in the perovskite framework seems to involve the disappearence of one Cu(III). This can explain the result obtained for x = 0.45 where δ was found to be negative (δ = -0.08) and with a deviation from zero too large to be included in the experimental error. If the iron content is higher than the solubility limit, the excess of iron (III) does not influence the Cu(III) content in the perovskite leading to an excess of oxygen by analysis. It results that the limit of composition should be around x = 0.40, this was confirmed by the X-ray diffraction study.

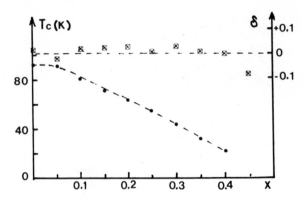

Fig. 9. Evolution of δ(⊠) and T_C (●) versus x in the oxides
$YBa_2Cu_{3-x}Fe_xO_{7-\delta}$.

X-ray diffraction patterns were performed over the full range of composition, using CuK_α radiation. Except for higher Fe contents, for which some extralines of very weak intensity are observed, all the diffraction patterns can be attributed to a single-phase. From the evolution of the cell parameters (Fig. 10), it can be seen that the c parameter decreases slightly and monotonously as x increases and that introduction of iron influences dramatically the symmetry of the cell, which from orthorhombic, becomes tetragonal (within the precision of our measurements), as soon as 5 % of iron are substituted for copper in $YBa_2Cu_3O_7$, in agreement with Maeno et al.[5c] and Roth et al.[13].

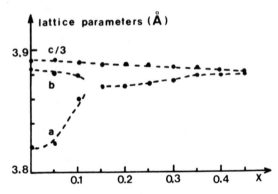

Fig. 10. Evolution as a function of x of the lattice parameters of the oxides $YBa_2Cu_{3-x}Fe_xO_{7-\delta}$.

Temperature-dependent dc resistivity measurements were carried out on sintered bars by a standard four-probe method using silver paste or indium contacts. Data were recorded from 300K down to 4.2K with measuring current densities of about 0.5 A/cm². At room temperature, the resistivity of the samples increases as the iron content increases ; for example ρ ∼ 5×10^{-4}, 8×10^{-4}, 1.7×10^{-3} Ω cm for x = 0.05, 0.15, 0.30 respectively. Except the composition corresponding to x = 0.45, which was found to be a semiconductor, superconducting properties were observed for all the samples. The variation of T_C (at midpoint of the resistive transition) versus x is plotted on Fig. 9. As it was stated before by several authors[5c,12], the critical temperature is not dramatically affected by introduction of iron in the copper sublattice and a smooth decrease of T_C is observed as x increases.

One interesting feature is that these materials remain superconductors with high critical temperature even if their cell symmetry is tetragonal. This is in contrast to the results obtained for the oxide $YBa_2Cu_3O_{7-\delta}$ for which superconducting properties were only observed in the orthorhombic form. This tetragonal cell has indeed been observed for $YBa_2Cu_{2.85}Fe_{0.15}O_7$ as well by X-ray diffraction on single crystals as by neutron powder diffraction[13]. This is confirmed by the systematical E.D. investigation which shows that less than 2 percent of the crystals are orthorhombic and twinned.

The determination of the structure of the oxide $YBa_2Cu_{2.85}Fe_{0.15}O_7$ by X diffraction on single crystal, and neutron powder diffraction shows that the Fe^{3+} ions are distributed over the two copper sites with a preference for the Cu(1) site : among the 0.15Fe per cell, 0.05 are located on the two Cu(2) positions and 0.10 on the Cu(1) site. The results are in agreement with the study by Mossbauer spectroscopy, of a sample of composition $YBa_2Cu_{2.82}Fe_{0.18}O_7$, by Qiu et al.[14] who found that Fe is substituted for both copper sites. Another important feature of this structural study lies in the fact the Cu(1) site is underoccupied. Both single-crystal X-ray diffraction and powder neutron diffraction show that this site appears to be deficient by about 2 %. The oxygen vacancies are not either ordered in rows running along [010] but are randomly distributed at the Cu(1) level over the 0, 1/2, 0 and 1/2, 0, 0 positions as it was previously observed on tetragonal compounds like $YBa_2Cu_3O_{6.25}$[15], $LaBa_2Cu_3O_{6.7}$[1b,16]. In fact this statistic distribution of the oxygen atoms over two sites give only an average structure and does not reflect the local structure. In fact the high resolution electron microscopy study of the 5 % Fe doped sample reveals a structure modulation with orthorhombic microdomains of about 20-30 Å diameter corresponding to a local order (Fig. 11).

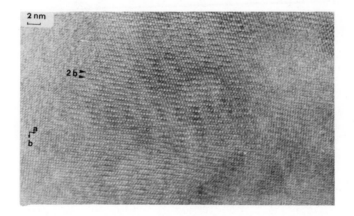

Fig. 11. [001] HREM image showing superstructures × 2b and modulations of the contrast.

The situation in the tetragonal Fe-doped $YBa_2Cu_3O_7$ oxides is quite different from that observed for the tetragonal undoped $YBa_2Cu_3O_{6.25}$. In the pure copper tetragonal oxide, the $[CuO_2]_\infty$ chains are interrupted or non-existent owing to the presence of great amount of copper (I) in twofold coordination. In the Fe-doped tetragonal oxides, as δ remains close to zero, the amount of copper (I) can be considered as negligible. Moreover, in this latter oxide iron can take a pyramidal coordination on the Cu(2) site and either a pyramidal or a tetrahedral coordination on the Cu(1) site. Thus, in these compounds the $[CuO_2]_\infty$ chains are probably formed of short segments linked together by means of iron polyhedra inducing a change in the orientation of the segments. This change of orientation is to be compared to that induced by twins ; but, contrary to microtwinning, it would appear in an aleatory way and over short distances, implying a metrically tetragonal symmetry.

REFERENCES

1. Mixed-valence copper oxide high T_C superconductors in Blue booklets.
    a) 1980 to May 1987, Caen, Mixed-valence copper perovskite and relatives.
    b) High T_C superconductors in Caen (CRISMAT), Grenoble (CRTBT), Bordeaux (LCS), Mai 1987-August 1987.
2. J.G. Bednorz and K.A. Müller, Z. Phys. B Condensed Matter, 64:189 (1986).
3. M.K. Wu, J.R. Ashburn, C.J. Torng, P.H. Hor, R.C. Meng, L. Gao, Z.T. Huang, Y.G. Wang and C.W. Chu, Phys. Rev. Lett., 58:908 (1987).
4. J. Labbé and J. Bok, Europhysics Letters, 3:1225 (1987).
5. Novel Superconductivity, Ed. A. Wolf and V.Z. Kresin, Plenum Press, N.Y. 1987. Proceeding of the International Workshop on Novel Super-conductivity, Berkeley, June 1987 ((a), Batlogg et al. p. 653, (b) Tarascon et al., p. 705, (c) Maeno et al. p. 1073).
6. P. Monod, M. Ribault, F. d'Yvoire, J. Jegoudez, G. Collin and A. Revcolevschi, J. Phys., 48:1367 (1987).
7. M. Tokumoto, H. Ihara, T. Matsubara, M. Hirabayashi, N. Terada, H. Oyanagi, K. Murata and Y. Kumura, Jpn J. Appl. Phys., 26:L1565 (1987).
8. H. Oyanagi, H. Ihara, T. Matsubara, M. Takumoto, T. Matsuhita, MM. Hirabayashi and K. Murata, Jpn. J. Appl. Phys., 26:L1561 (1987).
9. F. Baudelet, G. Collin, E. Dartige, A. Fontaine, J.P. Kappler, G. Krill, J.P. Itie, J. Jegoudez, M. Maurer, Ph. Monod, A. Revcolev-schi, H. Tolentino, G. Tourillon, M. Verdaguer. Orsay high T_C superconductors preprints, Vol. 4, October 1987.
10. Y. Lepage, W.R. McKinnon, J.M. Tarascon, L.H. Greene, G.W. Hull and D.M. Hwang, Phys. Rev., B35:7245 (1987).
11. J.M. Tarascon, M. Hervieu, B.. Domengès, B. Raveau, in preparation.
12. T.J. Kistenmacher, W.A. Bryden, J.S. Morgan, K. Moorgani, Y.W. Du, Z.Q. Giu, H. Tang and J.C. Walker, Phys. Rev., B36:8877 (1987).
13. G. Roth, B. Renker, G. Heger, V. Caignaert, M. Hervieu and B. Raveau, Z. Phys., submitted.
14. Z.Q. Qiu, Y.W. Du, H. Tang, J.C. Walker, W.A. Bryden and K. Moorjani, J. Magn. Mater., in press.
15. B. Domengès, M. Hervieu, V. Caignaert and B. Raveau, J. Spect. Microsc. Elect., in press.
16. J. Nakai, K. Imai, T. Kawashima and R. Yoshizaki, Jpn. J. Appl. Phys., 26:L1244 (1987).
17. H.W. Zandbergen, G. Van Tendeloo, T. Okabe and S. Amelinckx, Phys. Stat. Sol. (a), 103:45 (1987).
18. E. Hewat, M. Dupuy, A. Bourret, J.J. Capponi and M. Mazezio, Nature, 327, 4 june 1987.

CRITICAL OVERVIEW OF THEORIES FOR HIGH TEMPERATURE SUPERCONDUCTORS

C. M. Varma

AT&T Bell Laboratories
Murray Hill, New Jersey 07974

ABSTRACT

   In this overview I will discuss the principal theoretical ideas proposed for the mechanism of superconductivity in the oxide metals, their chief shortcomings and some of the experimental results which relate one way or the other to them.

INTRODUCTION

   Since the great excitement on the Copper-Oxide superconductors, which following the discovery by Bednorz and Müller[1] started last year, over 3,000 papers on this subject have appeared. I will review the principal theoretical questions and ideas that have been proposed to explain their properties. The materials belong to the class of ionic perovskite type materials; at the same time, with the conduction band half-filled, they are antiferromagnetic insulators. Because of the proximity of the appropriate copper and oxygen ionization levels they are also quite co-valent at the self-consistent ionic charge distribution. On the metallic side, a crucial question is: Are the properties consistent with a Fermi-liquid? On the superconducting side, a crucial question is: Does the order parameter have the conventional BCS symmetry? Unfortunately, the experimental results are not quite unambiguous on even such basic issues. Absent a clear guidance from the experiments two principal varieties of models have been developed with different prejudices about the answers to the above questions. One variety of models relies exclusively on magnetic fluctuations - the two principal sub-varieties of these rely on (i) exchange of spin fluctuations instead of phonons, (ii) the more radical point of view popularly known as Resonant Valence Bonds (RVB) in which superconductivity is due to condensation of pre-existing singlet pairs in the normal state. The other variety of models relies on the ionic interactions in these materials and the charge polarization fluctuations dressing the electrons for the high temperature superconductivity.

   It is now generally agreed that the conventional mechanism of pairing via electron-phonon interaction won't do. T_c is simply too high. Given the electronic structure of these materials[2,3] T_c for these materials has been estimated[4] from the strong-coupling theory (Eliashberg) to be less that 30K for the electron-phonon coupling constant $\lambda \to \infty$. The bipolaron mechanism can only do worse. In the range of λ where bipolaron mechanism may be valid the Eliashberg theory should give a higher T_c than the bipolaron mechanism. I will not discuss the phonon related mechanisms anymore except to note that the isotope effect measurements[5] have clinched this issue. It should be noted though that in concert with other mechanisms which give conventional s-wave pairing, phonons could help towards a higher T_c-higher, I estimate, by no more than about 10K.

I will first discuss a phase diagram[6] for the new superconductors and then discuss the magnetic and the charge polarization mechanisms.

PHASE DIAGRAM OF THE OXIDE SUPERCONDUCTORS:

A. Sleight draws a phase diagram for the oxide superconductors, including $BaPb_xBi_{1-x}O_3$, which I find, very revealing. With minor modifications this phase diagram is shown in Fig. (1). The axis is the metal-oxygen covalency. In the copper-oxide superconductors, the increase of covalency between copper and oxygen is achieved by substituting divalent atoms Ba, Sr etc. for trivalent La as in $La_{2-x}Sr_xCuO_4$ or simply by increasing the amount of oxygen as in going from $Y_1Ba_2Cu_3O_6$ to $Y_1Ba_2Cu_3O_7$. (In $BaPb_xBi_{1-x}O_3$ increasing x does the same.) This is an important point - the variations not merely change the number of carriers, they also shift the relative position of the Cu and oxygen levels.

For low covalency, the materials are ionic insulators, which if the metal ion is paramagnetic are antiferromagnetic. For $BaPb_xBi_{1-x}O_3$ this region is

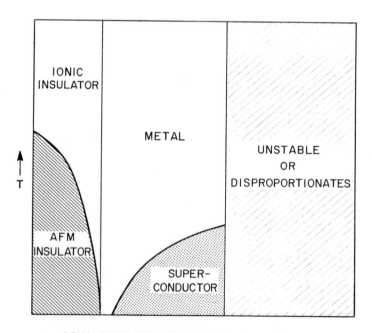

Fig. 1:   Generic phase diagram of the high temperature superconductors.

merely an insulator. With changing covalency a paramagnetic metallic phase results, which is itself unstable for further increase of covalency. This instability may be gentle and take the form of charge disproportionation as in $BaPb_xBi_{1-x}O_3$ or so severe that the compound simply does not form - for $La_{2-x}Sr_xCuO_4$, increase of Sr beyond 0.2 is not possible keeping the right composition of oxygen; for $YBa_2Cu_3O_x$, increase of oxygen content beyond $x \approx 7$ is not possible ordinarily.

The superconducting transition temperature T_c varies with change in covalency (and doping) in the general manner shown, (of course there will be subtler variations from compound to compound). The debate about the mechanisms for superconductivity couched in terms of this phase diagram is simply whether the physics of superconductivity is related to the magnetic instability or to the instability at higher covalency. We shall return to this phase diagram again.

THE MAGNETIC MECHANISMS:

With appropriate change in composition the oxide superconductors turn into insulating antiferromagnets.[7] (It is well however to remember that Cr, a high temperature antiferromagnet, turns by doping with its neighbor, V, to a conventional superconductor.) This is undoubtedly an effect of correlations. It leads one to think of residual antiferromagnetic correlations in the metallic state and the possibility that they promote superconductivity.[8,9] This idea has been invoked with some success in heavy Fermions.[10] The simplest version of the idea is to exchange spin fluctuations instead of phonons to produce attractive interactions between electrons near the Fermi surface. This has the following problems:

(a) As for the heavy Fermions,[10] the predicted state has a d-wave symmetry with a quasiparticle density of states which is linear in energy. (It is true that in purely two dimensional models with nearest neighbor hopping, extended s or combinations of s and d wave symmetrics become favorable for small deviations from 1/2 filling). Experimental results, such as are available, indicate conventional pairing. The clearest of these deduces the London penetration depth through μ-SR and finds its temperatures dependence given well by conventional s-wave quasiparticle spectrum and quite unlike that expected for other states. Tunneling experiments can in principle shed light on the quasiparticle spectrum but they have so far proved very difficult and irreproducible.

(b) Strong coupling theory puts strong limits on the maximum transition temperature which can be obtained through exchange of spin fluctuations.[12] Consider the phenomenological electron-electron interaction Hamiltonian

$$H_{int} = \sum_{k, k', q} J(q) c^+_{k'\uparrow} c_{k'-q\downarrow} c^+_{k+q\downarrow} c_{k\uparrow} \tag{1}$$

with

$$J(q) = J_o + J_1 \gamma(q) \tag{2}$$

where $\gamma(q)$ has the symmetry of the lattice. Now J_1 alone contributes to the attractive electron-electron interaction in unconventional channels but both J_o

and J_1, contribute to the effective mass. In strong coupling theory T_c is then given by

$$T_c \approx J_o \exp(-\frac{1+\rho J_o}{\rho J_1}) \tag{3}$$

assuming magnetic excitation exist to energy J_o. Here ρ is the electronic density of states at the Fermi-surface. To prevent an antiferromagnetic instability J_1/J_o must be small, $\leq \frac{1}{4}$, for instance, in a square lattice. Then even if the coupling constants ρJ_1, ρJ_o become infinite

$$T_c < J_o \exp(-4). \tag{4}$$

The model of course breaks down long before even this rather small T_c. Actually for $\rho J_o \to \infty$, $T_c \to 0$.

(c) There is really no evidence for any anomalous magnetic fluctuations in the superconducting state. Two magnon Raman scattering[13] in the antiferromagnet $YBa_2Cu_3O_6$ sees a peak in the magnetic fluctuations at $\omega \approx 3000$ cm^{-1}, which corresponds to $J \approx 1000\,°K$. But on introduction of more oxygen the peak decreases and is invisible at O_7. Of course a background signal remains but this is expected for an ordinary metal - the integrated spectral weight is equal to $NS(S+1)$ and is distributed almost up to the bandwidth.

There are other models,[9] which are of the local pairing via magnetic interaction variety, and which couple holes on oxygen atoms through localized spin excitations on the copper atoms. I think they would have the same difficulty with the above experiment. They would have similar difficulty with point (a) above and perhaps worse with point (b) since they bear the same relationship to Eliashberg theory of pairing via exchange of spin fluctuations that bipolaron ideas bear to Eliashberg electron-phonon theory. The Gutzwiller type calculations on the Hubbard model,[14] which find d-wave superconducting should give results similar to Eliashberg theory for T_c.

Some questions are occasionally raised about superconductivity in the AFM phase or in a strongly fluctuating state which may locally be regarded as AFM. Two points should be noted. If a mechanism, say electron-phonon interaction, of conventional s-wave superconductivity exists in the normal state, T_c in the AFM phase compared to the normal state suffers small differences due to transfer of spectral weight for pairbreaking from one region of (q, ω) space to another. This is a rather small effect.[15] The other effect is due to change of bandstructure. This is simply given by using the density of state near the Fermi-energy in the AFM phase rather than the normal phase in the expression for T_c.[15,16] If on the other hand, magnetic fluctuations in the normal state favor d or extended s-wave pairing, part of their spectral weight $<S(T)>^2$ out of $S(S+1)$ is lost on going to the AFM phase. So they are less effective and T_c is lower in the AFM phase (and also in a strongly AFM fluctuating regime) than would be in the normal phase. There is also again the change in the bandstructure which will generally reduce the density of states near the Fermi-surface.

RESONATING VALENCE BONDS (RVB):

The 1d spin $\frac{1}{2}$ antiferromagnetic Heisenberg model does not order even at

T = 0 due to quantum fluctuations.[17] Anderson[18] asked in 1973 whether the 2d spin $\frac{1}{2}$ triangular lattice also has no Néel order at T = 0. He proposed that the wave function is a linear combination of products of pair wave function each of which is a singlet

$$\psi = \sum_{\{ij\}} C_{\{ij\}} \prod_{i<j} (\alpha_i \beta_j - \beta_i \alpha_j) \tag{5}$$

Here $C_{\{ij\}}$ are coefficients for different configuration of singlet pairs. In 1987, motivated by early experimental results on La_2CuO_4 he suggested[19] that a 2d lattice has such a ground state as well rather than the Néel state modified by zero spin deviations.

The fascinating point of the idea is that if the BCS wave function

$$\Psi = \prod_k (u_k + v_k a^+_{k\uparrow} a^+_{-k\downarrow}) |0> \tag{6}$$

is projected to exactly N/2 pairs and then transformed to real space, it has exactly the functional form of eq. (5). The suggestion was then made that the 1/2 filled case, La_2CuO_4, would have a ground state like (5) and be insulating. As it is doped the number of pairs in any given region is not fixed and so it is possible that a state with a uniform phase (or a phase which is a well defined function of k) may have a lower energy. This corresponds to the phase of v_k/u_k in (5) independent of k and is therefore superconducting. The idea is that at high temperatures such singlet pairs already exist but without phase coherence and phase coherence is achieved at a temperature similar to the binding energy of pair i.e. the exchange energy. The pairs have of course constraints due to the strong-range repulsion underlying the model. The pair wave function must therefore vanish at the origin, or

$$\int \frac{v(k)}{u(k)} d^3k = 0. \tag{7}$$

This is of course the same reason which disfavors conventional s-wave superconductivity in liquid ^3He and in the heavy Fermions. Such a condition together with ferromagnetic spin fluctuations favors p-wave superconductivity in liquid ^3He and together with antiferromagnetic fluctuations favors "d-wave" type superconductivity in the heavy Fermions.

Since the RVB wavefunction with phase coherence has the same form as a BCS type wavefunction, one would expect the properties in the superconducting state to be similar to those of the conventional approaches. If the amplitude of the gap is small compared to E_F "d-wave" or "extended s-wave" type superconductivity would be favored; if the gap is comparable to E_F s-wave type pairing may also be imagined. (After all two hard-core atoms can bind in an s-wave channel to form molecules.)

The radical part of the idea is about the normal state. The assertion is that it is not a Fermi-liquid. The educated guess about the phase diagram of the 2d-Hubbard model is shown in fig. (2). Where in this phase diagram does RVB lie? Is there a distinct line separating it from a Fermi-liquid or does the Fermi-liquid at small U/t and substantially away from half-filling go continuously into a phase with zero quasi-particle amplitude?

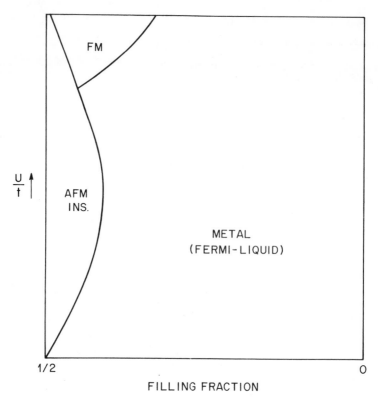

Fig. 2: Commonly accepted phase diagram (T=0) of the Hubbard model.

On the experimental side it has been discovered that both La_2CuO_4 (below 200° K) and $YBa_2Cu_3O_6$ (below 500° K) are Neél antiferromagnets.[7] The strong interactions are two dimensional with exchange energies of 0(1000° K) and the three-dimensional ordering is a parasitic effect due to the very long two dimensional correlations developed by $T \approx T_N^+$. The temperature dependence of the correlation lengths[20] is fitted[21] well by assuming that the 2d planes are headed for a Neél order at $T \to 0$ with spin deviations expected from spin wave theory and not to the disordered quantum RVB state. The dynamical scattering[20] observed above T_N is also consistent[22] with this.

On the theoretical side, the 2d Heisenberg model on a square lattice (and possibly on a triangular lattice as well) has a lower energy[23] for a ground state with AFM long range order than RVB wave functions with short bond lengths.[24] The energy of the latter is not much higher though but efforts to improve them by adding longer bonds lead to states with AFM long range order.[24] This is presumably due to the fact that general RVB states form an over complete set which will, given a chance, add up to approximate the actual ground state.

The argument was then made that while the half-filled case may be AFM, doping introduces RVB states locally near the holes because the hole may move freely in such a state compared to an AFM state.

So far we have discussed the original ideas for the RVB state. The actual further development has been along two main different lines. In one, approximations[25,26] on the Hubbard Hamiltonian

$$H = t \sum_{(ij),\,\sigma} C_{i\sigma}^+ C_{j\sigma} + U \sum_i n_{i\sigma} n_{i-\sigma} \tag{8}$$

in the $t/U \to 0$ limit to convert it to

$$H = \tilde{t} \sum_{<ij>,\,\sigma} c_{i\sigma}^+ c_{j\sigma} + 2t^2/U \left[\sum_{(ij)} (\vec{\sigma}_i \cdot \vec{\sigma}_j - n_i n_j/4) \right.$$

$$\left. + \sum_{(ijk)} c_{i\sigma}^+ c_{j\sigma} c_{j\sigma}^+ c_{k\sigma} \right] + \cdots \tag{9}$$

where $\tilde{t} = t<1-n>$. Mean field superconducting ground states are then sought. At this point this is no difference from superconductivity due to the exchange of AFM fluctuations[10] which we have already discussed, with all its attendant problems.

The other line of development springs from the remarks of Kivelson, Rokhsar and Sethna[27] that if a bond is broken in RVB and the two electrons taken far from each other, one will generate two spin $\frac{1}{2}$ neutral objects- called the spinons (obeying Fermi statistics) while on being doped a charged defect with spin zero called the holon (obeying Bose statistics) is generated. An electron is then a composite particle of a spinon and a holon. In Fermi-liquid theory the wave function of an electron is a linear combination of a quasiparticle (of charge e) and an incoherent combination of particle-hole pairs. In RVB the incoherent combination is being turned into a Bosonic charged quasiparticle. While in the original Hamiltonian there are constraints which keep the spinon and the holon related to each other, calculations in RVB are done for mean field Hamiltonians in which they are regarded[28] as independent. There are invariances in such Hamiltonians[29] which are absent in the original Hamiltonian. In one latest version of RVB[30] the holons are prevented from Bose condensing but pairs of them Bose condense by mutual attraction through exchange of interlayer electron exchange. The spinons stay non-interacting. All this is somewhat ad-hoc.

It is claimed that RVB explains several curious properties of the oxide superconductors - the linear specific heat, the linear temperature dependence of the resistivity, the behavior of the resistivity perpendicular to the planes, the curious tunneling curves, the tetragonal to orthorhombic transition[31] etc. In each instance such claims or the experimental results themselves appear to be on shaky grounds.[32]

A clear experimental test of RVB would consist in mapping the Fermi-surface. If the oxide superconductors in the metallic phase have a conventional Fermi-surface, which obeys Luttinger's theorem and looks even vaguely as that given by bandstructure calculations, RVB must be abandoned, at least, in the present context.

The RVB ideas have been so provoking that some beautiful and some interesting work[33] on quantum problems is being pursued. This sort of thing, I think, will be the enduring legacy of RVB — rather than an understanding of superconductivity in the oxide metals.

As a prelude to the next section, I mention here that Monte Carlo calculations on small lattices for the Hubbard model find[34] that the superconducting pair susceptibility for temperature $\gtrsim t/12$ lies below that for non-interacting electrons ($U = 0$). An Eliashberg calculation[35] for the pair susceptibility agrees with this result in this temperature region and predicts a $T_c \sim 0(10^{-3}t)$ for $U = 4t$. I believe the Eliashberg theory gives an upper limit to T_c.

SUPERCONDUCTIVITY THROUGH ELECTRONIC POLARIZATION

Consider the tight binding model for transition metal oxides:[36]

$$H = \sum_i (\epsilon_A n_{Ai} + \epsilon_B n_{Bi} + U_A n_{Ai\uparrow} n_{Ai\downarrow} + U_B n_{Bi\uparrow} n_{Bi\downarrow}$$

$$+ \sum_{<ij>} (V n_{Ai} n_{Bj} + t \sum_\sigma (c^\dagger_{Ai\sigma} c_{Bj\sigma} + h.c.)) \tag{10}$$

where A stands, say, for a metal ion d-orbital and B for an oxygen p orbital. For simplicity we consider only one orbital per atom. The parameter V represents the Madelung energy which stabilizes the basically ionic structure in these compounds. For $\epsilon_A - \epsilon_B$ large compared to t, U_A, U_B and V this model can be reduced to the Hubbard model but not otherwise.[4,36] The difference $\epsilon_A - \epsilon_B$ decreases to the right of the periodic table and even NiO[37] can not be studied in the framework of the Hubbard model. A useful quantity to define is $E_x (<(\epsilon_A - \epsilon_B))$, the energy cost to transfer a charge from B atom to A atom considering the mutual interaction in the new configuration. For $E_x \ll U$ and half-filling of the upper-most band, the material will again be an AFM insulator if $E_x \gg t$ with an exchange energy between the A atoms given by $J \approx (t^2_{A-B}/E_x)^2 (\frac{1}{E_x} + \frac{1}{U_A})$. For $\epsilon_A = \epsilon_B$, and $U_A = U_B = 0$, the model has a charge density wave transition away from half-filling, due to V. Weak-coupling calculations for general values of $\epsilon_A - \epsilon_B$, U_A and V and filling give[38] a spin density wave phase, a metallic phase and a charge density phase, which may be mapped to the phase diagram, fig. (1), discussed for the oxide superconductors.

The basic electronic structure feature common to the Cu-O superconductors are a Cu-O bonding band, a Cu-O antibonding band and an oxygen non-bonding band. These are shown in fig. (3). Tight binding fit to the bandstructure[2,3] give $\tilde{\epsilon}_A - \tilde{\epsilon}_B = 0$, the tilda refers to the mean-field renormalization of ϵ_A and ϵ_B due to the charge transfer as in a band-structure calculation. This is a situation for maximum covalency between Cu and O atoms at the self-consistent ionicity. The simplification of (10) to the Hubbard model is then not possible.[4,36]

Although the band structure is not a good starting point for discussing the insulating - AFM phase, it is assumed so for the metallic phase if this phase is a Fermi-liquid. Consider now the excitations with the bandstructure in fig. (3). Any low energy excitation near the Fermi-level changes the charge on Cu and oxygen atoms thereby activating the term proportional to V in (10). These low energy excitations then are not orthogonal to the charge transfer resonances arising from transitions between the non-bonding oxygen orbitals and the Cu

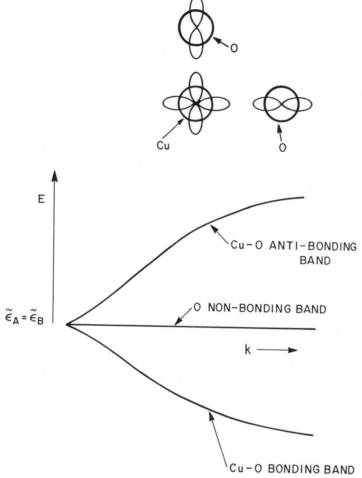

Fig. 3: Schematic bandstructure of the 2d Cu-O layers using $d_{x^2-y^2}$ Cu-orbital and p_x and p_y orbitals respectively for oxygen to the x and y direction from the Cu atom.

part of the hybrid orbitals near the Fermi-surface. Thus excitations near the Fermi-surface are dressed with Cu-O polarization fluctuations. It is proposed[4,36] that mutual interaction of electrons through such polarization fluctuations is attractive and can lead to high temperature superconductivity.

The main problem with this proposal is that it has never been shown that a sizable attractive interaction arises through such a mechanism. Consider jellium. The effective electron-electron interaction may be written

$$V_{eff}(q, \omega) = \frac{4\pi e^2}{q^2} \epsilon^{-1}(q, \omega). \tag{11}$$

To get attraction $\epsilon(q, \omega)$ must be negative. This will in general happen at small q for large ω (near plasmons in 3d). However to be useful for high temperature superconductivity $\epsilon(q, \omega)$ must be negative at low frequencies and over a substantial part of q space - especially large q. A fundamental property of a metal is that it perfectly screens any extra charge which means ϵ is positive in this range. It is for this reason that the various plasmon and acoustic plasmon ideas are useless for high temperature superconductivity.

It has been proposed[4,36] that the action of V, local field corrections and the specific electronic structure of the Cu-O superconductors change this and allow for a large attraction. Theoretical support for these ideas comes from the calculations of Hirsch and co-workers.[39] They have diagonalized small 2d lattices with the copper-oxide lattice structure in the superconducting compounds with the Hamiltonian (10). They find an effective attractive electron-electron interaction when V is large enough. The value of V required for attraction goes down as $|\epsilon_A - \epsilon_B|$ decreases. In a similar calculation for the Hubbard model they find no regime for attraction. This is consistent with their Monte-Carlo calculation of the superconducting pair-susceptibility.

If V is an important parameter, it should seriously affect the excitation spectra in the normal state. After some confusion, it is now generally agreed by most experimentalists[40] working in the infrared and optical spectra of the copper-oxide superconductors that a large broad non-Drude conductivity is observed stretching from the lowest frequencies to almost 1 eV. The conductivity is so large (absorption coefficient $> 2 \times 10^5 cm^{-1}$) that it can only come from transitions between oxygen -p states and copper -d states shifted down in energy, compared to the bandstructure, due to V. Such a broad and strong response is in fact found in model calculations[38] with the Hamiltonian (10). On increasing V or decreasing $\epsilon_A - \epsilon_B$, the calculations[38] give a charge density wave instability corresponding to a zone-boundary Cu-O breathing mode.

As remarked already, an association can be made of this instability with the rather violent instability (see fig. (1)) - the fact that the Cu-O superductors do not form beyond a certain doping which besides moving the material away from half-filling decreases $\epsilon_A - \epsilon_B$ and thereby increases covalency. In $BaPb_xBi_{1-x}O_3$ the instability is gentle — disproportionation of Bi^{3+} and Bi^{5+} and has the symmetry of the predicted CDW.

The electronic polarization mechanism associates superconductivity with the same basic interactions that engender the structural instability rather than those that engender antiferromagnetism. From the variation of T_c in fig. (1), one could argue in favor of this proposition.

The idea of high T_c through electronic polarization needs to be made convincing through detailed calculations. Also, the several anomalous properties of the oxide superconductors - the linear resistivity, the temperature dependence of the Hall coefficient, the nuclear relaxation rate, etc. need to be understood.

The polarization mechanism does not require magnetic ions. What it requires is low density of electrons and a not too large value of $(\epsilon_A - \epsilon_B)/t$ so that the metallic phase has important ionic fluctuation. The two dimensional Cu-O structure is helpful because of the proximity of the non-bonding levels to E_F.

A discovery of an oxide superconductor without magnetic metal ions and with high T_c would be a boost to the polarization mechanism as the only serious alternatives are the magnetic mechanisms. I think $BaPb_xBi_{1-x}O_3$ is already an example of this situation — a T_c of $12°K$ in this compound is unlikely with electron-phonon interactions.

ACKNOWLEDGEMENTS

It is a pleasure to thank my colleagues E. Abrahams, A. Aeppli, B. Batlogg, R. Cava, S. Coppersmith, P. Littlewood and S. Schmitt-Rink for innumerable discussions. This review of the theoretical ideas proposed is far from exhaustive. I have discussed only those that I am familiar with. My apologies to the proponents of other points of view.

REFERENCES

1. J. G. Bednorz and K. A. Müller, Z. Phys. *B64*, 189 (1986).

2. L. F. Mattheiss, Phys. Rev. Letters *58*, 1028 (1987); L. F. Mattheiss and D. R. Hamman, Solid State Comm. *63*, 395 (1987).

3. J. Yu et al., Phys. Rev. Lett. *58*, 1035 (1987); S. Massida et al., Phys. Lett. A 122, 198 (1987).

4. C. M. Varma, S. Schmitt-Rink and E. Abrahams, Solid State Comm. *62*, 681 (1987).

5. B. Batlogg et al., Phys. Rev. Lett. *58*, 2333 (1987); L. C. Bourne et al., Phys. Rev. Lett. *58*, 2337 (1987).

6. A. Sleight, (Preprint).

7. D. Vakinin et al., Phys. Rev. Lett. *58*, 2802 (1987); J. M. Tranquada et al., Phys. Rev. Lett. (to be published).

8. P. A. Lee and M. Read, Phys. Rev. Lett. *58*, 2691 (1987); J. E. Hirsch, Phys. Rev. Lett. *59*, 228 (1987); M. Cyrot, Solid State Comm. (1987); Y. Hasegawa and H. Fukayama, Jpn. J. Appl. Phys. *26*, L322 (1987); J. R. Schrieffer, X. -G Wen and S. C. Zhang, Phys. Rev. Letters *60*, 944 (1988); G. Kotliar and S. Liu (Preprint); K. Miyake et al., LTXVIII proceedings; and many many others.

9. V. J. Emery, Phys. Rev. Lett. *58*, 2794 (1987).

10. K. Miyake, S. Schmitt-Rink and C. M. Varma, Phys. Rev., *B34*, 6554 (1986); D. Scalapino, E. Loh and J. Hirsch, Phys. Rev. *B34*, 8590 (1986).

11. D. R. Harshman et al., Phys. Rev. (Rapid Comm.) *36*, 2386 (1987); G. Aeppli et al., Phys. Rev. B *35*, 7129 (1987). See also micro-wave absorption by E. M. Jackson et al., (Preprint) which fits s-wave and J. R. Cooper et al., Phys. Rev. B *37*, 638 (1988) which does not.

12. A. Millis, S. Sachdev and C. M. Varma, Phys. Rev. B April 1, 1988 (in press).

13. K. Lyon, P. Fleury and L. Schneemeyer, Phys. Rev. Letters *60*, 732 (1987).

14. C. Gros, J. Joynt and T. M. Rice, Z. Phys. (1987).

15. T. V. Ramakrishnan and C. M. Varma, Phys. Rev. *24*, 137 (1981).

16. E. Fenton, Solid State Comm. (1988).

17. H. Bethe, Z. Physik *71*, 205 (1931).

18. P. W. Anderson, Mater. Res. Bull. *8*, 153 (1973).

19. P. W. Anderson, Science V 235, 1196 (1987).

20. R. J. Birgeneau et al., Phys. Rev. Lett. *59*, 1329 (1987); G. Shirane et al., Phys. Rev. Lett. *59*, 1613 (1987); Y. Endoh (unpublished).

21. S. Chakravarthy, B. I. Halperin, and D. F. Nelson, Phys. Rev. Lett. *60*, 1057 (1988).

22. G. Reiter, Private Communication; A. Aeppli, Private Communication.

23. D. Huse and V. Elser, (Preprint).

24. S. Liang, B. Doucot and P. W. Anderson (Preprint).

25. G. Bhaskaran, Z. Zou and P. W. Anderson, Solid State Comm. *63*, 973 (1987).

26. A. E. Ruckenstein, P. J. Hirschfeld and J. Appel, Phys. Rev. *B36*, 857 (1987).

27. S. Kivelson, Rokhsar and J. Sethna, Phys. Rev. Rapid Comm. *B35*, 8865 (1987).

28. P. W. Anderson, G. Bhaskaran, Z. Zou and T. Hsu, Phys. Rev. Lett. *58*, 2790 (1987).

29. I. Affleck and J. B. Marston, to appear in Phys. Rev. (Rapid Communication); I. Affleck, Z. Zou, T. Hsu and P. W. Anderson, Phys. Rev. (Brief report) to be published.

30. J. Wheatley, T. Hsu and P. W. Anderson (to be published).

31. P. W. Anderson and Z. Zou, Phys. Rev. Lett. *60*, 132 (1988); P. W. Anderson et al., *58*, 2790 (1987).

32. (a) the calculation of the resistivity due to holon-spinon scattering yields $T^{3/2}$ in 2 dimensions rather than T, (C. Kallin and A. J. Berlinsky, Preprint).
    (b) the resistivity along the c-axis becomes increasingly parallel to that in the plane for better samples. See for instance, S. Martin et al. (Preprint).
    (c) the linear specific heat coefficient (as well as the slope of the linear resistivity) bear little relation to T_c. In one recent measurement (M Kato, Y. Maeno and T. Fujita, Physica *C152*, 116 (1988) the specific heat coefficient decreases in $La_{2-x}Sr_xCuO_4$ in going into the superconducting phase and is maximum in the "spin glass" phase (G. Aeppli, Private Communication). The thermal conductivity of many samples with linear specific heat is close to T^2 (Nunez-Ruggeiro et al. (Preprint), J. Graebner et al., (Preprint), E. Abrahams and C. M. Varma (Preprint)) suggesting a glassy explanation.
    (d) the tunneling curves in many instances show a BCS like spectrum. This is seen also in Andreev reflection measurements. See for instance, H. F. C. Hoovers et al. *C152*, 105 (1988).
    (e) the symmetry assumed for the gap function to explain the structural transition does not appear to be any of the solutions of the mean field theory, for example Affleck and Marston - Ref. 19.

33. F. D. Haldane, Phys. Rev. Letters, *60*, 635 (1988); S. Shastry, ibid, *60*, 639 (1988); I. Dzialoshinky, A. M. Polyakov and P. B. Wiegmann, to be published; P. B. Wiegmann, Phys. Rev. Letters *60*, 821 (1988); V. Kalmeyer and R. B. Laughlin, Phys. Rev. Lett. *59*, 2095 (1987).

34. J. Hirsch and H. Q. Lin, (Preprint).

35. G. Bickers, BAPS, *33*, 452 (1988).

36. C. M. Varma, S. Schmitt-Rink and E. Abrahams in *Novel Superconductivity*, Edited by V. Kresin and S. Wolf, Plenum Press, New York (1987).

37. J. Zaanen, G. A. Sawatzky and J. W. Allen, Phys. Rev. Letters, *55*, 418 (1985).

38. P. Littlewood, (Private Communication); A. Koyama and M. Tachiki (Preprint). Z. Tesanovic (Preprint).

39. J. Hirsch et al., to be published in Phys. Rev. Letters; C. A. Balseiro (Preprint).

40. K. Kamaras et al., Phys. Rev. Lett. *59*, 919 (1987); J. Orenstein et al., Phys. Rev. B *36*, 729 (1987); D. B. Tanner (Private Communication); I Timofeev (Private Communication).

SECTION 2: MATERIALS-STRUCTURE

SOME EFFECTS OF CHEMICAL SUBSTITUTION AND APPLIED PRESSURE ON HIGH T_c COPPER OXIDE SUPERCONDUCTORS

M. B. Maple, Y. Dalichaouch, E. A. Early, B. W. Lee, J. T. Markert,
M. W. McElfresh*, J. J. Neumeier, C. L. Seaman, M. S. Torikachvili**
K. N. Yang and H. Zhou

Department of Physics and
Institute for Pure and Applied Physical Sciences
University of California, San Diego
La Jolla, CA 92093   USA

ABSTRACT

Recent experiments on the effect of Na and Pr substitutions and applied pressure on high T_c copper oxide superconductors are briefly reviewed. The Na and Pr substituted systems that were investigated include $(La_{2-x}Na_x)CuO_{4-\delta}$, $(Y_{1-x}Na_x)Ba_2Cu_3O_{7-\delta}$, and $(Y_{1-x}Pr_x)Ba_2Cu_3O_{7-\delta}$, while the compounds studied under pressure were $YBa_2Cu_3O_{7-\delta}$ and $(Y_{0.5}Pr_{0.5})Ba_2Cu_3O_{7-\delta}$. The substitution of monovalent Na for trivalent La in $La_2CuO_{4-\delta}$ results in a continuous transition from semiconducting to metallic behavior, accompanied by the appearance of two superconducting components, one with $T_c \approx 20$ K and the other with $T_c \approx 40$ K, while the substitution of monovalent Na, and Pr, which appears to have a valence $\sim +3.9$, for trivalent Y in $YBa_2Cu_3O_{7-\delta}$ produces a continuous transition from metal to semiconductor and a gradual suppression of superconductivity. The onset of T_c of $YBa_2Cu_3O_{7-\delta}$ was found to increase from 95 K at 8 kbar to 107 K at 149 kbar, whereas applied pressure was observed to induce semiconducting behavior and suppress superconductivity in the compound $(Y_{0.5}Pr_{0.5})Ba_2Cu_3O_{7-\delta}$.

*Los Alamos National Laboratory, Los Alamos, NM 87545
**Department of Physics, San Diego State University, San Diego, CA 92182

INTRODUCTION

The discovery of high T_c superconductivity in the copper oxide compounds $(La_{2-x}M_x)CuO_{4-\delta}$ (M = Ca, Sr, Ba) with a maximum $T_c \approx 40$ K [1-5] and $RBa_2Cu_3O_{7-\delta}$ (R = Y or a lanthanide element, except for Ce, Pr, Pm and Tb) with $T_c \sim 95$ K [6-10] has generated a level of activity in the field of superconducting materials of unprecedented proportions. Considerable effort is being expended in characterizing the physical properties of the presently known high T_c copper oxide superconductors and in finding new oxides and related materials with high T_c's and other desirable superconducting and mechanical properties. On a fundamental level, there is a great deal of interest in determining the mechanism that is responsible for the high T_c superconductivity, while on an applied level, there is strong motivation to develop thin films and flexible conductors for technological applications. In an effort to obtain information concerning the mechanism of superconductivity in the high T_c copper oxide superconductors, we have been involved in an extensive investigation of the physical properties of these materials as a function of temperature T, magnetic field H, chemical substitution, and applied pressure P. In this paper, we briefly describe some of our recent experiments on the effect of chemical substitution and applied pressure on the normal and superconducting state properties of the high T_c copper oxide superconductors. Other aspects of our work on this problem have been described in the proceedings of several recent conferences and references to papers cited therein [7,11,12].

CHEMICAL SUBSTITUTION EXPERIMENTS

The $(La_{2-x}Na_x)CuO_{4-\delta}$ and $(Y_{1-x}Na_x)Ba_2Cu_3O_{7-\delta}$ Systems

The motivation for studying the $(La_{2-x}Na_x)CuO_{4-\delta}$ system was to compare the effect of monovalent Na substitutions with that of divalent Ca, Sr, and Ba substitutions. Electrical resistivity measurements on $(La_{2-x}M_x)CuO_{4-\delta}$ compounds with M = Ca [5], Sr [3-5], and Ba [1,2] have revealed a continuous transition with increasing x from semiconducting behavior for x = 0 to metallic behavior. The continuous semiconductor-metal transition is accompanied by the gradual development of superconductivity in which T_c exhibits a maximum as a function of x at x ≈ 0.15 with values of ~ 24 K, ~ 39 K and ~ 32 K for Ca [5], Sr [3-5], and Ba [1,2], respectively.

Magnetic susceptibility measurements on $La_2CuO_{4-\delta}$ compounds indicate antiferromagnetic ordering with a Néel temperature T_N that depends strongly on the oxygen vacancy concentration δ, ranging from ~290 K for $\delta \approx 0.03$ to ~ 0 K for $\delta = 0$ [13-15]. The Néel temperature T_N decreases rapidly as the concentration x of the alkaline-earth is increased. According to neutron diffraction experiments, the antiferromagnetic ordering is three dimensional with values of T_N that vary between 50 K and ~ 300 K and ordered magnetic moments between 0.2 μ_B and 0.5 μ_B [16-18]. A recent neutron diffraction study conducted by Shirane et al. [19] on a single crystal of La_2CuO_4 revealed three dimensional antiferromagnetic order below $T_N \approx 195$ K with the spins oriented parallel and antiparallel to the b-axis and a value of the ordered moment of 0.35 ± 0.05 μ_B. In addition, at 300 K a curious dynamical type of two dimensional antiferromagnetic behavior was found in which the spins are ordered *instantaneously* in two dimensions over distances exceeding 200 Å. No time-averaged staggered moment was observed in spite of the fact that the energy scale of the correlations is unusually large, corresponding to an effective dispersion of > 0.4 eVÅ. Since the structure factor is dynamical in character, Shirane et al. referred to this novel two dimensional magnetic state as a *quantum spin fluid* (QSF) state.

Recently, we observed bulk superconductivity in the $(La_{2-x}Na_x)CuO_{4-\delta}$ system for concentrations $0.11 \leq x \leq 0.41$[20]. Prior to this study, no *single* dopant other than M = Ca, Sr, or Ba had been found to produce superconductivity in the $(La_{2-x}M_x)CuO_{4-\delta}$ system. Exceptions exist for the cases of "doping" with lanthanum vacancies [21-24] or excess oxygen [25] in pure La_2CuO_4. The normalized electrical resistivity of $(La_{2-x}Na_x)CuO_{4-\delta}$, $\rho(T)/\rho(250$ K$)$, is shown in Fig. 1 as a function of temperature T for several concentrations x. For $x \leq 0.08$, semiconducting behavior was observed. The magnitude of the resistivity was largest for undoped La_2CuO_4, with $\rho(250$K$) \approx 4000$ mΩ-cm. The high temperature resistivity decreased from $\rho(250$ K$) \approx 18$ mΩ-cm for x = 0.06 - 0.08 to $\rho(250$ K$) \approx 4$ mΩ-cm for $x \geq 0.11$. Superconductivity was observed for all concentrations $0.11 \leq x \leq 0.41$. The resistive transition temperature T_c showed only a

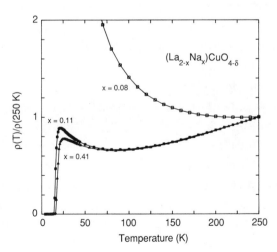

Fig. 1   Normalized electrical resistivity $\rho(T)/\rho(250$ K$)$ vs temperature for $(La_{2-x}Na_x)CuO_{4-\delta}$ compounds. The midpoint of the resistive superconductive transition occurs at $T_c = 16$ K (18 K) for x = 0.11 (0.41). After Ref. 20.

slight increase with x; the T_c onset of 21 K at x = 0.11 increased to 25 K at x = 0.41, while the T_c midpoint of 16 K at x = 0.11 increased to 18 K at x = 0.41. Transition widths (10% - 90%) between 3 and 4 K were observed for all superconducting samples. Magnetic susceptibility measurements revealed sizeable fractions of a Meissner effect indicative of bulk superconductivity, as well as evidence for bulk superconductivity with an onset of ~ 40 K for the sample with x = 0.41.

These experiments demonstrate the existence of bulk superconductivity in $(La_{2-x}Na_x)CuO_{4-\delta}$, the first group I alkali dopant in the $(La_{2-x}M_x)CuO_{4-\delta}$ system to display high T_c superconductivity. Superconducting properties were observed to increase with Na concentration up to x = 0.41. This is in contrast to what has been observed in $(La_{2-x}Sr_x)CuO_{4-\delta}$, where optimal behavior is found for x ≈ 0.15. The resistive superconducting transition temperatures reported here (~18 K) are lower than those observed for the Ca, Sr, or Ba substituted compounds (~24 K, ~39 K, and ~32 K, respectively). These results are unusual from the point of view of naive electronic arguments, where one might assume that Na^{1+} would be twice as effective as M^{2+} (M = Ca, Sr, or Ba) when substituting for La^{3+} and thus might incorrectly anticipate optimal behavior for x ≈ 0.075. However, from a structural point of view, the fact that the ionic radius of Na (0.97 Å) is closest to that of Ca (0.99 Å) of the alkaline earths (Sr - 1.12 Å, Ba - 1.3 Å) is consistent with the similarity between the T_c's obtained for Na and Ca.

In the $(La_{2-x}Na_x)CuO_{4-\delta}$ system, discussed above, increasing the value of x results in a transition from semiconducting to metallic behavior and the occurrence of superconductivity. In contrast, increasing x in the $(Y_{1-x}Na_x)Ba_2Cu_3O_{7-\delta}$ system produces a transition from metallic behavior towards semiconducting behavior and the eventual loss of superconductivity [26]. This is illustrated in Fig. 2 where the normalized electrical resistivity, $\rho(T)/\rho(300K)$, is plotted vs temperature for various Na concentrations $0 \leq x \leq$ 0.5. The data reveal that the onset of T_c is nearly constant within the range $0 \leq x \leq 0.3$, and then drops off and broadens for x ≈ 0.3. The ionic radius of Na^{1+} (0.970 Å) is comparable to that of many R ions (e.g.; Nd^{3+} - 0.995 Å, Sm^{3+} - 0.964 Å, Eu^{3+} - 0.950 Å) and is thus not expected to have a strong influence on the T_c of $YBa_2Cu_3O_{7-\delta}$. However, the depression of T_c could be the consequence of the monovalent state of Na for which fewer electrons are contributed to the conduction band. The feature that the two systems have in common is that over an appreciable range of Na concentration ($0.11 \leq x \leq 0.4$ for $(La_{2-x}Na_x)CuO_{4-\delta}$ and $0 \leq x \leq 0.3$ for $(Y_{1-x}Na_x)Ba_2Cu_3O_{7-\delta}$), the resistivity is metallic (i.e., the temperature coefficient of resistivity $\alpha \equiv (1/\rho)\,(d\rho/dT)$ is positive) and T_c is nearly constant, ~ 17 K in the former case and ~ 95 K in the latter case. For $(Y_{1-x}Na_x)Ba_2Cu_3O_{7-\delta}$, magnetic susceptibility χ vs T data reveal the gradual development of a magnetic moment with x that attains a value $\mu_{eff} \approx 3\ \mu_B$ per Na ion in the range $0.2 \leq x \leq 0.5$, although the Na^{1+} ions are presumably nonmagnetic.

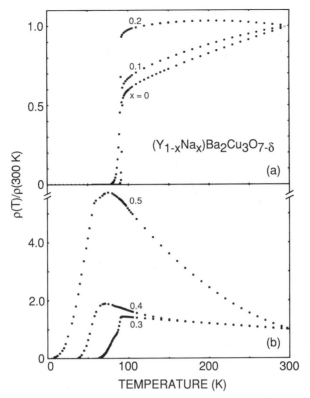

Fig. 2   Normalized electrical resistivity $\rho(T)/\rho(300\text{ K})$ vs temperature for $(Y_{1-x}Na_x)Cu_3O_{7-\delta}$ compounds. After Ref. 26.

The $(Y_{1-x}Pr_x)Ba_2Cu_3O_{7-\delta}$ System

Apart from $PmBa_2Cu_3O_{7-\delta}$, which has not been investigated because of the short half-life of Pm, the three $RBa_2Cu_3O_{7-\delta}$ compounds with R = Ce, Pr, and Tb do not exhibit superconductivity, probably because these R-ions are not trivalent in $RBa_2Cu_3O_{7-\delta}$. While $PrBa_2Cu_3O_{7-\delta}$ forms with the same orthorhombic perovskite-like crystal structure as superconducting $YBa_2Cu_3O_{7-\delta}$ (so called "123"-phase), the Ce and Tb counterparts form with a quite different crystal structure whose identity has not yet been reported. In an attempt to obtain information concerning the absence of superconductivity in the $RBa_2Cu_3O_{7-\delta}$ compounds with R = Ce, Pr and Tb, we have carried out experiments on $(Y_{1-x}R_x)Ba_2Cu_3O_{7-\delta}$ compounds with R = Ce, Pr and Tb [26]. While the Ce and Tb substitutions yield multiphase samples, single phase $(Y_{1-x}Pr_x)Ba_2Cu_3O_{7-\delta}$ compounds can be prepared throughout the entire range of Pr concentration ($0 \leq x \leq 1$).

Electrical resistivity $\rho(T)$ data for a series of $(Y_{1-x}Pr_x)Ba_2Cu_3O_{7-\delta}$ compounds are shown in Fig. 3. Similar $\rho(T)$ data for this system were recently reported by Soderholm et al. [27]. As x increases, the temperature coefficient of electrical resistivity α in the normal

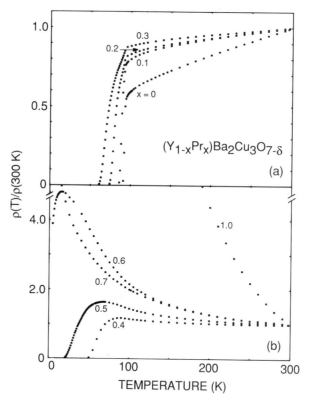

Fig. 3   Normalized electrical resistivity $\rho(T)/\rho(300\ K)$ vs temperature for $(Y_{1-x}Pr_x)Ba_2Cu_3O_{7-\delta}$ compounds. After Ref. 26.

state decreases continuously and changes from positive to negative at $x \approx 0.4$, while the resistive superconducting transition curves shift to lower temperatures and broaden. The magnitude of $\rho(T)$ was also found to increase with x; ρ (300 K) increased from 1.6 mΩ-cm at $x = 0$ to 42.7 mΩ-cm at $x = 1$. The superconducting transition temperature T_c, defined from the resistive superconducting transition curves as the temperature at which $\rho(T)$ dropped to 50% of its extrapolated normal state value, decreased monotonically with x from 93 K at $x = 0$ to 39 K at $x = 0.5$, as shown in Fig. 4. The vertical bars in Fig. 4 represent the width ΔT_c of the superconducting transition, defined as the difference in temperatures at which $\rho(T)$ dropped to 90% and 10% of its extrapolated normal state value.

Magnetic susceptibility measurements on the $(Y_{1-x}Pr_x)Ba_2Cu_3O_{7-\delta}$ compounds yield a Curie-Weiss contribution due to the Pr from which an effective magnetic moment μ_{eff} per Pr ion of 2.7 μ_B is obtained, indicating that the Pr ions have an intermediate valence of ~ 3.9.

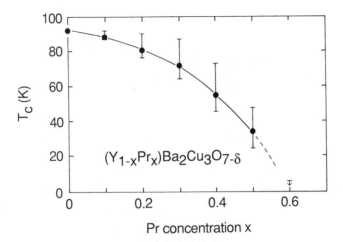

Fig. 4   Resistively measured T_c vs x curve for the $(Y_{1-x}Pr_x)Ba_2Cu_3O_{7-\delta}$ system. The solid line is a guide to the eye. Data points indicate transition midpoints and vertical bars represent the 10%-90% transition widths. After Ref. 26.

The transition of the normal state electrical resistivity $\rho(T)$ from metallic to semiconducting behavior and the monotonic decrease of T_c with increasing x in the $(Y_{1-x}M_x)Ba_2Cu_3O_{7-\delta}$ (M = Pr, Na) systems can be explained in terms of increasing oxygen deficiency in the structure, similar to what was observed in samples of $YBa_2Cu_3O_{7-\delta}$ when the oxygen content was decreased in a controlled manner [28]. Indeed, the higher than +3 valence state of Pr in this structure could lead to a situation equivalent to an oxygen deficiency. For the case of Na^{1+}, the smaller than trivalent state of Na in this structure could result in less oxygen concentration ($\delta > 0.1$) to balance the valence of the metal ions, which in turn could reduce more of the Cu from Cu^{3+} to Cu^{2+} and account for the development of a magnetic moment in the $(Y_{1-x}Na_x)Ba_2Cu_3O_{7-\delta}$ compounds.

EFFECT OF APPLIED PRESSURE

The Compound $YBa_2Cu_3O_{7-\delta}$ at Very High Pressure

Some of the dramatic increases of T_c to its present value of ~ 95 K involved pressure-induced increases to T_c onset values between ~ 40 K and ~ 50 K, depending on the definition of the T_c onset, for materials in the La-Ba-Cu-O system [29,30]. Measurements of T_c under pressure have also been reported for multiphase samples in the Y-Ba-Cu-O system

[31] and $RBa_2Cu_3O_{7-\delta}$ compounds, with R = Y, up to ~ 10 kbar [32], and R = Y, Gd, Er, and Yb, up to ~ 18 kbar [33]. In the study on the multiphase Y-Ba-Cu-O samples [31] ,T_c was found to increase from less than 90 K to about 91.5 K at 17.6 kbar, while for studies on the $RBa_2Cu_3O_{7-\delta}$ compounds [32,33], the largest increase in T_c occurred for the compound $YbBa_2Cu_3O_{7-\delta}$ in which T_c increased from 88 K at zero pressure to 92.5 K at 18.2 kbar.

Recently, we measured $T_c(P)$ of $YBa_2Cu_3O_{7-\delta}$ to ~ 149 kbar [11,34], using a clamped Bridgman anvil device. The sample was one of the earlier ones prepared in our laboratory and appears to be slightly deficient in oxygen, as evidenced by a slightly negative temperature coefficient of resistivity at atmospheric pressure and a rather large transition width $\Delta T_c \equiv T_c(90\%) - T_c(10\%) \approx 26$ K (see Fig. 5). At a pressure of 8 kbar, the resistive transition onset, 90%, 50% (midpoint), 10 % and 0% values occurred at 95 K, 94 K, 87 K, 68 K, and ~40 K, respectively. The resultant $T_c(P)$ curve, shown in Fig. 5, is nearly linear with the T_c onset increasing from ~ 95 K at 8 kbar to ~ 107 K at 149 kbar. To our knowledge, this is the highest value of T_c yet reported for a material known to exhibit a confirmed bulk superconducting transition. Furthermore, the results in Fig. 5 show no evidence of a maximum in $T_c(P)$ or a decrease in dT_c/dP, suggesting that even higher T_c's could be obtained at higher applied pressures.

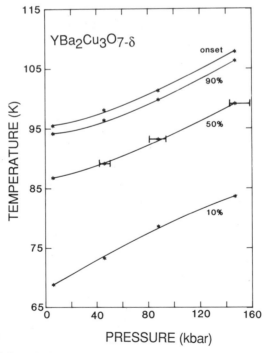

Fig. 5   Onset, 90%, 50% (midpoint), and 10% values of T_c vs pressure for $YBa_2Cu_3O_{7-\delta}$. After Ref. 34.

The rather large positive dependence of T_c on P in both the $(La_{2-x}M_x)CuO_{4-\delta}$ (M = Ba and Sr) and $RBa_2Cu_3O_{7-\delta}$ systems suggests that a common mechanism is responsible for the high T_c superconductivity of these two phases. Moreover, such a strong positive dependence of T_c on P is unusual among superconducting materials, in general.

Based on the pressure dependence of T_c of the $(La_{2-x}M_x)CuO_{4-\delta}$ (M = Ba and Sr) and $RBa_2Cu_3O_{7-\delta}$ compounds, as well as the small isotope effect in $RBa_2Cu_3O_{7-\delta}$ compounds with R = Y and Eu, Griessen [35] has recently argued that resonating valence bond (RVB) models [36] as treated by Fukuyama and Yosida [37] and by Cyrot [38] appear to be capable, in principle, of describing the superconductivity of these materials. Kaneko et al. [39] have shown that there is a linear relationship between the pressure dependence of T_c or the Néel temperature T_N of various superconducting, antiferromagnetic, and ferrimagnetic oxides and sulfides and $\kappa_v T_c$ or $\kappa_v T_N$, where κ_v is the volume compressibility, and suggest that the RVB models appear to be appropriate for the high T_c copper oxide superconductors.

The $(Y_{1-x}Pr_x)Ba_2Cu_3O_{7-\delta}$ System under Pressure

Recent measurements of $T_c(P)$ on the $(Y_{1-x}Pr_x)Ba_2Cu_3O_{7-\delta}$ system [40] have revealed rather interesting behavior. This is illustrated in Fig. 6 where the electrical resistivity ρ, normalized to its value at 300 K, is plotted vs temperature for several pressures between 0 and 18.5 kbar for a sample with the composition $(Y_{0.5}Pr_{0.5})Ba_2Cu_3O_{7-\delta}$. In

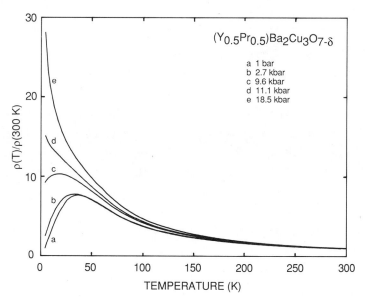

Fig. 6   Normalized electrical resistivity $\rho(T)/\rho(300\ K)$ vs temperature at various pressures for $(Y_{0.5}Pr_{0.5})Ba_2Cu_3O_{7-\delta}$. After Ref. 40.

marked contrast to the situation with $YBa_2Cu_3O_{7-\delta}$ where pressure renders the compound more metallic and increases T_c, the effects of applied pressure are to make the material less metallic and to suppress the superconductivity. This behavior is consistent with the expected pressure-induced increase of the valence of the Pr ions in $(Y_{0.5}Pr_{0.5})Ba_2Cu_3O_{7-\delta}$ towards +4, and reminiscent of that observed in superconductors containing Ce impurities where dramatic variations in T_c are induced under pressure due to corresponding changes in the Ce valence [41].

ACKNOWLEDGMENTS

The research at UCSD was supported by the U. S. Department of Energy under Grant No. DE-FG03-86ER45230 and the U. S. National Science Foundation-Low Temperature Physics-Grant No. DMR84-11839 (MST and BWL). Research at LANL was carried out under the auspices of the U. S. Department of Energy.

REFERENCES

1. J. G. Bednorz and K. A. Muller, Z. Physik **B64**, 189 (1986).
2. S. Uchida, H. Takagi, K. Kitazawa, and S. Tanaka, Jpn. J. Appl. Phys. **26**, L1 (1987).
3. R. J. Cava, R. B. van Dover, B. Batlogg, and E. A. Reitman, Phys. Rev. Lett. **58**, 408 (1987).
4. J. M. Tarascon, L. H. Greene, W. R. McKinnon, G. W. Hull, and T. H. Geballe, Science **235**, 1373 (1987).
5. K. Kishio, K. Kitazawa, S. Kanbe, I. Yasuda, N. Sugii, H. Takagi, S. Uchida, K. Fueki, and S. Tanaka, Chem. Lett. (1987), p. 429.
6. M. K. Wu, J. R. Ashburn, C. J. Torng, P. H. Hu, R. L. Meng, L. Gao, Z. J. Huang, Y. Q. Wang, and C. W. Chu, Phys. Rev. Lett. **58**, 908 (1987).
7. See, for example, M. B. Maple, Y. Dalichaouch, J. M. Ferreira, R. R. Hake, B. W. Lee, J. J. Neumeier, M. S. Torikachvili, K. N. Yang, H. Zhou, R. P. Guertin, and M. Kuric, Physica **148B**, 155 (1987).
8. See, for example, E. M. Engler, V. Y. Lee, A. I. Nazzal, R. B. Beyers, G. Lim, P. M. Grant, S. S. P. Parkin, M. L. Ramirez, J. E. Vasquez, and R. J. Savoy, J. Am. Phys. Soc. **109**, 2848 (1987).
9. See, for example, S. Hosoya, S. Shamoto, M. Onoda, and M. Sato, Jpn. J. Appl. Phys. **26**, L325 (1987).
10. See, for example, Z. Fisk, J. D. Thompson, E. Zirngiebl, J. L. Smith, and S. -W. Cheong, Solid State Commun. **62**, 743 (1987).

11. M. B. Maple, Y. Dalichaouch, J. M. Ferreira, R. R. Hake, S. E. Lambert, B. W. Lee, J. J. Neumeier, M. S. Torikachvili, K. N. Yang, H. Zhou, Z. Fisk, M. W. McElfresh, and J. L. Smith, in *Novel Superconductors*, S. A. Wolf and V. Z. Kresin, eds. (Plenum, New York, 1987) pp. 839 - 853.

12. M. B. Maple, to appear in *Proc. 1987 Symp. Low Temp. Phys.*, National Tsing Hua University, Hsinchu, Taiwan, R.O.C., Sept. 7 - 8, 1987, Chinese Journal of Physics (in press).

13. R. L. Greene, H. Maletta, T. S. Plaskett, J. G. Bednorz, and K. A. Muller, Solid State Commun. **63**, 379 (1987).

14. D. C. Johnston, J. P. Stokes, D. P. Goshorn, and J. T. Lewandowski, Phys. Rev. B **36**, 4007 (1987)..

15. Y. Maeno, Y. Aoki, H. Kamimura, J. Sakurai, and T. Fujita, Jpn. J. Appl. Phys. **26**, L402 (1987).

16. D. Vaknin, S. K. Sinha, D. E. Moncton, D. C. Johnston, J. M. Newsam, C. R. Safinya, and H. E. King, Jr., Phys. Rev. Lett. **58**, 2802 (1987).

17. T. Freltoft, J. E. Fischer, G. Shirane, D. E. Moncton, S. K. Sinha, D. Vaknin, J. P. Remeika, A. S. Cooper, and D. Harshman, Phys. Rev. B **36**, 826 (1987).

18. S. Mitsuda, G. Shirane, S. K. Sinha, D. C. Johnston, M. S. Alvarez, D. Vaknin, and D. E. Moncton, Phys. Rev. B **36**, 822 (1987).

19. G. Shirane, Y. Endoh, R. J. Birgeneau, M. A. Kastner, Y. Hidaka, M. Oda, M. Suzuki, and T. Murakami, Phys. Rev. Lett. **59**, 1613 (1987).

20. J. T. Markert, C. L. Seaman, H. Zhou, and M. B. Maple, to appear in Solid State Commun.

21. P. M. Grant, S. S. P. Parkin, V. Y. Lee, E. M. Engler, M. L. Ramirez, J. E. Vazquez, G. Lim, R. D. Jacowitz, and R. L. Greene, Phys. Rev. Lett. **58**, 2482 (1987).

22. K. Sakizawa, Y. Takano, H. Takigami, S. Tasaki, and T. Inaba, Jpn. J. Appl. Phys. **26**, L840 (1987).

23. Y. Nishihara, M. Tokumoto, K. Murata, and H. Unoki, Jpn. J. Appl. Phys. **26**, L1416 (1987).

24. See, e.g., B. D. Cullity, *Elements of X-ray Diffraction* (Addison-Wesley, Reading, Massachusetts, 1956) pp. 396–398.

25. J. M. Tarascon, L. H. Greene, B. G. Bagley, W. R. McKinnon, P. Barboux, and G. W. Hull, in *Novel Superconductivity*, S. A. Wolf and V. Z. Kresin, eds. (Plenum, New York, 1987) pp. 705 - 724.

26. Y. Dalichaouch, M. S. Torikachvili, E. A. Early, B. W. Lee, C. L. Seaman, K. N. Yang, H. Zhou, and M. B. Maple, Solid State Commun. **65**, 999 (1988).

27. L. Soderholm, K. Zhang, D. G. Hinks, M. A. Beno, J. D. Jorgensen, C. U. Segre, and I. K. Schuller, Nature **328**, 604 (1987).

28. D. E. Morris, U. M. Scheven, L. C. Bourne, M. L. Cohen, M. F. Crommie, and A. Zettl, preprint.

29. C. W. Chu, P. H. Hor, R. L. Meng, L. Gao, Z. J. Huang, and Y. Q. Wang, Phys. Rev. Lett. **58**, 405 (1987).

30. C. W. Chu, P. H. Hor, R. L. Meng, L. Gao, and Z. J. Huang, Science **235**, 567 (1987).

31. P. H. Hor, L. Gao, R. L. Meng, Z. J. Huang, Y. Q. Wang, K. Forster, J. Vassilious, C. W. Chu, M. K. Wu, J. R. Ashburn, and C. J. Torng, Phys. Rev. Lett. **58**, 911 (1987).

32. H. Yosida, H. Morita, K. Noto, T. Keneko, and H. Fugimori, Jpn. J. Appl. Phys. **26**, L862 (1985).

33. H. A. Borges, R. Kwok, J. D. Thompson, G. L. Wells, J. L. Smith, Z. Fisk, and D. E. Peterson, Phys. Rev. B **36**, 2404 (1987).

34. M. W. McElfresh, M. B. Maple, K. N. Yang, and Z. Fisk, to be published.

35. R. Griessen, Phys. Rev. B **36**, 5284 (1987).

36. P. W. Anderson, Science **235**, 1196 (1987).

37. H. Fukuyama and K. Yosida, Jpn. J. Appl. Phys. **26**, L371 (1987).

38. M. Cyrot, Solid State Commun. **63**, 1015 (1987).

39. T. Kaneko, H. Yoshida, S. Abe, H. Morita, K. Noto, and H. Fujimori, Jpn. J. Appl. Phys. **26**, L1374 (1987).

40. J. J. Neumeier, M. B. Maple, and M. S. Torikachvili, to be published.

41. M. B. Maple, Applied Physics **9**, 179 (1976).

DEFECT CHEMISTRY OF OXIDE SUPERCONDUCTORS

K.Fueki, K.Kitazawa, K.Kishio and T.Hasegawa

Department of Industrial Chemistry
Faculty of Engineering, University of Tokyo

INTRODUCTION

The discovery of La-Ba-Cu-O superconductors by Bednorz & Müller[1] has stimulated many investigators to study high-T_c oxide superconductors. Following the 40 K-class of superconductors with K_2NiF_4 structure[2], the 90 K-class of oxygen-deficient perovskites[3] was discovered, and very recently, superconductors of the Bi-Sr-Ca-Cu-O system, with an onset critical temperature of 120 K, have been reported. High-T_c superconducting oxides are complicated oxide systems. This paper aims to discuss high-T_c oxide superconductors from the viewpoint of defect chemistry.

GENERAL FEATURES OF PEROVSKITE-TYPE OXIDES

In Table 1, high-T_c oxide superconductors are summarized, together with the preceding oxide superconductor $Ba(Pb_{1-x}Bi_x)O_3$. Three kinds of oxide superconductors, BPBO, LSCO, and BYCO, are perovskite-related oxides.

Table 1. High-T_c oxide superconductors

Oxide	T_c/K	Crystal structure	
$Ba(Pb_{1-x}Bi_x)O_3$	13	perovskite	3-D
$(La_{1-x}M_x)_2CuO_{4-\delta}$		K_2NiF_4	2-D
M = Sr	40		
= Ba	30		
= Ca	20		
$Ba_2RCu_3O_{7-\delta}$	90	oxygen deficient	2-D
R = Y,Nd,Sm,Eu,		perovskite	
Gd,Dy,Ho,			
Er,Tm,Yb			
Bi-Sr-Ca-Cu-O	120		

This fact indicates that the perovskite ABO_3 is the most hopeful oxide candidate for superconductors. To begin with, we will briefly review the general features of perovskites. The "ideal" perovskite structure illustrated in Fig. 1(a), is cubic, with the B ion surrounded by 6 oxide ions and the A ion by 12 oxide ions. Each octahedron is connected with six neighbouring octahedra by sharing the corner oxide ions. Since the A-site hole is larger than the B-site one, cations of larger ionic radii prefer the A-site to the B-site.

Solid solutions of wide compositional range can be formed by substituting a B-site ion by another kind of cation. Similarly, solid solutions are formed by A-site substitution. When the B-sites are occupied by two kinds of cations, B' and B", and the B' to B" ratio is a simple integer one, such as 1:1, and 1:2, an ordered distribution of B' and B" is often observed. Similarly, two kinds of cations, A' and A", on the A-sites take a regular arrangement when the A' to A" ratio is a simple integer one. When the oxygen vacancy concentration is low, the vacancy distribution is random, but when the oxygen vacancy concentration is high, the vacancies are distributed regularly. Such an ordered distribution is well realized by the regular arrangement of coordination polyhedra shown in Fig. 2. In the case of $Ba_2YCu_3O_{7-\delta}$, the Ba-O plane and the Y-O plane are stacked regularly, as shown in Figs. 1(c) and 1(d). Moreover, BYCO has a large oxygen deficiency. In Fig. 2, the oxygen distribution in $Ba_2YCu_3O_7$ is represented by a combination of (b) and (c) and that in $Ba_2YCu_3O_6$ by a combination of (b) and (e).

In the "ideal" perovskite structure, AO and BO planes are stacked alternately. If one more AO plane is inserted, and shear is exerted in the [110] direction, the compound A_2BO_4 of K_2NiF_4 structure is obtained, as shown in Fig. 1(b).

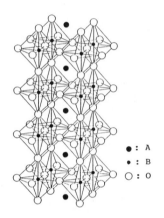

● : A
● : B
○ : O

Fig.1 (a)

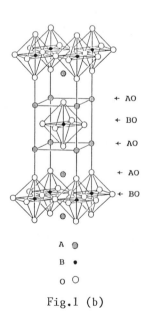

← AO
← BO
← AO

← AO
← BO

A ◉
B ●
o ○

Fig.1 (b)

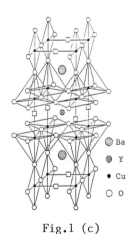

◉ Ba
◉ Y
● Cu
○ O

Fig.1 (c)

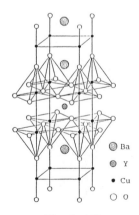

◉ Ba
◉ Y
● Cu
○ O

Fig.1 (d)

Structure 1(a) is three-dimensional, whereas structures 1(b), 1(c) and 1(d) are two-dimensional. It is often said that high-T_c is related to low dimensionality. The two-dimensional character of $(La_{1-x}Sr_x)_2CuO_4$ and $Ba_2YCu_3O_{7-\delta}$ structures could be responsible for the high T_c of these oxides.

AVERAGE VALENCE OF COPPER IONS AND ITS CONTROL

Perovskite-type oxides containing 3d transition metal ions on B-sites, exhibit electrical conduction due to those electrons originating from the transition metal ions. Conducting perovskites can be classified into two groups. In the first group, itinerant electrons are responsible for conduction. In the second group, transition metal ions are in mixed valence states. Conduction is caused by the hopping of electrons or electron holes. And some kinds of oxides exhibit superconductivity. Therefore, the control of a mixed valence state will be mentioned.

In perovskites containing 3d transition metal ions on B-sites, mixed valence states appear if A-site ions are substituted by allovalent ions. Let us consider $La_{1-x}Sr_xCrO_{3-\delta}$[4]. The valence of Cr in $LaCrO_3$ is +3. If La is partially replaced by Sr, the positive charge on the A-site decreases by the amount of x per mole. The decrease in the positive charge has to be compensated by an increase in the average valence of Cr ions and/or a decrease in the negative charge due to the removal of oxide ions. The electroneutrality condition gives:

$$3(1 - x) + 2x + (3 + z) = 2(3 - \delta),\tag{1}$$

where $3 + z$ is the average valence of the Cr ion. A simplification of Eq. (1) gives:

$$x = z + 2\delta,\tag{2}$$

Fig. 3 shows the plot of δ vs. $\log P_{O_2}$, for $La_{0.7}Sr_{0.3}CrO_{3-\delta}$. In region (I), where the oxygen partial pressure is low, $z = 0$ and $2\delta = 0.3$. In region (III), where the oxygen partial pressure is high, $\delta = 0$ and $z = 0.3$. In the middle oxygen pressure region, δ increases with the decrease in $\log P_{O_2}$. For $x = 0.3$, Eq. (2) gives:

$$z + 2\delta = 0.3,\tag{3}$$

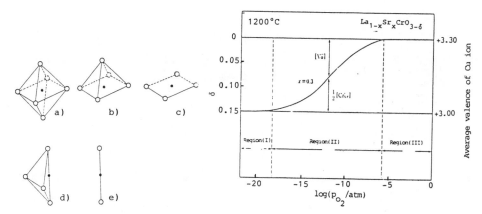

Fig. 2. Coordination polyhedra

Fig. 3. Oxygen deficiency and average valence of Cr ions as a function of $\log P_{O_2}$. $[La_{1-x}Sr_xCrO_{3-\delta}]$

43

We can therefore calculate z values which correspond to δ. The average valence, $3 + z$, of Cr ions is plotted on the ordinate on the righthand side.

Fig. 4 shows the plots of δ vs. log P_{O_2}, at constant temperatures for $(La_{0.92}Sr_{0.08})_2CuO_{4-\delta}$. The shape of the curves is similar to that shown in Fig. 3. It is noteworthy, that within this temperature range, where $P_{O_2} = 1$ atm, δ is quite small. This result indicates that the average valence of copper ions is mainly determined by the amount of Sr dopants. Figure 5 gives the relation between the average valence of Cu ions and the amount of Sr dopants. The samples annealed at 850 $^\circ$C at $P_{O_2} = 1$ atm were employed for chemical analysis. The solid line shows the experimental result. The electroneutrality condition is:

$$2x = z + 2\delta \qquad (4)$$

If $\delta = 0$, $2x$ is equal to z. This relation is shown by the broken line. The coincidence of the two lines is good. It points out clearly that the average valence of Cu ions is mainly determined by the amount of Sr dopants, as long as $(La_{1-x}Sr_x)_2CuO_{4-\delta}$ is annealed in 1 atmosphere of oxygen. Since $Ba_2YCu_3O_{7-\delta}$ has an exact metal ratio Y:Ba:Cu = 1:2:3, the average valence of the Cu ions is only controlled through the oxygen composition 7-δ. To determine the oxygen content is therefore essential. Figure 6 is a plot of δ, and the average valence of Cu ions against log P_{O_2}, at constant temperatures[5]. A wide homogeneous phase with δ ranging from 0.07 to about 0.9 is observed. Beyond the lower border-line, $Ba_2YCu_3O_{7-\delta}$ decomposes into Y_2BaCuO_5 and Ba-Cu-oxide. The average valence of copper ions is given on the ordinate on the righthand side.

The plot of δ vs. log P_{O_2} for $Ba_2GdCu_3O_{7-\delta}$ is given in Fig. 7.

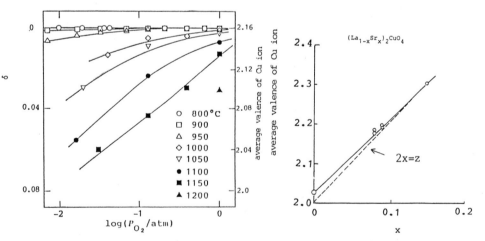

Fig. 4. Oxygen deficiency and average valence of $(La_{0.92}Sr_{0.08})_2CuO_{4-\delta}$.

Fig. 5. Relation between the average valence of Cu ions and the Sr concentration. $[(La_{1-x}Sr_x)_2CuO_{4-\delta}]$.

CONDUCTIVITY AND AVERAGE VALENCE

Figure 8 shows the relationship between the conductivity and the average valence of Cr ions, at constant Sr concentration[4]. The average valence was controlled by changing the partial pressure of oxygen. A good proportionality between conductivity and Cr^{4+} concentration indicates that the electron hole of Cr^{4+} is the majority charge carrier of electrical conduction. The slope of the plot gives the apparent mobility, which increases with the increase in Sr concentration. The Sr dopant plays the role of increasing the conductivity as a result of the increase in carrier concentration and mobility.

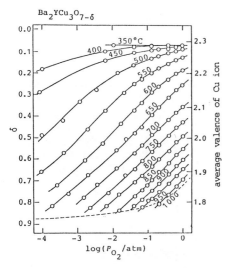

Fig. 6. Oxygen deficiency and average valence of Cu ions as a function of log P_{O2}. [$Ba_2YCu_3O_{7-\delta}$]

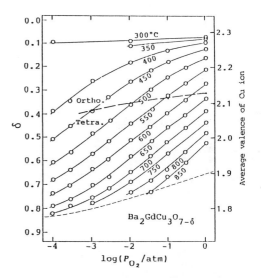

Fig. 7. Oxygen deficiency and average valence of Cu ions as a function of log P_{O2}. [$Ba_2GdCu_3O_{7-\delta}$]

The relationship between the average valence of Cu ions and conductivity for $(La_{1-x}Sr_x)_2CuO_{4-\delta}$ is given in Fig. 9. Data from Michel and his co-investigator[6] were used for the calculations. The logarithm of conductivity depends almost linearly on the average valence of Cu ions. As mentioned above, the average valence is mainly controlled by the Sr concentration, if the specimens are annealed and quenched in 1 atomosphere of oxygen. For $Ba_2RCu_3O_{7-\delta}$, the average valence of the copper ions is controlled by the oxygen partial pressure and by temperature. Figure 10 shows a plot of the logarithm of conductivity against the average valence of Cu ions for $Ba_2YCu_3O_{7-\delta}$. A linear relation is seen when the average valence is greater than 2.0. A similar plot for $Ba_2GdCu_3O_{7-\delta}$ is given in Fig. 11. Irrespective of temperature, all the data fall on a single straight line, when the average valence is greater than 2.0. The exponential increase in conductivity with Cu^{3+} concentration, suggests that doping increases the mobility remarkably.

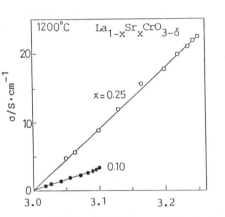

Fig. 8. Relation between σ and
the average valence of
Cr ions for $La_{1-x}Sr_xCrO_{3-\delta}$
(x = 0.1 and 0.25).

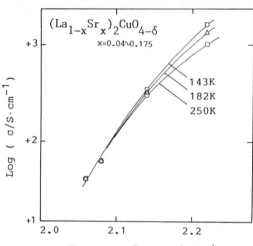

Fig. 9. Relation between σ and
the average valence of
Cu ions for $(La_{1-x}Sr_x)_2$-
$CuO_{4-\delta}$.

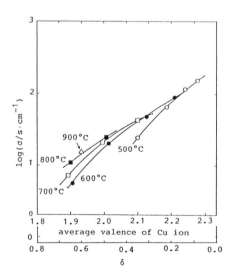

Fig. 10. Relation between log σ
and the average valence
of Cu ions for
$Ba_2YCu_3O_{7-\delta}$.

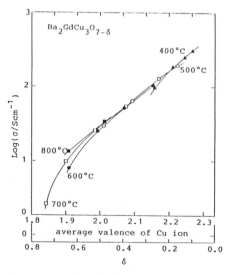

Fig. 11. Relation between log σ
and the average valence
of Cu ions for
$Ba_2GdCu_3O_{7-\delta}$.

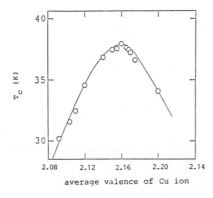

Fig. 12. Relation between T_C and the
average valence of Cu ions.
$[(La_{1-x}Sr_x)_2CuO_{4-\delta}]$

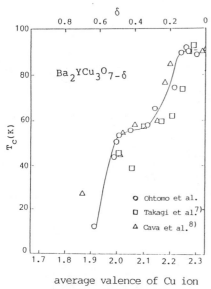

Fig. 13. Relation between T_C and
the average valence of
Cu ions. $[Ba_2YCu_3O_{7-\delta}]$

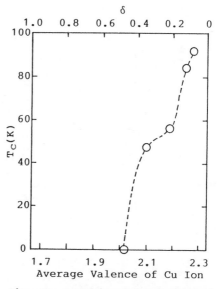

Fig. 14. Relation between T_C and
the average valence of
Cu ions. $[Ba_2GdCu_3O_{7-\delta}]$

CRITICAL TEMPERATURE AND AVERAGE VALENCE

Figure 12 shows the relation between the average valence of copper
ions and the critical temperature for $(La_{1-x}Sr_x)_2CuO_4$. The critical
temperature T_C increases initially with the increase of average valence,
reaches a maximum and decreases. The maximum T_C appears at an average
valence of 2.16.

The critical temperature of $Ba_2YCu_3O_{7-\delta}$ was measured using samples
annealed at the desired temperature and oxygen pressure and then quenched
in liquid nitrogen. The average valence was calculated from the annealing
condition. In Fig. 13, T_C is plotted against the average valence[7], and,
for comparison, data from another investigator[8] are also plotted. T_C
increases with the increase of average valence, and superconductivity

vanishes at an average valence of around 2.0. The T_c-average valence relation for $Ba_2GdCu_3O_{7-\delta}$ is given in Fig. 14.

SUPERCONDUCTIVITY AND TEMPERATURE DEPENDENCE OF CONDUCTIVITY

Figure 15 shows the Arrhenius plots for the conductivity of perovskite-type oxides. Perovskites with high conductivity, such as ReO_3 and $SrVO_3$, show metallic conduction. The temperature coefficient of conductivity is negative ($d\sigma/dT < 0$). On the other hand, perovskites with low conductivity have a positive temperature coefficient ($d\sigma/dT > 0$). such oxides are semiconductors or insulators. In the middle region, the temperature coefficient is nearly zero ($d\sigma/dT = 0$). It is noteworthy that all the lines meet at around log δ = 2 ~ 3 on the ordinate, if the lines

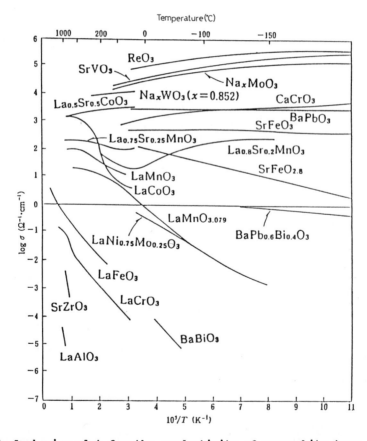

Fig. 15. Arrhenius plot for the conductivity of perovskite-type oxides.

are extended to 1/T = 0. Figure 16 gives the Arrhenius plot of high
temperature conductivity for $Ba(Pb_{1-x}Bi_x)O_3$[9] which exhibits
superconductivity at low temperatures. Data from Tanaka and his
co-investigators were used in Fig. 16. The temperature coefficient is
nearly zero and the absolute σ value is about 10^2 Scm^{-1}.

Similar plots for superconducting $(La_{1-x}Sr_x)_2CuO_{4-\delta}$ and $Ba_2GdCu_3O_{7-\delta}$
are given in Figs. 17, 18, and 19, respectively. Data from Michel and his
co-investigator were used for Fig. 17. All these plots show that
$d\sigma/dT \sim 0$ and $\sigma = 10^2 \sim 10^3$ Scm^{-1}, which suggests that superconductivity
is related to metal-insulator transitions.

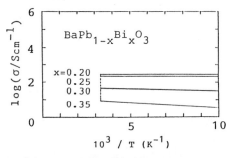

Fig. 16. Conductivity of the
superconducting oxide
$BaPb_{1-x}Bi_xO_3$.

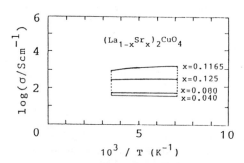

Fig. 17. Conductivity of the
superconducting oxide
$(La_{1-x}Sr_x)_2CuO_{4-\delta}$.

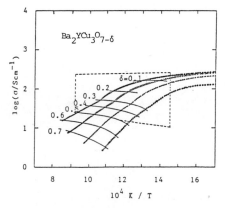

Fig. 18. High temperature
conductivity of the
superconducting oxide
$Ba_2YCu_3O_{7-\delta}$.

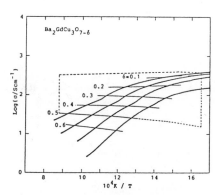

Fig. 19. High temperature
conductivity of the
superconducting oxide
$Ba_2GdCu_3O_{7-\delta}$.

Since the critical temperature, T_c, of $Ba_2YCu_3O_{7-\delta}$ increases with oxygen content, oxygen annealing at 500 to 550 °C for several hours, or slow cooling from high temperatures is usually carried out to incorporate oxygen into the oxide. Most of the sintered samples were porous, so that oxygen gas could easily penetrate and the oxygen incorporation was therefore completed within a short length of time. However, compact polycrystalline oxides and single crystals seem to need a long time to incorporate oxygen. Slow diffusion could pose serious problems for the fabrication process. We therefore carried out the following diffusion experiment.

Specimens were prepared as follows. A mixed aqueous oxalate solution of yttrium, barium and copper was mixed with ethanol. The precipitate was calcined for one day at about 800 °C in air to obtain BYCO powder. The average particle size was 0.2 ~ 0.7 μm. The pellet, prepared by pressing the powder, was sintered at 1020 °C under 1 atmosphere of oxygen partial pressure. The porosity was about 2%. Figure 20(b) is a photograph of the polished surface of the compact pellet. For comparison, the polished surface of a porous pellet prepared by the conventional method is shown in Fig. 20(a).

Two compact pellets, the thicknesses of which were 0.9 mm and 2.0 mm, were used for the diffusion measurements. A specimen was suspended from a beam of a thermo-microbalance and equilibrated with a predetermined oxygen pressure. Then the partial pressure of oxygen was changed stepwise and the weight change followed.

An example of the weight change is shown in Fig. 21. The solution of the diffusion equation for a semi-infinite plate gives

$$\frac{w(t)-w(\infty)}{w(0)-w(\infty)} = \frac{8}{\pi^2} \sum_{j=0}^{\infty} \frac{1}{(2j+1)^2} \exp\left[-\frac{(2j+1)^2\pi^2}{h^2}\tilde{D}t\right] \qquad (5)$$

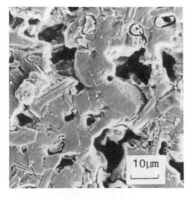

Fig. 20(a).

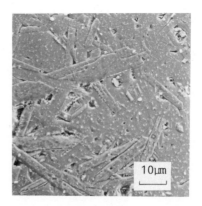

Fig. 20(b).

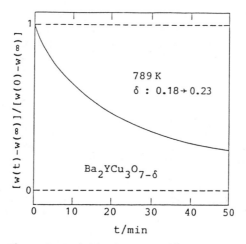

Fig. 21. Weight change - time curve.

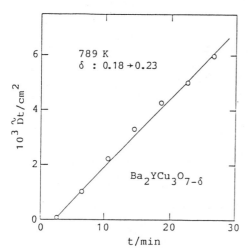

Fig. 22. Determination of D from
the plot of Dt vs. t.

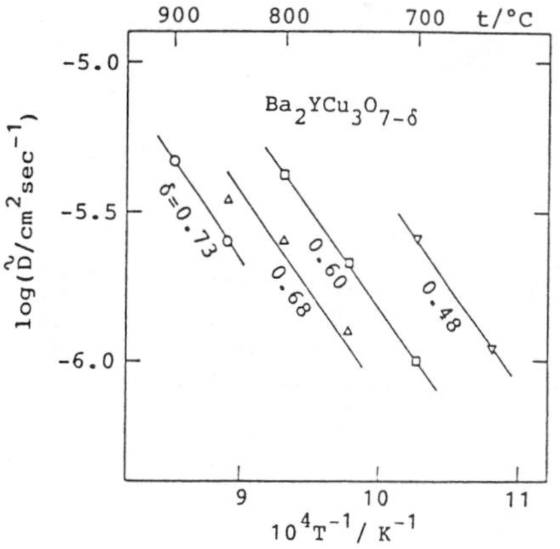

Fig. 23. Arrhenius plot for D.

where $w(0)$, $w(t)$, and $w(\infty)$ are the sample weights at times zero, t and infinity, respectively, j is an integer, D is the chemical diffusion constant, and h is the thickness. Accordingly, D can be determined from a set of data for the weight change. In order to check whether the process is diffusion-controlled or not, the measurement was made using two samples with different thickness. If the process is diffusion-controlled, D should be the same, irrespective of thickness. However, the thicker one gave slightly smaller D values. From the anisotropic crystal structure of BYCO, it is expected that oxygen diffuses along the Cu(I)-O planes preferentially.

The small difference in the D values between the two samples may be due to the difference in the average degree of grain orientation. It was found that D is a function of δ. The Arrhenius plot of D at constant δ values is given in Fig. 23. The activation energies determined from the slope ranged from 1.2 to 1.3 eV, which is fairly close to 0.9 - 1.3 eV observed for typical perovskite-type oxides. It is notable that D increases with the increase in oxygen content. With the increase in oxygen content, the O(I)-Cu(I)-O(I) distance decreases. The oxygen diffusion would probably be enhanced by a decrease in the O(I)-Cu(I)-O(I) distance.

REFERENCES

1. J.G.Bednorz and K.A.Müller, Z.Phys. B64, 189 (1986)
2. K.Kishio, K.Kitazawa, S.Kanbe, I.Yasuda, N.Sugii, H.Takagi, S.Uchida, K.Fueki and S.Tanaka, Chem.Lett. 429 (1987)
3. M.K.Wu, J.R.Ashburn, C.J.Torng, P.H.Hor, R.J.Meng, L.Gao, Z.J.Huang, Y.Q.Wang and C.W.Chu, Phys.Rev.Lett. 58, 908 (1987)
4. J.Mizusaki, S.Yamauchi, K.Fueki and A.Ishikawa, Solid State Ionics 12, 119 (1984)
5. K.Kishio, J.Shimoyama, T.Hasegawa, K.Kitazawa and K.Fueki, Jpn.J.Appl. Phys. 26, L1228 (1987)
6. C.Michel and B.Raveau, Rev.Chim.Minerale 21, 407 (1984)
7. H.Takagi, S.Uchida, H.Iwabichi, H.Eisaki, K.Kishio, K.Kitazawa, K.Fueki and S.Tanaka, Physica 148B, 349 (1987)
8. R.J.Cava, B.Batlogg, C.H.Chen, E.A.Rietmann, S.M.Zahurak and D.Werder, Phys.Rev. B36, 5719 (1987)
9. S.Uchida, K.Kitazawa and S.Tanaka, "Phase Transition", Gordon and Breach Science Publishers, New York, London, Paris, Montreux, Tokyo (1987)

DEVIATION FROM LINEARITY OF THE Cu-O CHAINS IN ORTHORHOMBIC YBa$_2$Cu$_3$O$_7$ BY HIGH-RESOLUTION NEUTRON POWDER DIFFRACTION

M. François, A. Junod, K. Yvon and A.W. Hewat[*]

Université de Genève, CH 1211 Genève, Switzerland and
[*]Institut Laue-Langevin, 156X, 38042 Grenoble, France

ABSTRACT

The structure of orthorhombic YBa$_2$Cu$_3$O$_{6.91}$ (T$_c$=90K) was refined from high-resolution data (sin$\theta/\lambda)_{max}$=0.91Å^{-1}) in the temperature range between 5K and 320K. The oxygen atoms on the so-called linear Cu-O chains occupy potential minima which are displaced from the chain axis along [100] by at least 0.08(1)Å at all temperatures. The Cu-O-Cu bond angle of the resulting Cu-O zig-zag chains, α=175(1)°, does not vary significantly as a function of temperature below T$_c$.

INTRODUCTION

The structure of orthorhombic, superconducting YBa$_2$Cu$_3$O$_7$ has been characterized by numerous neutron powder diffraction experiments (for a review see[1]). According to the structure refinements reported[2-7] the O4 oxygen atoms are located on linear Cu-O chains running along b, i.e. at the site 1e (0, 1/2, 0) of space group Pmmm. The occupancy of this site is lower than unity (usually between 0.60 and 0.93, depending on the oxygen content and thermal history of the sample[6,8,9], and its displacement amplitudes at low temperature (in particular those perpendicular to the chain axis) are anomalously large ($<$u$>^2$=0.012(1)Å^2 at 5K) compared to those of the other O atom sites in the structure ($<$u$>^2$=0.0040(2)Å^2)[10]. We propose here that the Cu-O chains in YBa$_2$Cu$_3$O$_7$ are not linear but zig-zag.

EXPERIMENTAL

A sample of about 10g was prepared by citrate pyrolysis[11]. Its X-ray Guinier powder pattern (CuKα radiation) was indexed on the known ortho-rhombic cell. It showed no trace of impurity phases. The sample was further characterized with respect to its physical properties (for results see ref[12]), and then placed into a cylindrical vanadium container (diameter=7mm). The neutron diffraction measurements were performed on the new high-resolution powder diffractometer D2B at ILL (Grenoble) in the temperature range between 5K and 320K. In order to obtain high spatial resolution of the structure a relatively short wavelength of λ=1.051Å was used, and the measurements were extended to scattering angles up to 2θ=145° (sinθ/λ=0.91Å^{-1}). The measurements were performed at 16 different increasing temperatures. None of the patterns showed evidence for diffraction peaks which could not be indexed on the known orthorhombic cell. Thus 3-dimensional long-range ordering of oxygen vacancies on the Cu-O chains can be excluded.

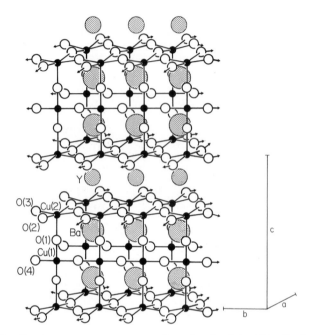

Fig.1. Average structure of orthorhombic $YBa_2Cu_3O_7$ with linear Cu(1)-O(4) chains running along b (from Ref.[4])

The structure refinements were performed with a version of a Rietveld program[13] which permits refinement of the background. Except for O4 (see below) the structure model used was that refined previously[2]: space group Pmmm (No47)[14], Ba in 2t (1/2, 1/2,±z; z=0.185), Y in 1h (1/2, 1/2, 1/2), Cu1 in 1a (0, 0, 0), Cu2 in 2q (0, 0,±z; z=0.355); O1 (called O4 in ref[2]) in 2q (0, 0,±z; z=0.159); O2 in 2s (1/2, 0,±z; z=0.377); O3 in 2r (0, 1/2,±z; z=0.378). The O4 atoms were assumed to occupy sites 2k (±x, 1/2, 0; point symmetry 2mm) which derive from site 1e (0,1/2,0; point symmetry mmm) by displacements of ±x(O4) perpendicular to the Cu-O chain axis (Fig.1). Statistical occupancy of 0.5 and isotropic temperature factors were assumed. This so-called "split-atom" model was previously tested with success on neutron diffraction data of much lower resolution ($\sin\theta/\lambda=0.45(1)Å^{-1}$)[15]. Preliminary structure refinements on our data showed that the occupancy factor for the O4 split-atom sites was lower than 0.50 and did not vary significantly as a function of temperature. Thus it was fixed during the final refinements at its average value of 0.455(5) , corresponding to the sample composition $YBa_2Cu_3O_{6.91(1)}$. The occupancy of another possible oxygen site in the structure (O5 at site 1b: 1/2,0,0) was tested during the refinements. The results showed that this site was not significantly occupied at any temperature.

Table 1: Structural parameters and agreement indices for $YBa_2Cu_3O_{6.91}$ at 5K and 95K

cell parameters(Å):

5K   a = 3.80922(6), b = 3.87886(6), c = 11.6365(2)

95K   a = 3.8091(1), b = 3.8789(1), c = 11.6435(3)

atomic parameters:

atoms		Ba	Y	Cu1	Cu2	O1	O2	O3	O4
x	5K	-	-	-	-	-	-	-	0.0247(14)
	95K	-	-	-	-	-	-	-	0.0209(21)
z	5K	0.1843(1)	-	-	0.3553(1)	0.1599(1)	0.3780(1)	0.3780(1)	-
	95K	0.1846(2)	-	-	0.3551(1)	0.1595(1)	0.3782(1)	0.3773(2)	-
B(Å²)	5K	0.27(2)	0.24(2)	0.26(1)	0.19(1)	0.42(2)	0.33(2)	0.35(2)	0.20(4)
	95K	0.30(2)	0.31(2)	0.32(2)	0.35(2)	0.50(2)	0.38(2)	0.37(2)	0.39(4)

agreement indices:

5K R_f=2.19, R_p=4.62, R_i=4.09, R_{wp}=6.23, R_{exp}=3.47

95K R_f=2.22, R_p=5.85, R_i=3.99, R_{wp}=7.53, R_{exp}=5.94

form of isotropic temperature factors: $\exp\{-B\sin^2\theta/\lambda^2\}$

for a definition of the various R values see [13]

RESULTS AND DISCUSSION

    All refinements (29 (14 atomic) parameters, 680 hkl values) converged satisfactorily. Partial results are summarized in Table 1. Table 1 contains the results of the structure refinement at 5K and 95K, and Fig. 2 shows the temperature dependence of the structural parameters involving the O4 split-atom position. Complete results are published elsewhere together with results obtained on a more reduced sample of composition $YBa_2Cu_3O_{6.86}$[10]. The agreement indices and estimated standard deviations of the atomic parameters are lower by a factor of about two, on the average, than those reported previously. The facts that the displacement parameter of the oxygen chain atom, x(O4), refines to non-zero values at all temperatures and that the thermal vibration parameter of that atom does not much differ from those of the other oxygen atoms justify the validity of the present (isotropic) split-atom model over the previous (anisotropic) unsplit-atom model which also has one more atomic parameter. As can be seen on Fig.2a, the displacement parameter is not quite independent of temperature as the distance of O4 from the Cu-O chain axis, δ=a.x(O4), decreases with temperature from 0.136(5)Å at 300K to 0.080(8)Å at about 90K. However this distance does not extrapolate to zero within that temperature interval and, more importantly, remains approximately constant below 90K. Thus the O4 atoms do not lie on the Cu-O chain axis as was assumed previously[2,7], but occupy potential minima at a distance of at least 0.080(8)Å on either side of the Cu-O chain at all temperatures. This new model implies that the Cu-O chains are no longer linear but zig-zag shaped and that the Cu1-O4 bond lengths, O4-O1 contact distances, and Cu1-O4-Cu1 bond angles (Cu1-O4=1.9412(3)Å, O4-O1=2.6863(8)Å, Cu1-O4-Cu1=175.3(3)° at 95K) differ significantly from those calculated for a linear chain model (Cu1-O4=1.9395(1)Å, O4-O1=2.6851(8)Å, Cu1-O4-Cu1=180°). As shown in Fig 2.b-d the temperature dependence of the former reflects that of the displacement amplitude δ=a.x(O4).

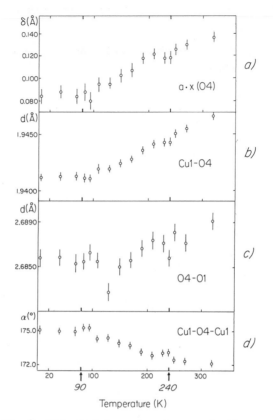

Fig.2. Structural parameters of the Cu1-O4 zig-zag chain in $YBa_2Cu_3O_{6.91}$ as a function of temperature. a) atomic displacements of O4, $\delta=a.x(O4)$, perpendicular to the Cu-O chain axis; b) Cu1-O4 bond lengths; c) O4-O1 contact distances; d) Cu1-O4-Cu1 bond angles, α. Arrows at 90K and 240K indicate possible anomalies.

A question which remains to be answered is the exact shape of the zig-zag chains and their three-dimensional correlation in the structure. According to our data no 3-d long range order exists at any temperature. However, there could be short-range order, either statically or dynamically, or two-dimensional order. Such effects are beyond the limit of detectability of the present powder data. Evidence for structural correlations at room temperature was recently obtained from electron diffraction experiments on $YBa_2Cu_3O_7$, and interpreted in terms of oxygen vacancy ordering[16]. The interpretation of our data differs in that a possible order does not concern the oxygen vacancies which are quite distant from each other at about 9 unit cells, on the average, along b in $YBa_2Cu_3O_{6.91}$, but the split atom positions O4 on either side of the chain axis. Possible zig-zag patterns for ordered but uncorrelated Cu-O chains in the basal plane are represented in Fig. 3. Whether or not such patterns actually occur in the structure should be investigated by a careful analysis of the diffuse scattering intensity, preferentially on well characterized and

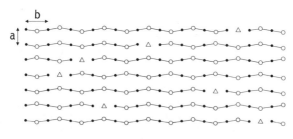

• Cu1   ○ O4   △ vacancy

Fig.3. Structural model of short range-ordered Cu1-O4 zig-zag chains within a single domain in $YBa_2Cu_3O_{6.91}$. For clarity the deviation of the Cu1-O4-Cu1 bond angles from linearity (about 5°) has been exaggerated.

stoichiometric single crystals. Notice that the directions of the zig-zag chains in twinned crystals change by about 90° at each twin boundary.

Another question which remains to be answered is wether or not order-disorder transitions occur, and if so, at what temperatures. In view of the close proximity of the two (symmetry equivalent) potential minima (≈0.2Å across the chain) ordering phenomena such as a condensation of soft modes on the Cu-O chains are likely to occur even at very low temperature. They could occur, for example, at the superconductive transition tempera-ture as a result of electron-phonon interactions and play a role in elec-tron pairing. Again, careful analyses of the diffuse scattering as a func-tion of temperature would clarify this hypothesis. Evidence for a structu-ral change near 90K was recently obtained from high-resolution X-ray work by plotting the orthorhombic lattice distortion parameter (b-a)/(b+a) as a function of temperature[17]. Our experiments (shown elsewhere[10]) confirm this result and in addition show possible anomalies near 240K. Anomalies at these two temperatures also occur for certain physical properties such as elasticity[18], bulk-modulus[19], specific heat[20,21], thermal expansion[22] and resistivity[21].

Acknowledgements

We thank B. Künzler for her help with the drawings. This work was supported by the Swiss National Science Foundation.

References

1. K. Yvon, Z. Physik (1988), to be published
2. M.A. Beno, L. Soderholm, D.W. Capone II, D.G. Hinks, J.D. Jorgensen, I.K. Shuller, C.U. Segre, K. Zhang and J.D. Grace, Appl. Phys. Lett. **51**, 57-59 (1987)
3. J.J. Capponi, C. Chaillout, A.W. Hewat, P. Lejay, M. Marezio, N. Nguyen, B. Raveau, J.L. Soubeyroux, J.L. Tholence and R. Tournier, Europhysics Lett. 3, 1301-1307 (1987)
4. M. François, E. Walker, J.L. Jorda, P. Fischer and K. Yvon, Sol. State Commun. **63**, 1149-1153 (1987).
5. J.E. Greedan, A.H. O'Reilly and C.V. Stager, Phys.Rev. **B35**, 8770-8773 (1987)
6. F. Beech, S. Miraglia, A. Santoro and R.S. Roth, Phys. Rev. **B35**, 8778-8781 (1987)

7. W.I.F. David, W.T.A. Harrison, J.M.F. Gunn, O. Moze, A.K. Soper, P. Day, J.D. Jorgensen, D.G. Hinks, M.A. Beno, L. Soderholm, D.W. Capone II, I.K. Schuller, C.U. Segre, K. Zhang and J.D. Grace, Nature 327, 310-312 (1987)

8. A.W. Hewat, J.J. Capponi, C. Chaillout, M. Marezio and E.A. Hewat, Sol. State. Commun. 64, 301-307 (1987)

9. J.D. Jorgensen, M.A. Beno, D.G. Hinks, L. Soderholm, K.J. Volin, R.L. Hitterman, J.D. Grace, Ivan K. Schuller, C.U. Segre, K. Zhang and M.S. Kleefisch, Phys. Rev. B36 3608-3616 (1987)

10. M. François, A. Junod, K. Yvon, P. Fischer, J.J. Capponi, P. Srobel, M. Marezio and A.W. Hewat, Sol.State Commun. (1988), to be published

11. D.H.A. Blank, H. Kruidhof and J. Flokstra, J. Phys. D 21, 226 (1988)

12. A. Junod, A. Bezinge, D. Cattani, M. Decroux, D. Eckert, M. François, A.W. Hewat, J. Muller and K. Yvon, submitted to Helvetica Physica Acta (1988)

13. D. B. Wiles and R. A. Young, J. Appl. Cryst. 14, 149-151 (1981)

14. International Tables for Crystallography Vol.A, D. Reidel publishing company (1983)

15. Z. Jirak, F. Pollert, A. Triska and S. Vratislav, Phys. Stat. Sol. (a) 102, K61-K66 (1987)

16. C. Chaillout, M.A. Alario-Franco, J.J. Capponi, J. Chenavas, J.L. Hodeau and M. Marezio, Phys. Rev. accepted

17. P.M. Horn, D.T. Keane, G.A. Held, J.L. Jordan-Sweet, D.L. Kaiser, F. Holtzberg and T.M. Rice, Phys. Rev. Lett. 59, 2772-2275 (1987)

18. Y. Wang, H. Shen, J. Zhu, Z. Xu, M. Gu, Z. Niu and Z. Zhang, J. Phys. C: Sol. State Phys. 20, L665-L668 (1987)

19. D.J. Bishop, A.P. Ramirez, P.L. Gammel, B. Batlogg, E.A. Rietman, R.J. Cava and A.J. Millis, Phys. Rev B36, 2408-2410 (1987)

20. T. Laegreid, K. Fossheim, E. Sandvold, O. Traetteberg, S. Julsrud, T. Helgesen, C.E. Gough and J.S. Abell, submitted to Phys. Rev. B, (1987)

21. R. Calemczuk, E. Bonjour, J.Y. Henry, L. Forro, C. Ayache and M.J.G.M. Jurgens, to be published (1988)

22. Y. Ruan, L. Li, X. Hu, X. Mao, G. Meng, D. Peng, J. Hu and Y. Zhang, Submitted to Chinese Phys. Lett. (1987)

ELECTRON IRRADIATION EFFECTS IN $YBa_2Cu_3O_{7-\delta}$ SINGLE CRYSTALS*

M.A.Kirk, M.C.Baker, J.Z.Liu and D.J.Lam
Materials Science Division, Argonne National Laboratory
Argonne, IL 60439, U.S.A.

and

H.W.Weber
Atominstitut der Österreichischen Universitäten
A-1020 Wien, Austria

Defect structures in $YBa_2Cu_3O_{7-\delta}$ produced by electron irradiation at 300 K, were investigated by transmission electron microscopy. Threshold energies for the production of visible defects were determined to be 152 keV and 131 keV (± 7 keV) in directions near the a- and b-axes, respectively (b > a, both perpendicular to c, the long axis in the orthorhombic structure). During above - threshold irradiations in an electron flux of 3 x 10^{18} $cm^{-2}s^{-1}$, extended defects were observed to form and grow to sizes of 10 - 50 nm over 15 minutes, in material thicknesses varying between 20 and 200 nm. Upon irradiation between the a- and b-thresholds, movement of twin plane boundaries and shrinkage of twinned volume were observed. All these findings suggest oxygen atom displacements in the basal plane with recoil energies near 20 eV. Above - threshold irradiations also show the collapse of c-axis long range order into a planar faulted defect structure with short range order peaks at 1.2 c and 1.07 c, depending on the irradiation direction.

INTRODUCTION

Several neutron and ion irradiation experiments (cf., e.g., references 1-3) on single- and polycrystalline samples of $YBa_2Cu_3O_{7-\delta}$ have been performed. They have shown that both the primary superconductive properties, such as the transition temperature T_c, as well as the intrinsic defect states responsible for flux pinning are affected by radiation. Whereas the transition temperature T_c immediately starts to decrease with radiation dose, thus indicating adverse effects on the pairing mechanism in these superconductors, the critical current densities j_c were observed to increase significantly at lower fluences (thus indicating the formation of new defects capable of flux pinning) and to decrease only at higher radiation doses. Of course, an investigation of the mechanisms responsible for these property changes is of fundamental interest.

* Work supported by the U.S.Department of Energy, BES-Materials Sciences, under contract W-31-109-Eng-38.

The simplest defects that can be produced in a metal, are those formed by electron irradiation. The threshold energy for the production of stable Frenkel pairs, self-interstitial atoms and vacancies, as well as the structure and properties of these defects, are essential for the understanding of many macroscopic properties. In the present paper, we report on experiments with single crystalline pieces of $YBa_2Cu_3O_{7-\delta}$ aimed at this task and made in an electron microscope with variable electron energy.

EXPERIMENTAL PROCEDURES

Samples suitable for electron microscopy were prepared from the same stock as those subjected to neutron irradiation[2,4]. Tiny single crystals formed during the growing procedure described in ref. 3 were crushed, deposited onto carbon films and then placed onto standard copper grids. The samples were either kept in the vacuum of the electron microscope (~10^{-6} mbar) or in desiccated air.

Samples having the correct orthorhombic phase were selected by pole diffraction patterns, either in a Philips 420 or in the AEI EM7 High Voltage Electron Microscope (HVEM). The imaging conditions were chosen to be G = 200 or 020, bright field and dark field. For the electron irradiation experiments, the beam energy was varied between 100 and 160 keV, the electron flux density was 3.10^{18} $cm^{-2}s^{-1}$. Electron dosimetry was made by in-situ Faraday cup measurements. The samples were kept at a temperature of 15 $^\circ$C; the influence of beam heating is not known accurately, but estimated to result in sample temperatures below 50 $^\circ$C.

Experimental investigations usually proceeded immediately following electron irradiation at a particular energy. However, when defect reactions after above-threshold irradiation were noted for the first time, waiting times of up to 24 hours had to be included.

RESULTS

The results of ambient temperature electron irradiation experiments are summarized in Figs. 1 - 4. Specific directions for the incident beam were selected from pole diffraction patterns. However, in order to avoid anomalous transmission and electron channeling effects, the beam was directed 5 - 10° off major crystallographic axes.

Fig. 1 illustrates the results of irradiation near the b-direction, which contains the oxygen atom in the basal plane. The upper series of 3 micrographs demonstrates the existence of a threshold energy between 123 and 139.5 keV, the formation of extended defects during irradiation with an electron energy of 139.5 keV, and a highly defective structure formed within 24 hours after irradiation. The lower series shows bright field micrographs of the defect production during above-threshold irradiation. The defects are found to grow to appreciable sizes (10 - 50 nm); their nature could not yet be determined, partly because of the heavy second defect reaction already occuring within a period of 24 hours.

From the results shown in Fig. 2 we will firstly discuss the threshold effect in the a-direction. In these micrographs, images along the a-direction are represented by the gray areas, whereas the darker areas inbetween results from dynamical contrast caused by twin boundaries. The data show clearly, that upon irradiating near the a-direction no extended defects are formed during irradiation, whereas a highly defective state is produced following irradiation at 159.8 keV and a waiting time of 24 hours. Hence, the threshold for damage production in the a-direction is determined to be (152 ± 8) keV. Secondly, important information on the behavior of the twin structure can be deduced from Fig. 2. Irradiation at 146.5 keV clearly demonstrates that the volume of twinned material (dark areas) shrinks considerably. If we assume, that the twinned material has exactly the same crystal structure, but is rotated by 90° about the c-axis with (110) boundary planes, then an irradiation along the a-direction corresponds to an irradiation along the b-direction in the twinned volume and is, therefore, *above* the threshold energy for defect production as determined from Fig. 1. Hence, the contrast features of these micrographs confirm the differences in threshold energies for the two crystal directions in the basal plane and are consistent with current ideas on the nature of twinning.

Some details of the highly defective state formed by above-threshold irradiation along the a-direction and after a waiting time of 24 hours are illustrated in Fig. 3. Amorphous appearing and planar faulted regions can be seen. The lattice fringes, corresponding to local planar configurations perpendicular to the c axis, are seen to include segmented sections and some very local bending. A relatively undamaged area in the very thinnest part of the material is also noted, which may perhaps be due to a loss of defects to very near surfaces.

Electron diffraction patterns taken from the highly defective structures formed following above-threshold irradiation and after a

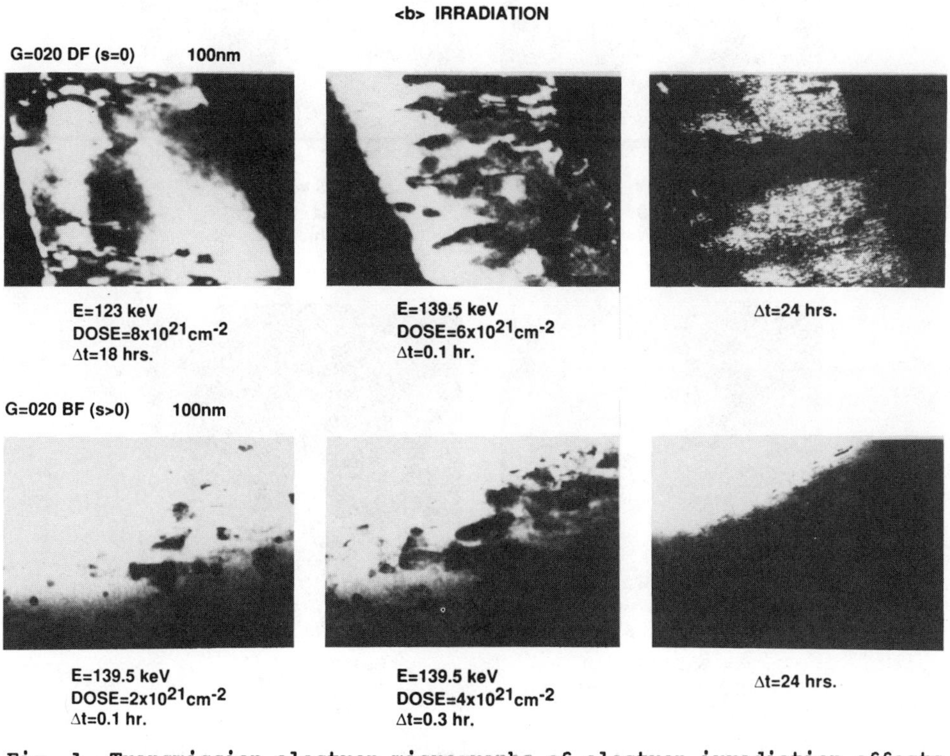

<b> IRRADIATION

G=020 DF (s=0)   100nm

E=123 keV
DOSE=8x10^{21}cm^{-2}
Δt=18 hrs.

E=139.5 keV
DOSE=6x10^{21}cm^{-2}
Δt=0.1 hr.

Δt=24 hrs.

G=020 BF (s>0)   100nm

E=139.5 keV
DOSE=2x10^{21}cm^{-2}
Δt=0.1 hr.

E=139.5 keV
DOSE=4x10^{21}cm^{-2}
Δt=0.3 hr.

Δt=24 hrs.

Fig. 1. Transmission electron micrographs of electron irradiation effects in the b-direction.

61

waiting time of 24 hours are displayed in Fig. 4. The data for both irradiation directions indicate an almost complete collapse of long range order in the c-direction and the formation of short-range order peaks with extended lattice spacings. These short-range order peaks occur at real lattice spacings of 1.39 nm for irradiation near the b-direction, and at 1.24 nm for irradiation near the a-direction. Within an accuracy of about 1 %, the a and b lattice parameters remain unchanged. Also the (103) type reflections are preserved (although broadened), and are especially strong following irradiation in the b-direction.

The diffraction patterns in Fig. 4 are consistent with recent observations of defect structures in this material obtained using high resolution techniques in electron microscopes at 200-400 kV. One observation[6], where electron beam damage was produced and followed, showed the expansion in the c-direction to take place, at least some of the time,

<a> IRRADIATION

G=200 BF (s=0)   100nm

E=131.5 keV
DOSE=4x10^{21}cm^{-2}
Δt=0.2 hr.

E=146.5 keV
DOSE=4x10^{21}cm^{-2}
Δt=0.2 hr.

E=146.5 keV
DOSE =4x10^{21}cm^{-2}
Δt=16 hrs.

E=159.8 keV
DOSE=4x10^{21}cm^{-2}
Δt=24 hrs.

Fig. 2. Transmission electron micrographs of electron irradiation effects in the a-direction.

in the basal plane of the original structure at the interface of a growing amorphous-appearing region. A defect structure in this same plane was observed by Zandbergen, et al.[7]. They proposed a model for the defective structure that includes an extra Cu-O "basal" plane adjacent to the original basal plane and displaced by a/2 or b/2. These $(CuO)_2$ double layers, as they call them, inserted randomly throughout a volume would produce a diffraction pattern exactly the same as for the b-irradiation in Fig. 4, including the approximate preservation of the (103) reflections and short-range order peaks at 1.39 nm.

From all these results, it seems highly probable that atoms are displaced by electron irradiations, at energies exceeding the respective thresholds. Applying relativistic kinematics[8], the maximum recoil energies for oxygen and copper atoms can be calculated. The results for the a- and b-directions (thresholds: 152 and 131 keV) are 23.9 and 20.3 eV for oxygen, 6.0 and 5.1 eV for copper, and 15 and 13 eV for copper displaced by recoil oxygen atoms. The observed motion of twin boundaries is consistent with the displacement of oxygen atoms and probable motion of an oxygen atom defect. The formation of an extra CuO layer over 24 hours suggests the additional displacement of Cu atoms, most likely by recoil oxygen atoms displaced near threshold, expecially in the b-direction irradiation. This resulting highly defective state would necessarily include high concentrations of oxygen and copper atom vacancies to produce the extra planes. The long time following irradiation needed to form this structure could be related to the long diffusion times of oxygen at room temperature[9].

<a> IRRADIATION

3 BEAM, (001) LATTICE FRINGES

 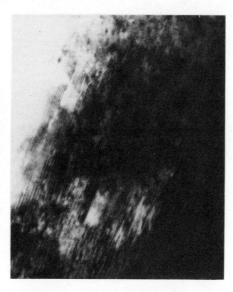

UNIRRADIATED   IRRADIATED (159.8 keV) + 24 hrs.

Fig. 3. Defect production by above-threshold irradiation and after a reaction period of 24 hours, a-direction.

Electron microscopy of single crystalline pieces of $YBa_2Cu_3O_{7-\delta}$ at various electron beam energies and as a function of electron fluence, has provided the following information on microstructural effects in this high-T_c superconductor:

- The threshold for defect production is anisotropic and depends on the crystal direction aligned parallel to the incident beam. The threshold energies were determined to be 152 and 131 keV (± 7 keV) for a- and b-axis irradiation, respectively. Preliminary data suggest, that the c-axis threshold exceeds 124 keV.

<b> IRRADIATION

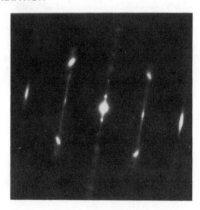

UNIRRADIATED   IRRADIATED (139.5 keV) + 24 hrs.

<a> IRRADIATION

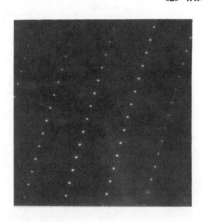

UNIRRADIATED   IRRADIATED (159.8 keV) + 24 hrs.

Fig. 4. Electron diffraction patterns of the planar faulted state formed by above-threshold electron irradiation and defect reaction within 24 hours.

- Extended defects are observed to form and grow during b-axis irradiation above the threshold energy. Above-threshold irradiation in the a-direction does not lead to visible defect production *during* irradiation.
- In both cases, slower defect motions occur over periods of 2 to 24 hours leading to highly defective states characterized by loss of long-range order in the c-direction and the appearance of short-range order peaks.
- These highly defective state is consistent with the formation of another CuO plane adjacent to the original basal plane and displaced by a/2 or b/2. The formation of this extra plane implies the displacement of Cu atoms by above-threshold electron irradiation.
- Twin boundary motion and shrinkage of twinned volume occurs during irradiation with electron energies between the two thresholds for the a- and b-directions. This effect implies a net movement of oxygen in the basal plane, in agreement with calculated recoil energies and structural considerations on the nature of twinning.
- As a practical conclusion, we wish to point out, that conventional transmission electron microscopy and high resolution electron microscopy *free* of electron beam effects, can only be carried out at beam energies ≤120 keV.

ACKNOWLEDGEMENT

One of us (H.W.Weber) wishes to express his gratitude for the hospitality, support and stimulating atmosphere experienced during an extended stay in the Materials Science Division, Argonne National Laboratory.

REFERENCES

1. H.Küpfer, I.Apfelstedt, W.Schauer, R.Flükiger, R.Meier-Hirmer, H.Wühl, H.Scheurer: Z.Physik 69 (1987) 167
2. A.Umezawa, G.W.Crabtree, J.Z.Liu, H.W.Weber, W.K.Kwok, L.H.Nunez, T.J.Moran, C.H.Sowers, H.Claus: Phys.Rev. B36 (1987) 7151
3. B.Egner, J.Geerk, H.C.Li, G.Linker, O.Meyer, B.Strehlau: Jap.J.Appl.Phys. 26-3 (1987) 2141
4. H.W.Weber, G.W.Crabtree, A.Umezawa, J.Z.Liu, L.H.Nunez: this volume, p.
5. J.Z.Liu, G.W.Crabtree, A.Umezawa, L.Zongquan: Phys.Lett. A121 (1987) 305
6. M.Chisholm, IBM, Yorktown Heights, private communication, March, 1988.
7. H.W.Zandbergen, R.Gronsky, K.Wang, and G.Thomas, preprint (submitted to Nature, January, 1988).
8. J.W.Corbett: "Electron radiation damage in semiconductors and metals", Academic Press, New York, 1966, p.13.
9. K.Fueki, this volume, p.

PHYSICAL PROPERTIES AND STRUCTURAL CHEMISTRY OF $YBa_2Cu_3O_x$

PREPARED BY A DEFINED OXYGEN SORPTION PROCEDURE

P.Meuffels, B.Rupp and H.Wenzl

Institut für Festkörperforschung
KFA Jülich
D 5170 Jülich, F.R.G.

ABSTRACT

Samples of $YBa_2Cu_3O_x$ ($6.0 < x < 7.0$) were prepared by a UHV desorption - absorption procedure. Employing this technique it is possible to obtain controlled and reproducible oxygen contents in order to study physical and structural properties correlated to the oxygen stoichiometry. Cell constants, phase transformations and superconducting transition temperatures are given as a function of oxygen content and/or temperature. Thermal desorption spectra of differently treated samples are compared and evolution of surface adsorbates and structurally bonded oxygen is discussed. The importance of in situ reloading of samples used for surface sensitive experiments is pointed out.

INTRODUCTION

The physical and structural properties of the high temperature superconducting ceramics $YBa_2Cu_3O_x$ ($6.0 < x < 7.0$) depend strongly on the oxygen content of the specimens. Moreover, in addition to the total amount of absorbed oxygen, the distribution of the oxygen atoms on the lattice sites is of considerable influence on the correlated properties, i.e. superconducting transition temperatures. For a reliable analysis of the material's properties it is thus necessary to establish a reproducible preparation technique covering the whole oxygen composition range from 6.0 to 7.0. Quenching is difficult to standardize if performed in different laboratories and does not yield equilibrium samples nor does it provide ordered oxygen distribution. The absorption - desorption procedure described below in this first report enables one to produce material with defined oxygen contents under clean conditions combined with in situ resistance measurements while loading or deloading the samples.

EXPERIMENTAL

For all of our experiments a high vacuum system described elsewhere[1] was employed. Thermal desorption experiments were performed under dynamic vacuum with a heating rate of 5K/min up to about 1040 K. The desorption products werde monitored with a quadrupole mass spectrometer. After completion of the desorption a pressure of 1.2 bar pure oxygen was adjusted in the sample chamber. The specimen was then cooled down to room temperature at 2 K/min in order to reload the specimen with oxygen. Subsequently a new desorption experiment was started in order to investigate the changes induced by this desorption- absorption procedure.

A series of $YBa_2Cu_3O_x$ alloys with varying oxygen content $0<X<1$ was prepared in a similar manner. Powdered starting material of $YBa_2Cu_3O_x$, prepared by a standard sintering procedure[1], was degassed in vacuum at 2 K/min up to 1040 K, evacuated for one hour at 1040 K and subsequently cooled to room temperature. Under this conditions an oxygen content of 6.00(3) was obtained as verified by the analysis of our desorption experiments described below. At room temperature a given amount of oxygen was admitted to the sample chamber and the temperature was raised to 1040 K again. Under slow cooling to room temperature (0.75 K/min) all oxygen was absorbed. The oxygen content of the sample at a given O_2-pressure and temperature was evaluated by ideal gas law calculations. Finally the powder was pressed into pellets and analyzed by X-ray diffraction. The superconducting transition was examined by an inductive method[1].

In situ high temperature resistance measurements were employed during these degassing and reloading experiments in order to observe phase transitions as a function of oxygen content and temperature. A thin disk of $YBa_2Cu_3O_x$ from the same material as the powder was placed close to the Pt-crucible containing the powder. Pt contact wires were bounded to four edges of this sample using Pt-cement. The resistance was measured by a four point method. One single sample was employed during all experiments. As the resistance showed the same value after each degassing run a deterioration of the sample could be excluded.

X-ray analysis (monochromatized Co $K\alpha_1$ radiation, transmission geometry, 6N Ge added as an internal standard) confirmed the structure of the samples, the amount of impurity phases detectable by X-Ray diffraction was found to be less than 3%. Calculated profile intensities based on atomic position parameters[2] and on refined cell constants were in excellent agreement with the observed spectra. The increase of the cell parameter c_0 with decreasing oxygen concentration was used as a sensitive measure do determine the actual oxygen content.

RESULTS AND DISCUSSION

Desorption studies

Thermal desorption measurements were performed in order to characterize $YBa_2Cu_3O_x$ samples and to demonstrate the importance of clean atmosphere handling and in situ reloading of samples with oxygen. Fig.1 shows the desorption spectrum of O_2, CO_2 and H_2O desorbing from a sample which has been stored in air for a few months. This sample was evacuated for 24 hours before measuring the spectrum. It was analyzed by means of X-ray diffraction indicating furthermore an inhomogenous oxygen distribution within the sample.

As can be seen in Fig.1 oxygen is not the only desorption product observed. Similar results have been reported by Keller et al.[3]. The considerable amount of water desorbing between 400K and 600K is due to chemisorbed water on the sample surface. The two maxima indicate chemisorbed water to exist in at least two states with different binding energies, most probably bound to unreacted barium oxide formed due to incomplete sintering. From 900 to 1000K CO_2 is released. We conclude that in air

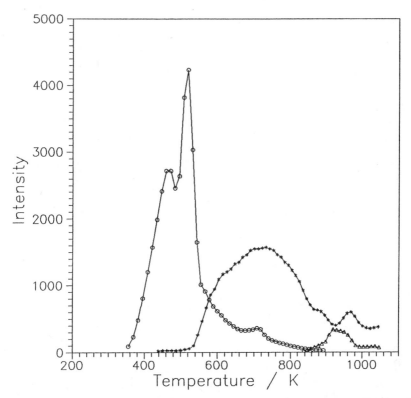

Fig. 1. Temperature programmed thermal desorption spectra (TDS) of
O_2 ($*$), H_2O (O) and CO_2 ($\triangle$) evolving from $YBa_2Cu_3O_x$
which has been stored in air for a few months (rate 5 K/min).

incompletely reacted parts of the sample and/or decomposition products on
the sample surface (most probably BaO), react with atmospheric CO_2. After
completion of this desorption run the sample was reloaded with oxygen as
described above.   X-ray diffraction analysis revealed that material produced from low quality starting material employing this method is perfectly orthorhombic with an oxygen content of 7.0. The desorption spectrum of
O_2 desorbing from this optimized sample is presented in Fig.2. Only negle-
gible amounts of H_2O and CO_2 are released, the measured intensities being
in the range of the rest gas background of the apparatus.

    Our high temperature resistance measurements during degassing in
vacuum reveal a pronounced maximum in the resistance R(T) around 950K.
Above this maximum the sample shows semiconducting behavior. Below the
maximum an abrupt change in slope of R(T) at about 890K is observed. This
kink, however, is not correlated to the orthorhombic/tetragonal (O-T)
phase transition which under these conditions should occur at temperatures
far below. This fact is evident from an extrapolation of the data given in
fig.1 of ref.[4], too. Sageev Grader et. al.[5] attributed the observed maximum to a metallic/semiconducting transition at an oxygen content of about
6.2. We wish to emphasize that the resistance in these types of experiments is governed by both the oxygen concentration dependence, dR/dX, plus
the temperature dependence, dR/dT. As a consequence, one should be cau-
tious in attributing semiconducting or metallic behavior to the various
slopes of the curves, if the concentration dependence of R(T) is not
known. We suggest that the observed abrupt change in slope of R(T) at 890K

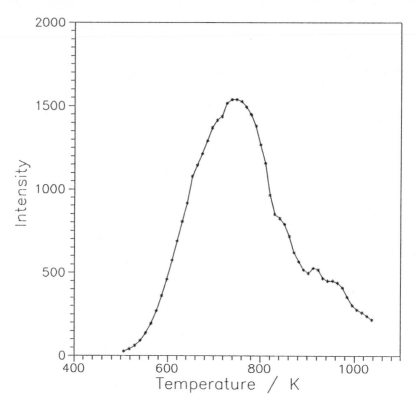

Fig. 2. TDS of O_2 desorbing from in situ reloaded $YBa_2Cu_3O_x$

(under vacuum) indicates the metallic/semiconducting transition and we
assume that this transition occurs at an oxygen content of 6.2 to 6.3 in
the sample. The further increase of R(T) up to 950K is in our opinion
governed by the oxygen loss in the sample, i.e. the increase of R(T) due
to the decrease in carrier concentration, $dR(T)/dX > 0$, overcompensates the
semiconducting behavior, $dR(T)/dT < 0$, within this temperature range.

By integrating the area under the desorption curve the total amount
of oxygen desorbed at a given temperature can be determined. Thus, as the
sample had an oxygen content of 7.0 before desorption, we can calculate a
value of about 0.75 mole oxygen atoms removed from one mole sample up to
890K. Up to about 750K, the center of the desorption peak, 0.45 mole
oxygen atoms per mole of sample are released. At this point, 6.5-6.6 mole
oxygen atoms remain in the sample. Within this concentration range the O-T
phase transition is observed by our resistance measurements while reloading the samples with oxygen.

The total amount of oxygen removed until the end temperature of 1040K
sums up to 0.9 mole of oxygen atoms per mole of sample. As a consequence,
subsequent to this degassing procedure, samples used for the preparation
of a series of $YBa_2Cu_3O_x$ alloys with varying oxygen content were held at
1040 K for another hour under vacuum in order to achieve specimens with a
reproducible oxygen content (6.0) as verified by X-ray analysis.

Summarizing, our desorption measurements show that every examination
of $YBa_2Cu_3O_x$ with surface sensitive methodes requires the removal of

surface adsorbates by degassing to obtain reliable results. Samples for this kind of investigations like XPS, UPS amd ELS should be reloaded in situ leading to a dramatic improvement of both the bulk properties and the specimen surface. A first successful application of this in situ reloading procedure is described by Sen and coworkers[6]. Fig.2 shows that oxygen can be removed readily at temperatures well below 600 K. Thus, a gentle reloading treatment which is especially required for the examination of sensitive epitaxial thin films by surface specific techniques, can be applied. On the other hand, this easy release of oxygen from the samples has to be taken into consideration when local heating under low pressures takes place (TEM, SAD).

Observation of the O-T phase transition

Measurements of the resistance of the $YBa_2Cu_3O_x$ superconductor during degassing or reloading with oxygen provide a good method to determine the high temperature order-disorder transition [4,5,7,8]. In the present investigation such measurements were employed during the preparation of the series of $YBa_2Cu_3O_x$ alloys in order to obtain the dependence of the O-T phase transition on the oxygen content. A typical result of one of these measurements is presented in Fig.3. The abrupt change in slope of the resistance, R(T), coincides with the distinct kink in the oxygen pressure curve during oxygen uptake by the sample. The abrupt slope change of R(T) can be correlated to the O-T transition as has been assumed by other groups[4,5,7,8]. Freitas and Plaskett[4] concluded from their measurements that this transition is second order in nature. The corresponding change in slope of the oxygen pressure supports the assumption, that the tetragonal and orthorhombic phases have different enthalpies for oxygen evolution or uptake[9].

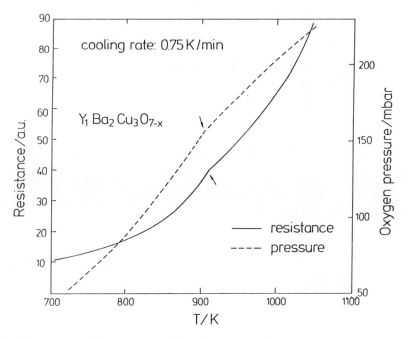

Fig. 3.   Typical results for the temperature dependence of the resistance and oxygen pressure during loading a $YBa_2Cu_3O_x$ sample with oxygen at 0.75 K/min.

The oxygen content of the samples at the O-T transition temperatures could be calculated by ideal gas law calculations, because the sample mass, sample chamber volume, pressure difference due to oxygen uptake and the effective gas temperature were known. Our results show that between 800 and 950 K the O-T transition occurs nearly at the same oxygen content of 6.55 to 6.6, i.e. we expect an almost straight line in the pseudobinary phase diagram of $YBa_2Cu_3O_x$ (temperature vs. oxygen content). The transition temperature, decreasing with decreasing oxygen pressure, is nearly independent of the oxygen content. Similar results were obtained by Jorgensen et al.[10] and O'Bryan and Gallagher[11].

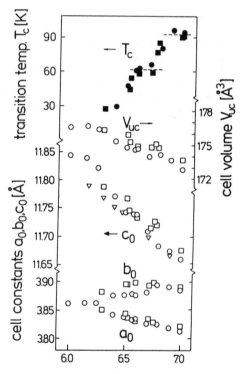

Fig. 4. Cell constants (empty symbols) and superconducting transition temperature (full symbols) as a function of oxygen content in $YBa_2Cu_3O_x$: (O): this work, (□): Cava et al.[12], (▽): Ono[13].

At the present state of the investigations, we cannot decide conclusively whether this assumption holds for lower temperatures. Cell constants based on the orthorhombic structure cell including data from Cava et al.[12] and compiled data from Ono[13] are presented in Fig.4. Despite fair agreement with the data reported in the literature, we observe deviations from smooth or continous variation of the cell constants vs. oxygen content.

These deviations are significantly larger than expected experimental errors. Moreover, profile simulations show that in some regions (6.3-6.5, appr. 6.7) there are some ambiguities as far as calculated vs. observed intensities are concerned, reflecting some inconsistency in the observed patterns. How far, and if at all, one can attribute this to a kind of phase separation, remains the subject of a further study[14]. In any case, we like to encourage multiphase profile refinements of neutron diffraction data on samples in the composition range stated above. In addition to the ambiguities in the powder patterns, rather broad superconducting transitions are observed in our work as well as by Cava et al.[12] in the same concentration regions. We confirm the existence of two distinct regions of superconducting transition temperatures of 94 and 62 K, respectivly, in good agreement with data reported previously[12].

LITERATURE

1. P.Meuffels, E.Pörschke and B.Rupp, Physica C, to be published.
2. A.W.Hewat, J.J.Capponi, C.Chaillout, M.Marezio and E.A.Hewat, Sol.State Commun., 64(3) (1987) 310.
3. S.W.Keller, K.J.Leary, A.M.Stacy and J.N.Michaels, Mat.Lett. 5(9) (1987) 375.
4. P.P.Freitas and T.S.Plaskett, Phys.Rev.B 36(10) (1987) 5723.
5. G.Sageev Grader, P.K.Gallagher and E.M.Gyorgy, Appl. Phys. Lett. 51(14) (1987) 1115.
6. P.Sen, B.Dauth, B.Rupp, Ch.Ziegler, W.Göpel and W.Gudat 8th General Conference of the Cond.Matter Div. of the European Physical Society, July 8-15th, 1988,Budapest, Hungary.
7. B.Fisher, E.Polturak, G.Koren and A.Kessel, Solid State Commun. 64(1) (1987) 87.
8. M.Gurvitch and A.T.Fiory, Phys.Rev.Lett. 59(12) (1987) 1337.
9. G.Sageev Grader and P.K.Gallagher, Adv.Ceram.Mater. 2(3B) (1987) 649.
10. J.D.Jorgensen, M.A.Beno, D.G.Hinks, L.Soderholm, K.J.Volin, R.L.Hitterman, J.D.Grace and I.K.Schuller, Phys.Rev.B 36(7) (1987) 3608.
11. H.M.O'Bryan and P.K.Gallagher, Adv.Ceram.Mater. 2(3B) (1987) 640.
12. R.J.Cava, B.Batlogg, C.H.Chen, E.A.Rietman, S.M.Zahurak and D.Werder, Phys.Rev.B 36(10) (1987) 5719.
13. A.Ono, Jap.J.Appl.Phys. 26(7) (1987) L1223.
14. B.Rupp, in progress.

CORRELATION OF T_c WITH THE RARE EARTH SIZE IN $(Ln)Ba_2Cu_3O_{6+x}$ SUPERCONDUCTORS (Ln = La, Nd, Sm, Eu, Yb " COCKTAILS") : SINGLE CRYSTAL X-RAY DIFFRACTION AND ELECTRON MICROSCOPY STUDIES

Jean Galy*, Renée Enjalbert*, Patrice Millet*
Marie José Casanove** & Christian Roucau**

* Laboratoire de Chimie de Coordination CNRS, 205 route de Narbonne, 31077 Toulouse, France - ** Laboratoire d'Optique Electronique CNRS, 29 rue Jeanne Marvig, 31055 Toulouse , France

In a previous paper, structural systematics around non-stoichiometric perovskite derivatives, $YBa_2Cu_3O_{9-n}$ with n = 1, 2, 3, were developed implying Wadsley's defects for the non-quantified n values [1]. In this series, $YBa_2Cu_3O_{6+x}$ (with $0 \leq x \leq 1$) exhibits remarkable superconductive properties and, in agreement with other reported results, a more precise domain has been found with T_c above 90 K for $0.66 \leq x \leq 1$. This orthorhombic phase is associated with high T_c values. It is well known that at high temperature, it undergoes an orthorhombic $\rightarrow$ tetragonal phase transition accompanied by O_2 loss, this phenomenon being reversible.

It was tempting to modify chemically such compounds. In the present paper, we report the substitution of rare earths for yttrium and, more particularly , rare earth mixtures which were achieved, leading to the so-called superconductive $(Ln)Ba_2Cu_3O_{6+x}$ "cocktail" phases (Ln = mixture of rare earths). The task was to follow their structural evolution and eventually the steric influence upon T_c.

$(Ln)Ba_2Cu_3O_{6+x}$ phases as powders

The first "cocktail", a mixture of five rare earths (Y, Nd, Sm, Eu, Yb: ratio 1:1:1:1:1) was obtained according to the reaction: $0.1Y_2O_3 + 0.1Nd_2O_3 + 0.1Sm_2O_3 + 0.1Eu_2O_3 + 0.1Yb_2O_3 + 2 BaCO_3 + 3 CuO \longrightarrow (Ln)Ba_2Cu_3O_{6+x}$ with (Ln) = 0.2 Y + 0.2Nd + 0.2Sm + 0.2Eu + 0.2Yb and x = 0.85 [2].

The various powders were thoroughly mixed, ground and pelletized under 10 t.cm^{-2}. After heating in air at 950°C for 12 h, the sample was treated again as previously and the resulting pellet annealed under a stream of O_2 for 8 h prior to cooling at a rate of 2°C.min^{-1}. All the "cocktails" were similarly prepared in order to obtain a homogeneous series of samples.

All the samples were characterized by X-ray powder pattern technique (CuKα radiation) and their superconductivity checked by levitation test. The above mentioned $(Ln)Ba_2Cu_3O_{6.85}$ cocktail is isostructural with $YBa_2Cu_3O_{6.87}$ (Fig. 1) and crystallizes in the orthorhombic structure. It is to be noticed that the lines corresponding to 103, 013 and 110, whose typical evolution has been enlightened in a previous paper [1], are rather thin like in $YBa_2Cu_3O_{6+x}$ (Fig. 1), indicating that a statistical distribution of the rare earth "cocktail" has occurred in-between the $[Ba_2Cu_3O_7]_n^{3n-}$ layers (Fig. 2).

The critical temperature T_c was determined from four-probe conductivity measurements on the ceramic $(Ln)Ba_2Cu_3O_{6.85}$. The beginning of the transition occurs at $T_1 = 91.2$ K while below $T_2 = 88.8$ K, the resistivity is smaller than 10^{-5} Ω which is to be compared with that of $YBa_2Cu_3O_{6.87}$ starting from $T_1 = 93$ K with $T_2 = 88$ K under the same conditions.

The cell parameters of $(Ln)Ba_2Cu_3O_{6.85}$ are: a = 3.821(2) Å, b = 3.884(2) Å, and c = 11.684(6) Å. Such parameters correspond to a weighted value of the cell parameters of the five pure phases corresponding to Y, Nd, Sm, Eu, and Yb. Some sort of "pseudo" rare earth ionic radius can be extrapolated, r = 1.055 Å. It logically takes place in-between those of the pure superconductive phases of dysprosium and of gadolinium.

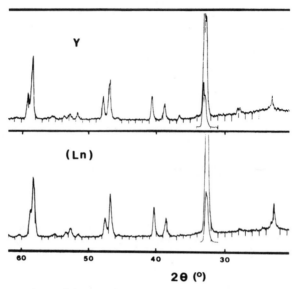

Fig .1. Comparison of the powder patterns of the $YBa_2Cu_3O_{6+x}$ (x=0.87) and $(Ln)Ba_2Cu_3O_{6+x}$ (x=0.85) phases .

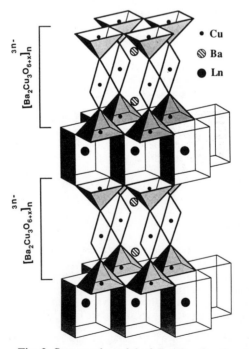

Fig. 2. Structural model of $(Ln)Ba_2Cu_3O_7$ phase

This possibility to simulate any ionic radius size or, in other words, to use "pseudo" rare earths, will be used later on.

(Ln)Ba$_2$Cu$_3$O$_{6+x}$ as single crystals

X-ray single crystal study. A single-crystal was grown using the flux method with an excess of CuO [3]. A black crystal with a prismatic shape (0.3x0.012x0.15mm) was isolated from the mass.

The crystallographic study revealed a "pseudo" tetragonal symmetry due to the already observed twinning. Actually, the orthorhombic cell, with a = 3.8748(7) Å, b = 3.8736(6) Å, c = 11.709(1) Å, could hardly be distinguished from a tetragonal one since the hkl and khl reflections coalesced. Careful numeric absorption corrections were applied on collected reflections observed up to θ = 75° for MoKα radiation. A set of some 1673 unique reflections observed with F≥3σ(F) was obtained with a good R$_{merge}$ value of 0.018 .

The structure was refined in the space group Pmmm pointed out by Capponi et al. [4] for YBa$_2$Cu$_3$O$_7$ using Fourier and Fourier difference synthesis and least-squares techniques. The final result is indicated in the following table.

Positional parameters, site occupancies
and temperature factors (Å^2) (x 10^4 for U)

(Ln)Ba$_2$Cu$_3$O$_{6.37}$

(Ln) = 0.2Y+0.2Nd+0.2Sm+0.2Eu+0.2Yb

a = 3.8748(7) Å
b = 3.8736(6) Å
c = 11.709(1) Å

Atom	occupancy	x	y	z	U11	U22	U33	U$_{eq}$.	B$_{eq}$.
(Ln)	1.	1/2	1/2	1/2	62(1)	59(1)	62(1)	61	0.48
Ba	1.	1/2	1/2	0.18793(3)	120(1)	117(1)	92(1)	110	0.87
Cu(1)	1.	0	0	0	145(4)	154(4)	99(3)	133	1.05
Cu(2)	1.	0	0	0.3567(1)	54(1)	54(1)	90(2)	66	0.52
O(1)	1.	0	0	0.1670(8)	171(25)	173(23)	621(65)	322	2.54
O(2)	1.	0	1/2	0.3773(4)	97(9)	57(7)	116(11)	90	0.71
O(3)	1.	1/2	0	0.3761(4)	68(8)	110(10)	128(12)	102	0.81
O(4)	0.23	0	1/2	0	91(51)	49(39)	60(47)	67	0.53
O(5)	0.14	1/2	0	0	84(81)	36(62)	103(101)	75	0.59

orthorhombic system
space group : Pmmm

MoKα radiation ; θ ≤75°
numeric absorption correction
R$_{merge}$ = 0.018
1673 unique reflections observed with F≥ 3σ (F)
34 parameters
R = 0.053   Rw= 0.059
w = 1/σ ^{2}(F)+0.010F^2
s = 0.79
g (ext II) = 1.09 x 10^{-6}

Selected interatomic distances (Å)

(Ln) - O(2)	2.412(3)	Cu(1) - O(1)	1.969(11)
(Ln) - O(3)	2.420(3)	Cu(1) - O(4)	1.937(1)
		Cu(1) - O(5)	1.937(1)
Ba - O(1)	2.749(5)		
Ba - O(2)	2.945(4)	Cu(2) - O(1)	2.208(11)
Ba - O(3)	2.933(4)	Cu(2) - O(2)	1.952(1)
Ba - O(4)	2.932(4)	Cu(2) - O(3)	1.951(1)
Ba - O(5)	2.913(3)		

The chemical result which can be extracted from this structural analysis is readily obtained. During the melting, some O$_2$ losses have occurred, giving a lowest value of x; i.e., (Ln)Ba$_2$Cu$_3$O$_{6.37}$. Consequently, the tendency to give a tetragonal phase similar to YBa$_2$Cu$_3$O$_6$ has drastically contracted the gap between a and b parameters.

Electron microscopy study. Electron microscopy observations have been performed in order to elucidate the local structure of the single crystals and, in particular, to verify their structural homogeneity. Small single crystals of (Ln)Ba$_2$Cu$_3$O$_{6+x}$ are fixed with silver paint on copper grids provided with a central hole. The specimens are then observed using a Philips EM400 microscope working at 120 kV and fitted with a field emission gun.

Two different phases have been found in these compounds, which is in good agreement with the X-ray studies. The first phase clearly presents an orthorhombic symmetry structure whereas the second phase has a tetragonal symmmetry structure, within the accuracy of the measurements.

I- Orthorhombic phase

A bright field image of the orthorhombic phase is displayed in figure 3. The micrograph shows evidence for high twinning. The corresponding diffraction pattern is displayed in figure 4 where the little white arrow indicates the direction of the consequent splitting of the Bragg reflection. The twin walls are parallel to the (110) or (-110) planes. Two domains separated by a wall are related by a rotation of $\pi/2$ around the c-axis.

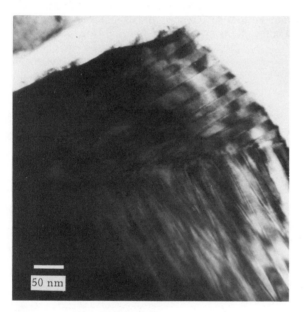

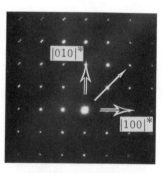

Fig. 3. Bright field image of the orthorhombic phase .

Fig. 4. Orthorhombic phase zone axis [001] .

II-Tetragonal phase.

A second set of single crystals mostly present a tetragonal symmetry structure which indicates a lower oxygen content. The micrograph displayed in figure 5 does not reveal the presence of any twin and there is no splitting of the Bragg reflections on the corresponding diffraction pattern (Fig. 6)

III-Occurrence of a superstructure in the tetragonal phase

A superstructure has been evidenced in a single crystal, at room temperature. The high resolution image of this region is displayed in figure 7. It mainly shows two sets of lattice planes belonging to the superstructure (-1/2,1/2,0) and (1/4,1/4,0) with respective interplanar distance of 5.48 Å and 10.96 Å, and the (1 0 0) planes of the tetragonal structure. All the spots have been indexed using the tetragonal cell ($a_t \cong 3.875$ Å). The extra spots define a supercell $a_t\sqrt{2} \times 2a_t\sqrt{2}$ in the (0 0 1) plane. Taking into account our experience on $YBa_2Cu_3O_{6+x}$ related compounds which also present superstructures when the oxygen content corresponds to $x \leq 0.5$, [5,6] it is reasonable to assume that this superstructure is due to the occurrence of oxygen vacancy ordering in this region of the crystal. It must be mentioned that the main part of this specimen has an orthorhombic structure. Some structural inhomogeneities have been therefore evidenced in this single crystal. The part played by possible rare earths ordering in this material is not known in spite of the good indications of the X-ray powder patterns for a statistical distribution.

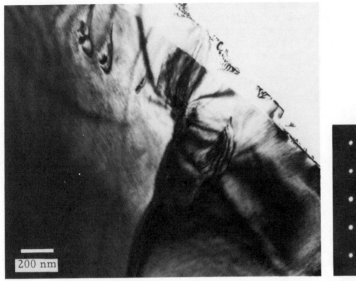
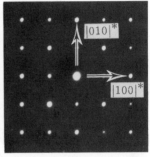

Fig. 5. Bright field image of the tetragonal phase .   Fig. 6. Tetragonal phase zone axis [001]

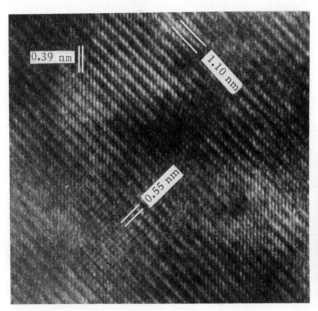
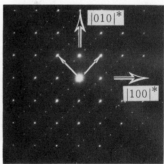

Fig.7. High resolution image of the superstructure   Fig. 8. Diffraction pattern corres-
occurring in the tetragonal phase .   ponding to figure 7 .

Rare earth "cocktails" and T_c

It has been shown that phases of rare earth "cocktails" of the type $(Ln)Ba_2Cu_3O_{6+x}$ behave very much like the pure compound typical of this superconductive family; i.e., $YBa_2Cu_3O_{6+x}$.

The behaviour of two simple "cocktail" families is briefly reported here: $(La_{1-y}Y_y)Ba_2Cu_3O_{6+x}$ and $(La_{1-y}Nd_y)Ba_2Cu_3O_{6+x}$.

The oxygen content is calculated from the weight change after each heat treatment. The pure lanthanum compound (y=0, x=0.7), prepared under the above chemical conditions, exhibits a T_c below the liquid nitrogen temperature (i.e.,71K). For the other compounds, the oxygen concentration is x=0.85± 0.05. It is readily seen in figure 9 that it is possible to improve the critical temperature T_c with a small amount of Nd or Y, and to raise it above 80K.

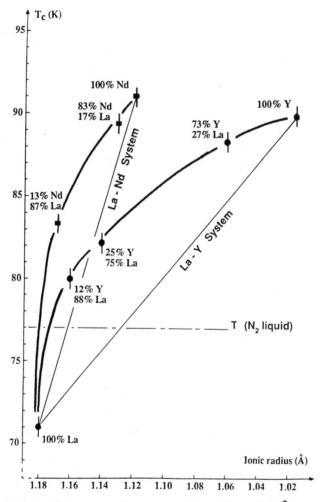

Fig. 9. T_c evolution versus the average ionic radii of the Ln^{3+}. La - Y and La - Nd systems. The midpoint superconducting transitions are indicated.

Such a result calls upon the influence of the ionic radius of the rare earths on T_c and shows that "pseudo" rare earths with the apparent size of cerium (III) and praseodymium (III) give superconductive phases.

[1] J. Galy, R. Enjalbert, P. Millet, C. Faulmann & P. Cassoux. J. Solid State Chem., in press.
[2] P. Millet, R. Enjalbert, J. Galy, C. Faulmann, P. Cassoux, H. Rakoto & S. Askénazy. C. R. Acad. Sci., in press.
[3] H. Takei, H. Takeya, Y. Iye, T. Tamegai & F. Sakai. Jpn. J. Appl. Phys. 26:L1425 (1987).

[4] J. J. Capponi, C. Chaillout, A. W. Hewat, P. Lejay, M. Marezio, N. Nguyen, B. Raveau, J. L. Soubeyroux, J. L. Tholence & R. Tournier. Europhys. Lett. 3 : 301 (1987).
[5] M. J. Casanove & C. Roucau. Europhys. Lett., submitted.
[6] C. Chaillout, M. A. Alario-Franco, J. J. Capponi, J. Chenavas, J. L. Hodeau & M. Marezio. Phys. Rev. (Brief Comm.) , in press.

ELECTRICAL RESISTIVITY, MAGNETIC SUSCEPTIBILITY, AND INFRARED SPECTRA OF

SUPERCONDUCTING RBa$_2$Cu$_3$O$_7$ WITH R = Y,Sc,Tm,Ho,Eu,Nd,Gd

[1]H.Gruber, [1]E.Krautz, [2]H.P.Fritzer, [2]K.Gatterer, [2]G.Sperka
[2]W.Sitte, and [3]A.Popitsch

[1]Institut für Festkörperphysik, [2]Institut für Physikalische
und Theoretische Chemie, Technische Universität; [3]Institut
für Anorganische Chemie, Universität Graz; A-8010 Graz
Austria

INTRODUCTION

Superconductivity in the metallic oxide system Ba-La-Cu-O at
temperatures above 30 K has first been reported in October 1986 by
J.G.Bednorz and K.A.Müller[1]. By partial substitution of Ba with Sr
C.W.Chu et al[2] obtained a compound with a transition temperature T$_c$ of
about 55 K. In February 1987 M.K.Wu, C.W.Chu et al[3] found a sensationally
high T$_c$ of about 93 K in the oxide system Y-Ba-Cu-O. The superconducting
phase in this system was identified by R.J.Cava et al.[4] as a layered
orthorhombic distorted, oxygen-deficient perovskite structure with the
composition YBa$_2$Cu$_3$O$_{9-d}$ where d = 2.1.

In this contribution we report results of measurements of the
electrical resistivity, magnetic susceptibility, and infrared transmission
spectra in the RBa$_2$Cu$_3$O$_{7-x}$ superconducting system with R = Y, Sc, Tm, Ho,
Eu, Nd, Gd, and Ti. Furthermore, the samples were characterized by X-ray
powder diffraction patterns. The data presented here are extensions of two
earlier contributions[5,6] in this exciting field. In addition, we report on
preliminary results of fluorine-doped samples.

EXPERIMENTAL

All samples were prepared from commercially available powders of the
rare earth elements as metals or metal oxides together with CuO and BaCO$_3$
in the desired composition. Each finely ground mixture was first calcined
for 6 hours at 920°C in a pure oxygen stream (99.99 % O$_2$), then ground,
pelletized and sintered again for 16 hours at 920°C in oxygen, and slowly
cooled down to room temperature. This treatment has been repeated twice
with a slow cooling rate down to 200°C in oxygen. All operations were
performed in platinum and alumina crucibles to avoid side reactions. The
composition and crystal structure of the compounds were confirmed by
electron microprobe analysis , X-ray powder diffraction patterns, and
infrared-spectra, respectively. The electrical resistivity measurements
were made by the standard four-point technique. The magnetic
susceptibility has been measured at various inductions and temperatures
with a modified Faraday method using a sensitive pendulum system with the
apparatus SUS 10 manufactured by H.Paar KG, Graz, Austria. Superpure KCl
(E.Merck, Darmstadt) has been used as calibration substance. The
vibrational transmission spectra were measured on a Perkin-Elmer grating
spectrometer (at 300 K and 10 K) using CsI or polyethylene pellets as
sample holders.

RESULTS AND DISCUSSION

Resistivity. Fig.1 shows the temperature dependence of the electrical resistivity of $RBa_2Cu_3O_{7-d}$ with R = Y, Eu, Ho, Tm, Gd. The samples containing Ho and Tm exhibit the lowest resistivity values.

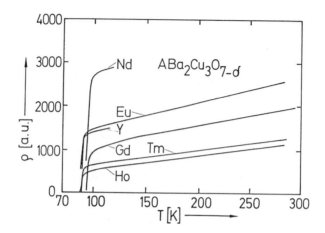

Fig.1 Temperature dependence of the resistivity for the densely sintered samples of $RBa_2Cu_3O_{7-d}$ with R = Y,Eu,Tm,Ho,Gd

Among all samples $NdBa_2Cu_3O_{7-d}$ shows the highest transition temperature and the highest resistivity. Partial substitution of the rare earth elements Ho and Tm by Sc has only a small influence[5,9] on the transition temperature whereas the values of electrical resistivity increased distinctly (Fig.2).

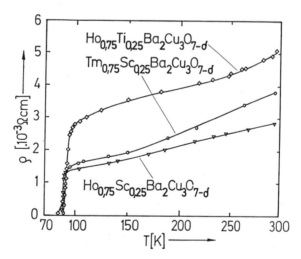

Fig.2 Temperature dependence of the resistivity of $RSc_{0.25}Ba_2Cu_3O_{7-d}$ with R = Ho,Tm and $Ho_{0.75}Ti_{0.25}Ba_2Cu_3O_{7-d}$

Further measurements[7] indicated that increasing Sc content leads to an increase of the resistivity. The repetition of the heat treatment in oxygen reveals a remarkable lowering of the resistivity. In case of partial substitution of Ho by Ti, for example for $Ho_{0.75}Ti_{0.25}Ba_2Cu_3O_{7-d}$, the onset transition temperature remains at 92 K, but the transition is less sharp and extends down to 85 K. Fig.3 shows the data for the F- and the Ag-doped compounds. The T_c-values are slightly increased to about 98 K for the Ag-doped sample.

X-ray diffraction. The main differences between the undoped $NdBa_2Cu_3O_7$ and the fluorine-doped compound occur in the (200), (020) and (006) region. The c-axis increases with the amount of F-. Similar differences were observed in the powder spectra of the Ag- and Ti-substituted compounds. We conclude that the substitution increases the c-axis rather than the a- or b-axis.

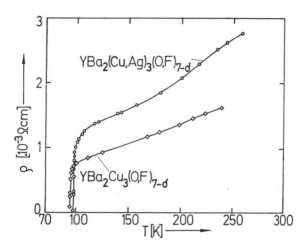

Fig.3   Temperature dependence of the resistivity of
$YBa_2(Cu,Ag)_3(O,F)_{7-d}$ and $YBa_2Cu_3(O,F)_{7-d}$

Magnetic susceptibility. Fig.4-6 show the temperature dependence of the magnetic gram susceptibility X and 1/X, respectively, for various representatives of the $RBa_2Cu_3O_{7-d}$ system.

For R = Ho, Tm, Sc (Fig.4) the susceptibility decreases with temperature due to the behaviour of the Pauli paramagnetism for the trivalent ions. Furthermore, the incorporation of Sc is related to a further decrease of X, probably as a result of the corresponding decrease of the Pauli and Van Vleck paramagnetism. The Tm,Sc-compound shows a strong increase of X with the magnetic induction B indicating a partial destruction of the superconducting phase.

Doping with Ti instead of Sc reveals a very pronounced field dependence (Fig.5). The paramagnetic contribution is already dominant at the small induction B of 1.05 T. It can be seen clearly that the T_c-values decrease with increasing B.

Fig.6 shows the magnetic behaviour of $YBa_2(Cu,Ag)_3(O,F)_{7-d}$ at various B-values. The striking feature is the sharp decrease of T_c with increasing values of B, which is not accompanied by an increase of paramagnetic contributions. The X-behaviour at temperatures below the various T_c-values seems to indicate a magnetic phase transition. It is not impossible that the superconducting sample changes to a normal conducting material.

Vibrational spectra. Infrared transmission spectra should be especially useful to provide experimental data in the region of the superconducting energy gap at about 200 cm-1. It is clear that the knowledge of the vibrational properties would help considerably to decide between the various proposed mechanism of superconductivity. Difficulties arise since,

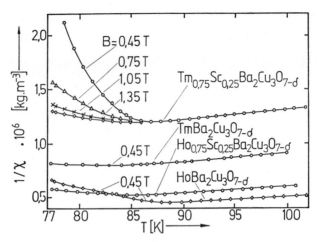

Fig.4 Temperature dependence of the inverse magnetic
susceptibility for $RBa_2Cu_3O_{7-d}$ with R = Ho,Tm and partial
substitution by Sc at different magnetic induction

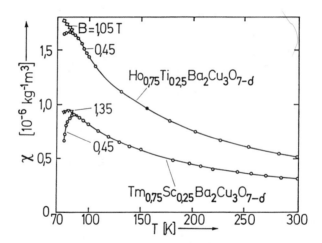

Fig.5 Temperature dependence of magnetic susceptibility for
$Ho_{0.75}Ti_{0.25}Ba_2Cu_3O_{7-d}$ and $Tm_{0.75}Sc_{0.2}Ba_2Cu_3O_{7-d}$ at
different magnetic induction

in general, infrared absorption bands tend to disappear when the ceramic
material becomes metallic. However, some vibrational features survive
(superimposed on the background of the spectrum) in the superconducting
phase despite the metallic behaviour. Since these materials are highly
anisotropic, this may be an explanation for the observation of vibrations
in these "metals".

Spectroscopic studies of some of our superconductors yield rather
surprising results as is shown in Fig.7. It is worth mentioning that the
undoped $YBa_2Cu_3O_{7-d}$ (D) shows very weak vibrational modes in the gap
region, i.e., around 200 – 300 cm[-1]. The observed modes between 250 and
350 cm[-1] (Fig.7) correspond to bond-bending vibrations of the O–Cu–O–Cu–O–
type where the superconducting pairing is likely to take place[10,11].
Despite the fact that this spectral region is strongly affected in the
doped compounds their superconducting properties remain.

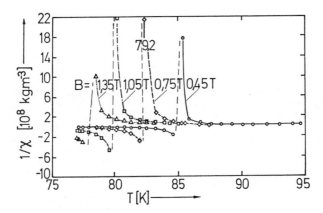

Fig.6　Temperature dependence of the inverse magnetic susceptibility of $YBa_2(Cu,Ag)_3(O,F)_{7-d}$ at increasing magnetic induction

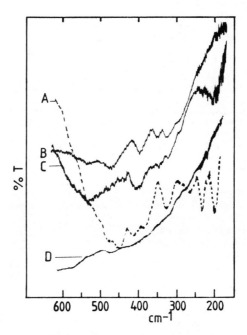

Fig.7　Transmission spectra of A = $YBa_2 Cu_3 O_6 F$, B = $Ho_{0.75}Sc_{0.25}Ba_2 Cu_3 O_{7-d}$, C = $Ho_{0.75}Ti_{0.25}Ba_2 Cu_3 O_{7-d}$, and D = $YBa_2 Cu_3 O_{7-d}$ in the wavenumber range from 200 cm^{-1} to 600 cm^{-1}

ACKNOWLEDGEMENTS

H.P.F.expresses his thanks to K.A.Müller (Zürich) and W.H.Stevens (Nottingham) for stimulating discussions at the IX.Jahn-Teller Symposium, Nottingham (UK), September 1987. W.S. expresses his thanks to J.Maier (Max-Planck-Institut für Festkörperforschung, Stuttgart) for valuable discussions at the 6th Int.Conference on Solid State Ionics, Garmisch-Partenkirchen (GFR), September 1987. We are also indebted to Firma A.Paar KG, Graz, Austria, for providing the SUS 10 magnetic equipment.

REFERENCES

1. J.G.Bednorz and K.A.Müller, Possible High-T_c Superconductivity in the Ba-La-Cu-O System, Z.Phys.B 64:189 (1986).
2. C.W.Chu, P.H.Hor, R.L.Meng, L.Gao, Z.J.Huang, and Y.Q.Wang, Superconductivity at 54,5 K in the Lanthanum-Barium-Copper-Oxide system, Phys.Rev.Lett.58:405 (1987).
3. M.K.Wu, J.R.Ashborn, C.J.Torng, P.H.Hor, R.L.Meng, L.Gao, Z.J.Huang, and C.W.Chu, Superconductivity at 93 K in a New Mixed-Phase Y-Ba-Cu-O Compound System at Ambient Pressure, Phys.Rev.Lett.58:908 (1987).
4. R.J.Cava, B.Batlogg, R.B.van Dover, D.W.Murphy, S.Sunshine, T.Siegrist, J.P.Remeika, E.A.Rietman, S.Zahurak, and G.P.Espinosa, Bulk Superconductivity at 91 K in Single-Phase Oxygen-Deficient Perovskite $Ba_2YCu_3O_{9-d}$, Phys.Rev.Lett. 58:1676 (1987).
5. H.Gruber, E.Krautz, K.Gatterer, and H.P.Fritzer, Temperaturabhängigkeit und Strukturabhängigkeit der elektrischen Leitfähigkeit und magnetischen Suszeptibilität von Metalloxiden, Annual Meeting of the Austrian Physical Society, September 23-25, 1987, Graz, Austria.
6. H.Gruber, H.P.Fritzer, K.Gatterer, G.Sperka, und W.Sitte, Einfluß der Struktur und Nichtstöchiometrie auf die physikalischen Eigenschaften von Metalloxiden, 7th Annual Meeting of the Austrian Chemical Society, November 3-6, 1987, Wien. Austria.
7. H.Gruber, H.P.Fritzer, K.Gatterer, W.Sitte, and A.Popitsch, Physical Properties and Chemical Doping of Superconducting Perovskites, to be presented at the International Conference on High-Temperature Superconductors and Materials and Mechanism of Superconductivity, February 29 - March 4, 1988, Interlaken, Switzerland.
8. L.Pauling, Influence of Valence, Electronegativity, Atomic Radii and Crest-Trough Interactions with Phonons on the High-Temperature Copper-Oxide Superconductors, Phys.Rev.Lett. 59:225 (1987).
9. M.Murakami, H.Teshina, M.Morita, S.Matsuda, H.Hamada, M.Matsuo, T.Innzuka, R.Suyama, M.Sugiyama, K.Sawano, H.Kubo, M.Abe, and M.Nagumo, Superconductivity in High-T_c Rare Earth Ba-Cu-Oxide, J.Appl.Phys.Japan 26:1061 (1987).
10. D.Cahen, Z.Moisi, and M.Schwartz, Effects of Ag/Cu Substitution in $YBa_2Cu_3O_7$ superconductors, Mat.Res.Bull.22:1581 (1987).
11. M.Cardona, R.Liu, C.Thomsen, M.Bauer, L.Genzel, W.König, A.Wittlin, U.Amador, M.Barahona, F.Fernandez, C.Otero, and R.Saez, Infrared and Raman Spectra of the New Super-conducting Cuprate Perovskites $MBa_2Cu_3O_7$, M = Nd,Dy,Er,Tm, Solid State Commun.65(1):71 (1987).

HEAT CAPACITY MEASUREMENTS ON SOME HIGH-T_c SUPERCONDUCTORS

M.Slaski*, B.D.Dunlap[#], M.V.Nevitt[#], and Z.Sungaila

*Instytut Fizyki, P.K. 30-084 Krakow, ul.Podchorazych 1
Poland
[#]Argonne National Laboratory, Argonne, IL 60439, USA

INTRODUCTION

The heat capacity has been obtained for superconducting samples of $La_{1.85}Sr_{.15}CuO_4$[1] and $REBa_2Cu_3O_{7-x}$[2,3] (RE = Gd,Dy,Ho,Er). The $\Delta C/\gamma T$ values of 2 to 5 for the $La_{1.85}Sr_{0.15}CuO_4$ compound places this material well into the strong coupling limit. From the data for $REBa_2Cu_3O_{7-x}$, the degeneracies of the crystal field ground state are determined and the magnetic moment in the ground state and the energy separation to the first excited crystal field state are estimated. The magnetic transitions are observed at T_m = 2.2, 0.95, 0.17, and 0.59 K for the Gd, Dy, Ho, and Er compounds, respectively. The antiferromagnetic ordering of the Gd ions is independent of the presence or the absence of superconductivity and is unchanged in both insulating and metallic systems.

PREPARATION, STRUCTURE AND SUPERCONDUCTING PROPERTIES

$La_{1.85}Sr_{.15}CuO_4$

Nitrates of the metals, citric acid and ethylene glycol were heated to 80 - 90 °C. The resultant solution was dried, calcinated, ball milled, pressed and sintered at 1150 °C. Finally, the material was annealed in oxygen at 500 °C for several hours. We obtained single-phase material as determined both by x-ray and neutron diffraction with a superconducting transition temperature of 31 K (from ac susceptibility and resistivity data).

$REBa_2Cu_3O_{7-x}$

Mixtures of the RE-sesquioxides, $BaCO_3$ and CuO, with the metals in a 1:2:3 mole ratio, were sintered in a pure flowing O_2 atmosphere for 10 hours at 975 °C, annealed at 650 °C for 6 hours and furnace cooled (100 °C/hour). Part of the $GdBa_2Cu_3O_{7-x}$ compounds was annealed in nitrogen at 850 °C for 12 hours to reduce the oxygen content and to change the crystal structure from orthorhombic to tetragonal. Samples prepared in this way give single-phase materials as determined by x-ray and neutron diffraction and superconducting transition temperatures of 92 - 93 K determined from ac susceptibility and resistivity data. Resistivity data show that the materials remain superconducting down to T = 0.6 K.

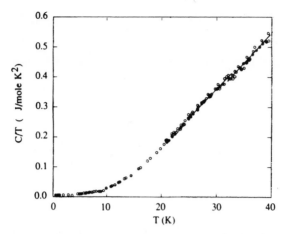

Fig. 1. Heat capacity of La$_{1.85}$Sr$_{.15}$CuO$_4$. The straight lines are drawn to show the change in C/T at T$_c$.

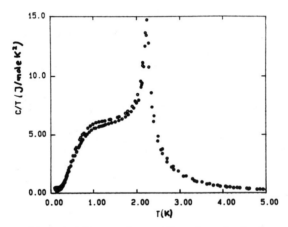

Fig. 2. Heat capacity of GdBa$_2$Cu$_3$O$_{7-x}$. The closed circles correspond to the superconducting compound and the open circles to the non-superconducting one.

RESULTS

La$_{1.85}$Sr$_{.15}$CuO$_4$

Heat capacity data have been obtained on two samples of 2.47 and 10.8 g in two different calorimeters, which have been described[4,5] previously. The absolute values, determined independently from the two sets of measurements, agreed very well and showed good reproducibility. In the vinicity of T$_c$ the heat capacity is very large due to phonon contributions, and the observation of an anomaly due to superconductivity is expected to be difficult, but inspection of the data shows that the effect is clearly present (Fig. 1). In individual temperature scans, the change of slope at T$_c$ becomes apparent and we estimate $\Delta C/T_c$ = 20±5 mJ/mole K^2.

Attempts to model the data with an exponential temperature dependence for the electronic contribution below T$_c$ and a Debye function for the phonons do not give reasonable results. It is, therefore, impossible to derive γ from these data at present. Taking γ values in the range 4.9 < γ < 7.3 mJ/mole K^2 from ref.6, we obtain $\beta = \Delta C/\gamma T_c$ = 2 - 5.

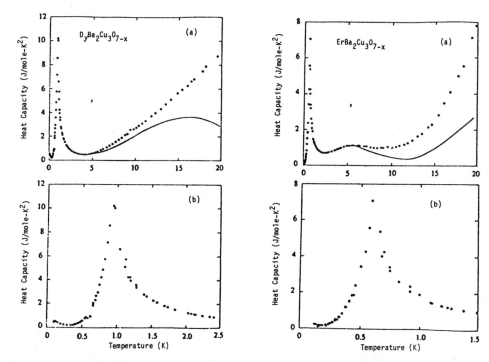

Fig. 3. Heat capacity of $DyBa_2Cu_3O_{7-x}$. The solid line shows the result of subtracting the heat capacity of $YBa_2Cu_3O_{7-x}$.

Fig. 4. Heat capacity of $ErBa_2Cu_3O_{7-x}$. The solid line shows the result of substracting the heat capacity of $YBa_2Cu_3O_{7-x}$.

$REBa_2Cu_3O_{7-x}$

Heat capacity data has been obtained for all compounds in the temperature range T = 0.1 - 20 K (Figs. 2 - 5). In order to correct for the lattice heat capacity contribution, we have made use of the available data for $YBa_2Cu_3O_{6.8}$[7]. The $YBa_2Cu_3O_{6.8}$ material has the same T_c as the other compounds and an identical Cu-O sublattice, but does not contain paramagnetic ions. We therefore assume that we may use the measured heat capacity of $YBa_2Cu_3O_{6.8}$ to obtain all contributions, which are not specifically related to the paramagnetic ions.

At the lowest temperatures a small upturn is seen in the heat capacity (for the Dy, Ho, and Er compounds) which is presumably the onset of a Schottky anomaly due to nuclear hyperfine splitting. At higher temperatures we observe the magnetic transitions. The magnetic ordering occurs at 2.2, 0.95, 0.17 and 0.59 K for the Gd (both compounds; semiconducting and superconducting), Dy, Ho and Er compounds, respectively.

The heat capacity data of $GdBa_2Cu_3O_{7-x}$ did not fall simply to zero below T_m, but showed a shoulder centered near 1 K. The origin of this additional feature is unknown but may arise from an incommensurate-commensurate transition in the magnetic ordering. This feature is very well reproduced (in both compounds) in the current data and should be considered to be an intrinsic feature of the material. Thus, the magnetic structure is unchanged in spite of the fact that the material has gone from metallic to semiconducting behaviour, and has at the same time lost its superconductivity. Since the magnetism is independent of the conductivity of the compound, it is clear that magnetic interactions mediated by conduction electrons (RKKY) are unimportant.

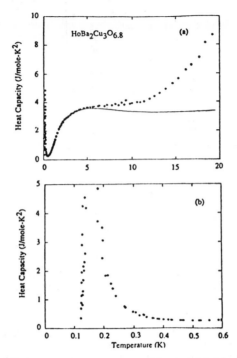

Fig. 5. Heat capacity of $HoBa_2Cu_3O_{7-x}$. The solid line shows the result of subtracting the heat capacity of $YBa_2Cu_3O_{7-x}$.

Table 1.

Properties of $REBa_2Cu_3O_{6.8}$

RE	T_c^a	μ_{eff}^b	T_m^b	Energy	Moment	Multi-plicity
	(K)	(μ_B)	(K)	(K)	(μ_B)	
Dy	93.5(1.4)	10.4(2)	0.95(2)	0	≈ 6	2
				40		
Ho	92.0(2.4)	10.2(3)	0.17(3)	0	≈ 0	1
				8		
Er	91.8(1.8)	9.1(1)	0.59(2)	0	≈ 4.5	2
				90		2

[a]Values given are for the midpoint of the 10 - 90% resistive transition, and the values in parantheses give the 10 - 90% width.
[b]Values in parantheses give estimated errors in the last significant figure.

CONCLUSIONS

$La_{1.85}Sr_{.15}CuO_4$

Heat capacity data show the appearance of bulk superconductivity. Analysis of the data gives $\Delta C/\gamma T_c = 2 - 5$ which places this material well into the strong coupling limit of superconductivity.

$REBa_2Cu_3O_{7-x}$

Magnetic transitions are observed at 2.2, 0.95, 0.17, and 0.59 K for the Gd, Dy, Ho and Er compounds, respectively (Table 1). It is argued that they are due predominantly to dipolar interactions. The magnetic state coexists with superconductivity in all cases. By integrating C/T vs. T for the Dy, Ho, and Er compounds (after subtracting the heat capacity of $YBa_2Cu_3O_{7-x}$), we can obtain the temperature dependence of the magnetic entropy.

For the $DyBa_2Cu_3O_{7-x}$ compound, we conclude that the magnetic ordering occurs within a crystal-field doublet ground state. The heat capacity at higher temperatures is in very good agreement with a Schottky peak due to a second doublet lying at 40 K above the ground state.

For $HoBa_2Cu_3O_{7-x}$ we assume that the ordering occurs out of a singlet ground state (combined nuclear-electronic ordering). A number of crystal field levels lie within 50 K of one another (the second singlet state lies approximately 8 K above the ground state).

For the $ErBa_2Cu_3O_{7-x}$ compound, we assume ordering to occur out of a doublet ground state. The second Kramers doublet is located 90 K above the ground state. The antiferromagnetic transition temperature for the Gd ions in the material is the same for both the semiconducting and the superconducting compounds. This provides direct evidence for the lack of magnetic interactions mediated by conduction electrons in these materials.

REFERENCES

1. B.D.Dunlap, M.V.Nevitt, M.Slaski, T.E.Klippert, Z.Sungaila, A.G.McKale, D.W.Capone, R.B.Poeppel and B.K.Flandermeyer, Phys.Rev. B.35 (1987) 7210
2. B.D.Dunlap, M.Slaski, D.H.Hinks, L.Soderholm, M.Beno, K.Zhang, C.Segre, C.W.Crabtree, W.K.Kwok, S.K.Malik, I.K.Schuller, J.D.Jorgensen and Z.Sungaila, J.Magn.Magn.Mat. 68 (1987) L139.
3. B.D.Dunlap, M.Slaski, Z.Sungaila, D.G.Hinks, K.Zhang, C.Segre, S.K.Malik and E.E.Alp, Phys.Lett. (submitted)
4. H.A.Kierstead, B.D.Dunlap, S.K.Malik, A.M.Umarji and G.K.Shenoy, Phys.Rev. B32 (1985) 135
5. R.J.Trainor, G.S.Knapp, M.B.Brodsky, G.J.Pokorny and R.B.Snyder, Rev.Sci.Instr. 46 (1975) 95
6. W.E.Kwok, G.W.Crabtree, D.G.Hinks, D.W.Capone, J.D.Jorgensen and K.Zhang, Phys.Rev. B35 (1987) 5343
7. M.V.Nevitt and T.E.Klippert, Phys.Rev. (submitted)

SPECIFIC HEAT OF THE SUPERCONDUCTING COMPOUND $YBa_2Cu_3O_{7-x}$

IN MAGNETIC FIELDS

G.K.Panova, M.N.Khlopkin, N.A.Chernoplekov, A.A.Shikov
and I.V.Kurchatov
Institute of Atomic Energy, Moscow, USSR
B.Fogarassy, L.Granassy, S.Pekker, and L.Mihaly
Central Research Institute for Physics, Budapest, Hungary

ABSTRACT

In the present work the specific heat of a $YBa_2Cu_3O_{7-x}$ specimen was studied experimentally in the temperature range from 3 to 85 K and in magnetic fields of 0 and 8 T. Several characteristic parameters were obtained, including the superconducting transition temperature T_c, the Debye temperature θ and the Sommerfeld parameter γ.

EXPERIMENTAL

The investigated specimen was sintered at the CRIP (Budapest) in the form of a disk, 10 mm in dia. and 2.2 mm thick. Its mass amounted to 0.628 g. The x-ray structure analysis has shown that the specimen is single-phase and has a orthorhombic structure with the parameters: a = 3.822 A; b = 3.891 A; c = 11.677 A. the specimen was porous, its resistivity amounted to 5 mOhm cm at 300 K and to 2.4 mOhm cm at 100 K. The temperature dependences of the electrical resistance at the superconducting transition, detected by noncontact and four-contact methods, are given on the right-hand side of Fig. 1.

RESULTS

Fig. 1 presents the specific heat of $YBa_2Cu_3O_{7-x}$ at the superconducting transition in the absence of a field and in a field of 8 T. There are jumps (marked by arrows in Fig. 1) observed in the temperature dependence of the specific heat, which correspond to the transition into the superconducting state (at H = 0) or the mixed state (at H = 8 T). In both cases the magnitude of the jump $\Delta C/T_c$ has a value of about 2 mJ/g-atom K^2. The critical temperatures of the superconducting transition, T_c, determined from the respective jumps of the specific heat, amount to: 79 K in zero field and 73 K in a field of 8 T. We can determine the magnitude of the derivative of the upper critical magnetic field at T_c: $dH_{c2}/dT = -1.4$ T/K. An estimate of the upper critical field $H_{c2}(0)$ from the relation $H_{c2}(0) = 0.7\ T_c|dH_{c2}/dT|$ yields for $H_{c2}(0) = 80T$.

Fig. 2 presents the specific heat of $YBa_2Cu_3O_{7-x}$ in the temperature range from 3 to 12 K using the coordinates C/T vs. T^2 without a magnetic field and in a field of 8 T. Below 10 K the temperature dependence of the specific heat is satisfactorily described by the law: $C = \gamma_H T + \beta T^3$, which corresponds to a straight line in the above mentioned coordinates. The magnitude of the coefficient β is practically independent of magnetic field and amounts to 0.034 mJ/g-atom K^2, which corresponds to a Debye temperature $\Theta = 384$ K. The coefficient γ_H depends to a significant extent on the magnetic field. It increases from 0.6 mJ/g-atom K^2 in zero field to 1.0 mJ/g-atom K^2 in a field of 8 T, allowing us to estimate the value: $d\gamma_H/dH = 0.05$ mJ/g-atom K^2 T.

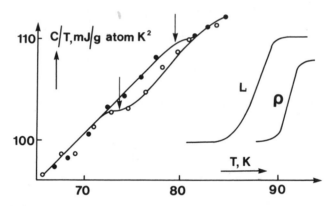

Fig. 1. Specific heat of $YBa_2Cu_3O_{7-x}$ at the superconducting transition in the absence of a magnetic field (·) and in a field of 8 T (o) and temperature dependence of the electrical resistance at the superconducting transition, detected by a noncontact (L) and a four-contact (ρ) method.

DISCUSSION

It has been shown theoretically[1] and experimentally[2], that in a type II superconductor in the mixed state, the coefficient γ_H depends linearly on the magnetic field H, reaching (at $H = H_{C2}(0)$) the magnitude of the Sommerfeld parameter γ, which characterizes the density of electron states at the Fermi level (renormalized by the electron-phonon interaction). It allows us to estimate the magnitude of γ from the relation: $\gamma = H_{C2}(0) \cdot d\gamma_H/dH$, which yields for the investigated specimen: $\gamma = 4$ mJ/g-atom K^2.

In the frame of the Bardeen-Cooper-Schrieffer theory of superconductivity, the values of the specific heat jump $\Delta C/T_c$ and of the Sommerfeld parameter γ, are related by the relation: $\Delta C/T_c = 1.43\ \gamma$. However, in $YBa_2Cu_3O_{7-x}$ the superconducting transition takes place at such high temperatures that, according to the estimates[3], the electron-

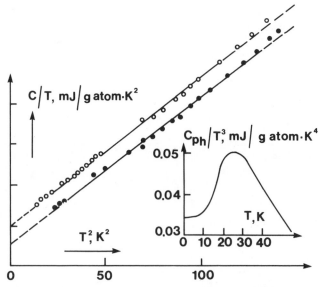

Fig. 2. The specific heat of $YBa_2Cu_3O_{7-x}$ using the coordinates C/T vs. T^2 without a magnetic field ($\cdot$) and in a field of 8 T (o). The inset shows the temperature dependence of the $YBa_2Cu_3O_{7-x}$ phonon specific heat in the C_{ph}/T^3 vs. T coordinates.

phonon interaction fails to renormalize the electronic specific heat. Assuming that in this case the BCS relation is satisfied, but only for a non-renormalized density of electronic states, the magnitude of the electron-phonon coupling constant λ can be estimated from the relation: $\Delta C/T_c = 1.43\ \gamma/(1+\lambda)$, which yields $\lambda \approx 2$ for $YBa_2Cu_3O_{7-x}$.

In the temperature range above 10 K the temperature dependence of the specific heat is observed to deviate from the law: $C = \gamma_H T \pm \beta\ T^3$. The inset of Fig. 2 shows the temperature dependence of the $YBa_2Cu_3O_{7-x}$ phonon specific heat plotted as C_{ph}/T^3 versus T. This dependence reaches a maximum at $T \approx 23$ K, which proves that an intensive peak exists in the density of phonon states at an energy of the order of 12 meV.

REFERENCES

1. K.Maki, Phys.Rev. 139, 702 (1965)
2. G.R.Stewart, B.L.Brandt, Phys.Rev.B 29, 3908 (1984)
3. V.Z.Kresin, G.O.Zaitsev, ZhETF 74, 1887 (1978)

EFFECT OF Cu/Zn SUBSTITUTION UPON SUPERCONDUCTIVITY IN

$YBa_2(Cu_{1-x}Zn_x)_3O_{7-\delta}$ AND $La_{1.85}Sr_{0.15}Cu_{1-x}Zn_xO_{4-\delta}$

K. Remschnig[*], P. Rogl[*], E. Bauer, R. Eibler
G. Hilscher, H. Kirchmayr and N. Pillmayr

[*]Institut f. Physikalische Chemie der Univ. Wien
A-1090 Wien, Währingerstraße 42
Institut f. Experimentalphysik der T.U. Wien
A-1040 Wien, Wiedner Hauptstraße 8-10, Austria

INTRODUCTION

The phenomenon of high-Tc superconductivity in the two families of ceramic Cu-oxides (LaBaCuO[1], YBaCuO[2]) generated a wide spectrum of theoretical models which indicate that experimental data of these materials are of prime interest. According to recent bandstructure calculations[3,4] the main contribution to the density of states at the Fermi energy comes from the antibonding combinations of the $Cu-d_{x^2-y^2}$ - orbitals with the p_x- and p_y-orbitals of the neighbouring O-atoms. Therefore we substituted Cu by Zn whose ionic size and orbital structure is close to Cu and present results on the structural chemistry, electrical resistivity and heat capacity of well characterized samples.

MATERIALS AND METHODS

Samples of average composition $YBa_2(Cu_{1-x}Zn_x)_3O_7$ and $La_{1.85}Sr_{0.15}Cu_{1-x}Zn_xO_4$ were prepared from appropriate amounts of Y_2O_3, $BaCO_3$, La_2O_3, CuO and ZnO, all 99.99% pure; the materials were mixed and pressed into pellets of 8mm diameter at $30kgcm^{-2}$. Then they were heated in an alumina boat in a tubular furnace in air to 940°C or 1100°C (for YBaCuO and LaSrCuO respectively) in 8hrs, held at the temperature for 24hrs, and cooled to room temperature, reground, recompacted, fired once more for 24hrs at 940/1100°C and LaSrCuO finally "quenched" to room temperature. Subsequently several samples as labeled in the figures were additionally heated for 24hrs at 880/900°C under a stream of oxygen and slowly cooled.

Lattice parameters and standard deviations were evaluated by a least squares extrapolation method from room temperature Guinier-Huber powder photographs, using monochromatized $CuK\alpha_1$ - radiation and employing an internal standard of 99.9999% pure Ge (a_{Ge} = 5.657906 Å). Ac-susceptibility measurements were performed with a standard lock-in technique at 81Hz with a field amplitude of 1 Oersted. For electrical resistivity measurements a standard 4-point technique was employed. The heat capacity of samples with a mass of about 2g was measured over the temperature range 1.5 – 60K in an automated adiabatic calorimeter employing the Nernst step heating method.

RESULTS AND DISCUSSION

In contrast to the rather extended solid solubility of Zn in $La_{1.85}Sr_{0.15}Cu_{1-x}Zn_xO_4$ (0.0 < x < 0.5), X-ray analysis of the samples $YBa_2(Cu_{1-x}Zn_x)_3O_{7-\delta}$ revealed the existence of a rather limited solid solution up to values of x < 0.1 (see fig. 1). For all samples x < 0.1 Guinier powder patterns were completely indexed on the basis of a primitive orthorhombic unit cell consistent with the structure type of $YBa_2Cu_3O_{7-\delta}$[5]. The variation of the unit cell dimensions as a function of the Cu/Zn – exchange shows an overall increase of the lattice parameters with increasing Zn-content. However, at small values of x = 0.01 the a and c parameters reveal a small but significant minimum. As far as the existence of this minimum is concerned there is no difference between samples

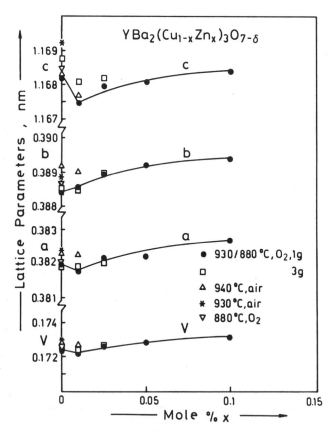

Fig.1. The lattice parameters a,b,c and the unit cell volume as a function of the Zn content x. The various annealing conditions are as labeled. □ corresponds to a 3g sample with the same condition as for □.

annealed and slowly cooled in air or under a stream of one bar of oxygen, the unit cell dimensions, however, being larger in air than after heat treatment in oxygen. This phenomenon has been recently observed[6] and was interpreted as due to a higher concentration of Cu^{3+} and thus a closer bonding between Cu^{3+}-O in the O-richer oxygen annealed material. Due to the close X-ray scattering factors of Zn and Cu no distinction can be made from X-ray analysis concerning the crystallographic substitutional mode of Cu/Zn replacement. Final decisions will be derived from neutron diffraction experiments whether a general statistical distribution of (Cu, Zn)-atoms or a preferential occupation by Zn of one of the two Cu sites is more appropriate. From the lattice parameter dependence as a function of Zn-content, however, a preferential or partially preferential occupation seems to be more likely in order to explain the occurrence of the minimum by two antagonistic influences:
i) the general increase of the unit cell volume by replacement of Cu by the larger Zn atoms
ii) a decrease of the unit cell dimensions by a change in the oxygen content and/or even more speculative by Cu^+/Zn^{2+} substitution.

Fig. 2 shows the temperature dependence of the electrical resistivity for $YBa_2(Cu_{1-x}Zn_x)_3O_7$ with $0 < x < 0.1$, while fig. 3 presents the ac-susceptibility of four samples with Zn concentrations as labeled as a function of temperature. The narrowest transition width (10%-90% ΔT_c) was obtained for the undoped sample of about 2.6 K whereas in Zn substituted specimens the transition width increases with raising Zn content from 7.5K to 11.5K. The resistive transitionwidths are in all cases somewhat smaller than those of the corresponding susceptibility measurements. It is obvious from fig. 2 that the normal state resistivity exhibits a linear temperature

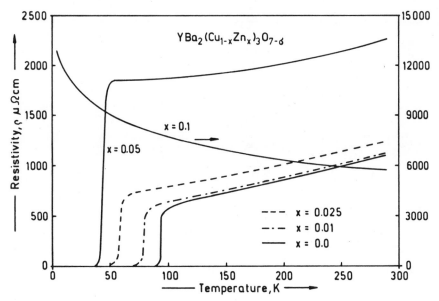

Fig.2. Temperature dependence of the electrical resistivity
for various x values x = 0.0, 0.01, 0.025, 0.05, 0.1.

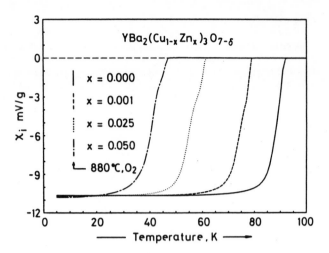

Fig.3. Temperature dependence of the ac-susceptibility for various x values.

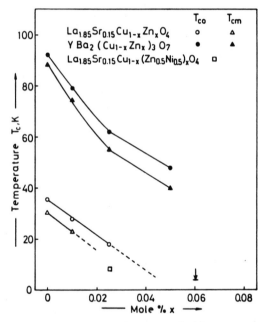

Fig.4. Concentration dependence of the superconducting
transition temperature Tc, Tco = Tc onset, Tcm =
Tc midpoint; ↓ means no transition observed down
to 4.2K

dependence. The slope R = $d\rho/dT$ in units of $(\mu\Omega cm/K)$
decreases: 2.58, 2.41, 2.16 for x = 0.0, 0.01, 0.025
respectively. In the LaSrCuO system we also observed this
linear temperature dependence of ρ but at higher temperatures
T > 150K. Zn substitution increases in both cases the

resistivity in the normal state whereby with raising Zn
content a negative temperature coefficient becomes of growing
importance. In this connection it is noteworthy that the high
temperature linear ρ (T) dependence, which is preserved up to
higher Zn concentrations in YBaCuO than in the LaSrCuO system
was used[7] as an argument that the electron-phonon enhancement
factor λ is rather small in both systems.

The initial decrease of the superconducting transition
temperature Tc on substituting Cu by Zn is for YBaCuO larger
than for LaSrCuO (dTc/dx = 11.5 and 8.5K/Znmol%). This is seen
in fig. 4, which plots Tc as a function of the Zn
concentration x. Superconductivity is suppressed below 4.2K in
the YBaCuO and LaSrCuO system for x = 0.06 and x = 0.1
respectively. The latter finding is in agreement with recent
investigations on the effect of transition metal substitution
on superconductivity of YBaCuO[8,9] and EuBaCuO[10]

Isoelectronic substitution of Cu by equal amounts of Zn
and Ni $(Ni_{0.5}Zn_{0.5})x$ causes a larger reduction of Tc than the
equivalent amount of Zn in LaSrCuO (x = 0.025 ΔTc-onset =
-10K) whereas just the opposite occurs in YBaCuO: for x =
0.025 we find Tc = 69K and 55K for the isoelectronic and Zn
substitution respectively giving ΔTc =+14K. In comparison with
other transition metal substitutions Zn was found to have the
largest ability to suppress superconductivity in YBaCuO, even
larger than the well known pair breaking effect of the
magnetic 3d-elements which was explained by the reduction of
N(Ef) as d-electrons fill up the antibonding dx^2-y^2 – band[8].
This effect should not occur on employing an isoelectronic
substitution of Cu and will be discussed together with the
heat capacity results later.

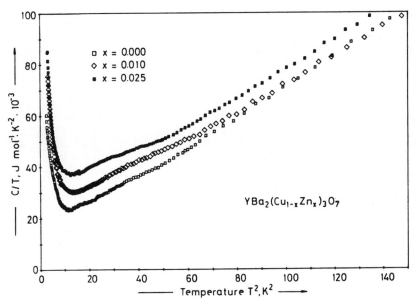

Fig.5. Low temperature specific heat in a C/T against T^2
representation for various x values.

The results of the low temperature specific heat measurements are displayed in fig. 5 in a conventional C/T against T^2 presentation. Below 3.3K the data exhibit a remarkable upturn which cannot satisfactorily be described in terms of a high temperature part of a Schottky anomaly ($C_{sch} \propto A/T^2$). From the nearly linear low temperature regime ($11K^2$–$60K^2$) of this plot we extrapolate as in the LaSrCuO system[11] a finite C/T value at $T = 0$, which we denote with γ^* in order to clearly distinguish between this term in the superconducting state and the normal state electronic specific heat coefficient γ. The γ^*-values in units of mJ/Cu-molK^2 increase: 5.6, 7.8, 10.6 for $x = 0.0$, 0.01, 0.025 respectively while β the coefficient of the T^3 term – correlated with the Debye temperature – decreases in the same sequence of Zn contents: 0.166, 0.158, 0.133 mJ/Cu-molK^4. (1Cu-mol$_*$ = 222g ,397.6g for YBaCuO and LaSrCuO respectively). The γ and β value of the parent compound are in good agreement with Ayache et al.[12] and Junod et al.[13]. However, using a high temperature ($T^2 > 60K^2$) extrapolation we end up with $\gamma^* = 3$mJ/Cu-molK^2. A remarkable low γ^*-value of about 0.83 but with a comparable β value as above has been obtained for a well characterized sample kindly supported by Schultz[14]. The significant difference between both samples with almost the same Tc and transition width ΔTc = 3K appears to be the low temperature behavior: for the latter specimen the low temperature upturn occurs already at about 40K^2.

Concerning the minimum of the lattice constants in fig. 1 at 1% Zn we may speculate that the high temperature slope of the C/T against T^2 plot reflects this trend qualitatively. In fig. 5 only the crossover of both curves can be seen at higher temperatures, however, β of $x = 0.01$ is lower than that of $x = 0.0$ and 0.025 for $T^2 > 150K^2$. A lower β value corresponds to a larger Debye temperature which is in a first approximation related with a smaller volume and is thus observed for $x = 0.01$.

On substituting Cu by Zn a general trend emerges for both the YBaCuO and the LaSrCuO systems: γ^* rises significantly, in particular for the latter system from 2.5 to 10.4 mJ/Cu-molK^2 for $x = 0.0$ and 0.025 respectively. Under the assumption of an almost linear increase of γ^* upon replacement of Cu by a transition metal in YBaCuO ($d\gamma^*/dx = 2.1$mJ/(Cu-molK^2.mol%Zn)) we estimate γ^* for a specimen with 10% Fe to be 26 mJ/Cu$_{0.9}$Fe$_{0.1}$-molK^2. The experimental value obtained for this particular composition – kindly supplied by Fellner[15]- amounts to 24 in the same units. This appears to be a rather systematic trend and cannot simply be attributed to a normal state electronic heat coefficient γ of an impurity phase, since then, the second phase though not detectable by X-rays should exhibit an unreasonable high γ value comparable to a heavy fermion system.

According to these observations we suppose that the isoelectronic substitution of Cu (by equal amounts of Zn and Ni) yields a similar effect upon γ^*, since the degradation of superconductivity is of the same order though not equivalent to Zn. As already mentioned Xiao et al.[16] explained the suppression of superconductivity in terms of filling the antibonding $d_{x^2-y^2}$ by valence electrons supplied from the alloyed Zn content. In our experiment, however, the mean valence electron concentration remains constant but superconductivity is degraded. This finding cannot be used as a direct argument against the rather convincing suggestion of Xiao et al., since on the local atomic scale the effect of bandfilling is presumably of major importance.

In conclusion we state that the results of these recent investigations[8,9,10,16] concerning the suppression of superconductivity by transition elements are largely consistent with those of the present study. The effect of filling up the antibonding d-band on a local scale appears to be of importance for the reduction of Tc rather than the influence of the mean valence electron concentration. The specific heat "linear" term γ^* in the superconducting state rises substantially with growing Cu/Zn substitution in both the YBaCuO and the LaSrCuO systems; its origin, however, is not yet resolved.

ACKNOWLEDGEMENT: P. R. wants to express his gratitude to Auer-Remy Vertriebsges. Seltene Erden, GmbH, Hamburg, BRD. Two of us (G.H., N.P.) are indebted to the Fonds zur Förderung der wissenschaftlichen Forschung in Österreich (Proj. Nr. 6104).

REFERENCES
1. J. G. Bednorz, and K. A. Müller, Z. Phys. B 64:189 (1986)
2. M. K. Wu, J. R. Ashburn, C. J. Torng, P. H. Hor, R. L. Meng, L. Gao, Z. J. Huang, Y. Z. Wang, and C. W. Chu, Phys. Rev. Lett. 58:908 (1987)
3. L. F. Mattheiss, Phys. Rev. Lett. 58:1028 (1987)
4. W. M. Temmermann, Z. Szotek, P. J. Durham, G. M. Stocks, and P. A. Sterne, J. Phys. F 17:L319 (1987)
5. J. B. Jorgensen, M. A. Beno, D. G. Hinks, L. Söderholm, K. V. Volin, R. L. Hittermann, J. D. Graze, I. K. Schuller, G. U. Segre, K. Zhang, and M. S. Kleefisch, Phys. Rev. B 36:3608 (1987)
6. J. M. Tarascon, W. R. McKinnon, L. H. Greene, G. W. Hull, and E. M. Vogel, Phys. Rev. B 36:227 (1987)
7. M. Gurvitch, and A. T. Fiory, Phys. Rev. Lett. 59:1337 (1987)
8. G. Xiao, F. H. Streitz, A. Gavrin, Y. W. Du, and C. L. Chien, Phys. Rev. B 35:8782 (1987)
9. P. Mandal, A. Poddar, P. Choudhury, A. N. Das, and B. Ghosh, J. Phys. C 20:L953 (1987)
10. S. B. Oseroff, D. C. Vier, J. F. Smyth, C. T. Salling, S. Schultz, Y. Dalichaouch, B. W. Lee, M. B. Maple, Z. Fisk, J. D. Thompson, J. L. Smith, and E. Zirngiebl, Sol. State Commun. 64:241 (1987)
11. G. Hilscher, N. Pillmayr, E. Bauer, R. Eibler, K. Remschnig, and P. r̄ gl, to be published
12. C. Ayache, B. Barbara, E. Bonjour, R. Calemczuk, M. Couach, J. H. Henry, and J. Rossat-Mignod, Sol. State Commun. 64:247 (1987)
13. A. Junod, A. Bezinge, D. Cattani, J. Cors, M. Decroux, Ø. Fischer, P. Genoud, L. Hoffmann, J. L. Jorda, J. Muller, and E. Walker, Japan. J. Appl. Phys. 26:1119 (1987)
14. L. Schultz, Siemens Forschungslabor Erlangen, FRG
15. I. Felner, Racah Institute, Hebrew Univ. Israel
16. S. X. Dou, N. Savvides, X. Y. Sun, A. J. Bourdillon, C. C. Sorrell, J. P. Zhou, and K. E. Easterling, J. Phys. C 20:L1003 (1987)

EFFECT OF COMPOSITION AND ATOMIC SUBSTITUTIONS ON STRUCTURE AND SUPERCONDUCTIVITY OF $Y_1Ba_2Cu_3O_x$

M.M.Oleksienko, N.N.Matyushenko, L.F.Verkhorobin
V.V.Derevyanko, V.N.Golovin, and S.D.Lavrinenko

Institute of Physics & Technology, the Ukrainian
Academy of Sciences, Kharkov 310108, USSR

In the present work, ceramic samples of Y-Ba-Cu-O of a varied composition were synthesized in order to identify a superconducting phase and to construct the phase diagram of the system. From the viewpoint of elucidating the nature of high temperature superconductivity, an important role is given to experiments studying the influence of the substitution for elements in different lattice sites of high temperature superconductors on crystal structure, electric and magnetic properties. The replacement of atoms in the starting compound $Y_1Ba_2Cu_3O_x$ by the atoms of elements differing in mass, size, charge and spin may reveal the role of different mechanisms responsible for high temperature superconductivity. In the present work, the influence of the total and partial atomic substitutions of rare-earth elements Dy, Ho, Er, Pr, etc. for Y, and Sr, Fe for Ba and Cu, respectively, has been investigated.

High-T_c superconducting samples were synthesized from powders of Y_2O_3, $BaCO_3$ and CuO using the usual ceramics technology. The final sintering of pellets was performed in air or oxygen atmosphere at temperatures of 900-975° C. After the sintering the samples were slowly cooled in the furnace. The procedure of preparing high-T_c superconductors with atomic substitution was essentially the same as previously described[1,2] . The crystal structure was investigated using the X-ray diffraction technique (DRON-UM1 diffractometer, CuKα radiation). The resistivity was measured by the usual four-probe method. The superconducting transition temperature was deduced from resistivity measurements or by the induction method using an a.c. bridge R 591.

The comparison of T_c values for ceramic samples of different compositions has shown that the critical temperature of the superconducting transition is observed in the samples near $YBa_2Cu_3O_y$. As X-ray diffraction data indicate, there exist, at least, three quaternary compounds in the range of compositions investigated (see Fig. 1). The simplest X-ray diffraction patterns have revealed $YBa_2Cu_3O_x$, $YBa_3Cu_2O_z$, Y_2BaCuO_y. The first two phases showed a black colour, and the last one - green. We have made indexing for these compounds, defining more exactly the lattice constants by the least-squares method, the reflection spectrum was analysed with a BESM-6 computer using the programs of ref.[3] . We have found that $Y_1Ba_2Cu_3O_x$ has an orthorhombic lattice with the parameters a = 3.826(2)A, b = 3.881(1)A, c = 11.699(2)A; Y_2BaCuO_y has an orthorhombic lattice with a = 12.189(5)A, b = 7.141(1)A, c = 5.660(1)A, and $YBa_3Cu_2O_z$ has a tetragonal lattice with a = 5.793(1)A and c = 8.320(1)A. Note that the phase diagram cross section shown in Fig. 1 for Y-Ba-Cu-O is not isothermal, since the cooling from the reaction temperature (950°C) was slow. Yet, in a number of samples, $YBa_2Cu_3O_x$ had a high-temperature (tetragonal) modification. In this case, the samples were not superconducting (down to 60 K). This is observed, in particular, for many samples involved

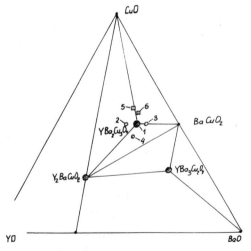

Fig. 1. Phase diagram of the system Y-Ba-Cu-O.

by composition into the triangle $CuO-YBa_2Cu_3O_x-CuBaO_2$ and
having a low melting temperature. In samples with apparent
traces of slight melting, the $YBa_2Cu_3O_x$ phase usually has
the tetragonal modification (shown by squares in Fig. 1).
The analysis of a superconducting transition in the samples
lying in different phase regions around $YBa_2Cu_3O_x$ has shown
the narrowest transition for pellets of a stoichiometric composition (Fig. 2, curve 1). The samples with a composition
approaching stoichiometry inside the triangle $CuO-YBa_2Cu_3O_4-$
Y_2BaCuO_y also demonstrate a rather high transition temperature
(Fig. 2, curve 2), whereas the lowest transition characteristics were observed in the samples lying in the triangle
$CuO-YBa_2Cu_3O_x-CuBaO_2$ (Fig. 2, curves 5,6). Thus, for the
sample being on the line $CuO-YBa_2CU_3O_x$ the transition is
strongly widened (Fig. 2, curve 5), and for the sample lying
inside the triangle the onset of the transition temperature
is essentially shifted to lower temperatures (Fig. 2, curve
6). X-ray diffraction studies indicated the phase $Y_1Ba_2Cu_3O_x$
to have a rhombic structure in this sample. However, the
lattice parameters and intensities of split peaks differ from
those of a stoichiometric sample. The substitution of heavy
rare-earth metals for Y brought no essential changes in T_c,
which is in agreement with the known data[1] . Neither have we
observed any change of the rhombic structure typical of
$Y_1Ba_2Cu_3O_x$. We succeeded in observing the transition to a
superconducting state in $PrBa_2Cu_3O_{7-\delta}$ at $T_c^{on} \sim 80K$(see Fig. 3).

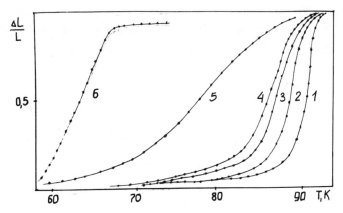

Fig. 2. T_c variation of different samples.
The composition of samples 1-6 is
presented in Fig. 1.

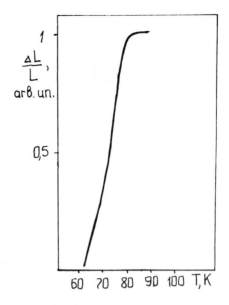

Fig. 3. Induction measurements of
T_c for the system Pr-Ba-Cu-O.

(The replacement of Y by Pr was considered previously to result in the loss of superconductivity[1]). Data on the existence of superconductivity in the system Pr-Ba-Cu-O with $T_c^{on} \sim 90$ K and $T_c^{on} \sim 69$ K, $\Delta T_c \sim 35$ K have been reported in ref.[4] and ref.[5], respectively.

The substitution of Sr atoms for a half of Ba atoms has led to some T_c decrease and an extended superconducting transition ($T_c^{on} \sim 85$ K, $\Delta T_c \sim 8$K). This result is inconsistent with the data on the increase of T_c up to 338 K[6] in $Y(BaSr)_2Cu_3O_{7-\delta}$. In our case, the compound has a clearly defined rhombic lattice, whereas in[6], a mixture of rhombic and tetragonal phases was observed (the latter phase probably accounted for the metastable state with zero resistance).

The data of most interest were obtained on investigating the system $YBa_2Cu_{3-x}Fe_xO_{7-\delta}$ ($0 \leqslant x \leqslant 0.24$), i.e., with 8% of Cu replaced by Fe atoms. Results of the induction measurements are illustrated in Fig. 4. With the substitution for 0.5-2.0% Cu, the critical temperature changes weakly, and the widening

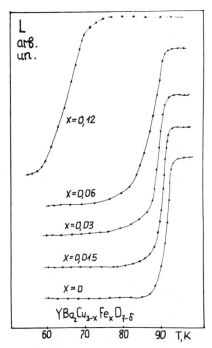

Fig. 4. Change in T_c of the system $YBa_2Cu_{3-x}Fe_xO_{7-\delta}$ with copper replaced by iron atoms.

of the superconducting transition is insignificant (these data are in good agreement with resistivity measurements). With 4% of Cu replaced by Fe atoms, T_c is sharply decreased, and with the 8% Fe substitution the compound remains in the normal state at $T \geqslant 50$ K. Characteristically, with a substitution for 2% Cu the rhombic structure of the compound changes into tetragonal, and the transition from one-dimensional superconductivity to the two-dimensional one is accompanied by an abrupt T_c decrease. As the Fe content grows, the metal-like behaviour of resistivity (nearly a linear dependence of $R(T)$) gradually turns into semiconductor one, which indicates the localization of charge carriers. Since the mass and size of Fe atoms are close to those of Cu atoms, it can be suggested that the different spin states of the atoms play a crucial role in the evolution of structure, kinetic and magnetic properties after the substitution of elements.

REFERENCES

1. S. Ohshima and T. Wakiyama, Superconducting and structural properties of the new $Ba_{1-x}Ln_xCuO_{3-y}$ compound system (Ln= La, Ce, Pr, Nd, Sm, Eu, Gd, Tb, Dy, Ho, Er and Yb). Jap. J. Appl. Phys. 26: L815 (1987).
2. J. Xiao, F.H. Streitz, A. Savrin, Y.W. Du, and C.I. Chiem, Effect of transition metal elements on the superconductivity of Y-Ba-Cu-O, Phys. Rev. B35: 8782 (1987).
3. V.P. Serykh, Kompleks programm dlya strukturnogo analiza polikristallov, Preprint KhFTI AN UkrSSR N85-8, Kharkov (1985).
4. M. Murakami et al., Superconductivity in high T_c Rare earth Ba-Cu-oxide, Jap. J. Appl. Phys. 26: 1061 (1987).
5. N.E. Alekseevskij et al., Problema vysokotemperaturnoj sverkhprovodimosti, Inform. Materialy, Ural. Otd. AN SSSR, Sverdlovsk. Pt. 1: 88 (1987).
6. H. Ihara, N. Terada, and M. Ia, Possibility of superconductivity at 65°C at Sr-Ba-Y-Cu-O system, Jap. J. Appl. Phys. 26: 169 (1987).

TRANSPORT AND SPECTROSCOPIC PROPERTIES OF UNDOPED AND Zn-DOPED CERAMIC SUPERCONDUCTORS

H. Kuzmany, M. Matus, and E. Faulques

Institut für Festkörperphysik, Universität Wien and
Ludwig Boltzmann Institut für Festkörperphysik in
Wien, Strudlhofg.4, A-1090 Vienna, Austria and

K. Remschnig and P. Rogl

Institut für Physikalische Chemie, Universität Wien
Währingerstr. 38, A-1090 Vienna, Austria

Abstract: Structural, transport, and spectroscopic properties of the ceramic superconductor $La_{2-x}Sr_xCuO_4$ and $YBa_2Cu_3O_{7-\delta}$ have been investigated as a function of a Cu/Zn-substitution. An increase of the in-plane lattice parameters in a(b) direction and a decrease of the parameter in c direction is correlated to changes in vibrational frequencies for the in plane and out of plane Cu-O stretching modes. T_c was found to decrease dramatically with increasing Zn-substitution. For the temperature dependence of the resistivity, Mott's model for hopping conduction could be applied for the metallic, as well as for the semiconducting temperature and doping range.

INTRODUCTION

Since the discovery of superconductivity at temperatures reasonably above the boiling point of liquid nitrogen in several layered perovskite type materials[1,2] the main interest focused on two series of compounds: the $La_{2-x}Sr_xCuO_4$ and the $YBa_2Cu_3O_{7-\delta}$ systems [3]. One possibility to learn about the mechanism of conduction in these materials is to study the influence of substitution of components. We report structural, transport and spectroscopic properties of Cu/Zn-substituted ceramics. The results for $YBa_2Cu_3O_{7-\delta}$ are compared with similar experiments on oxygen deficient samples [4] and the results from the conductivity measurements are related to predictions from a MOTT-HUBBARD model for transport properties [5].

EXPERIMENTAL

The substituted ceramics were prepared by repeated grinding and sintering stoichiometric quantities of the starting materials in air. Except for the highest Zn-substitution (16 %) in the Y-compound always single phase material was obtained.

Standard experimental techniques have been used for the X-ray powder diffraction analysis [6] and for the determination of the conductivity, the IR and the Raman spectra. Interatomic distances have been derived assuming atom positions as obtained in[7] from neutron diffraction experiments. For a random Cu/Zn substitution excellent agreement was found between calculated and observed x-ray intensities. The IR-experiments were performed with compressed KBr-pellets on a Fourier spectrometer.

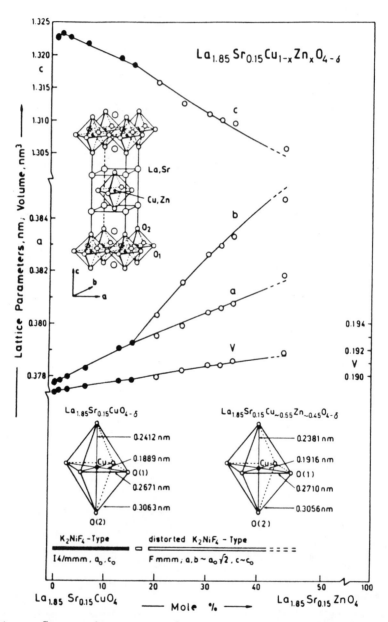

Fig.1   Structural parameters of $La_{1.85}Sr_{0.15}CuO_4$ for various doping concentrations with Zn

RESULTS

Structure:

Doping $La_{1.85}Sr_{0.15}CuO_4$ with Zn resulted in a characteristic change of the lattice parameters and consequently in a change of the Cu-O distances as shown in Fig.1. For up to 15 % Zn-substitution the b.c.t. K_2NiF_4-type structure is retained with increasing base plane unit cell parameters a (b), wheras c decreases. At Zn-concentrations higher than $\delta = 0.15$, a tetragonal to orthorhombic lattice distortion has been revealed ($Fmmm$ or $Cmca$), and beyond ca 45 % the single phase nature of the material is lost.

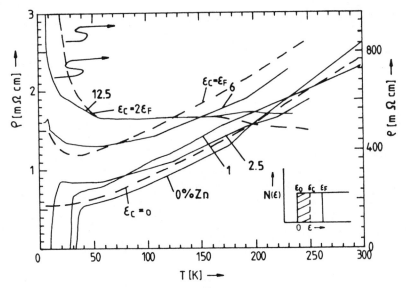

Fig.2   Specific resistivity of $La_{1.85}Sr_{0.15}Cu_{1-\delta}Zn_\delta O_4$ for various values of δ. The dashed curves are as calculated from a MOTT-HUBBARD model for transport in a system with localized electronic orbitals. Insert: schematic drawing of the two-dimensonal density of states; ϵ_o band edge, ϵ_c mobility edge, ϵ_F FERMI-level.

Transport:

The transport properties are dramatically changed by the Cu/Zn-substitution. Fig.2 shows the specific resistivity vs temperature for various Zn-concentrations. For the unsubstituted and weakly substituted material a metallic temperature dependence of the resistivity and a transition to superconductivity at low enough temperatures for Zn-concentrations less than 6 % is observed. With increasing substitution a transition to a semiconducting temperature dependence of the resistivity occurs particularly at low temperatures. Very similar results have been obtained for Zn substitution in $YBa_2Cu_3O_{7-\delta}$. Concentrations up to 8 % ($\delta = 0.24$) yielded single phases with a transition to superconductivity. Higher Zn-concentrations could not be incorporated in the same phase and led to a semiconducting behaviour. Fig.3 shows the progressive decrease of transition temperature with increasing Zn substitution for the two compounds in comparison with the behaviour of oxygen deficient $YBa_2Cu_3O_{7-\delta}$. The less dramatic quenching of the superconductivity in the Y-compound may be due to the distribution of the substituted atoms on two lattice positions (three atoms) from which only one is immediately involved in the process of superconductivity.

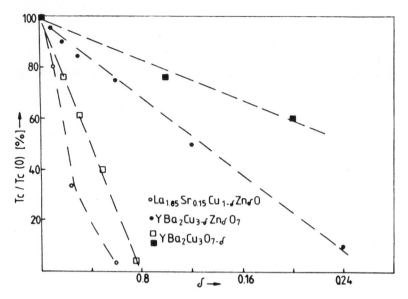

Fig.3   Relative change of transition temperature vs concentration of substitution or oxygen deficiency. The open and full squares correspond to an oxygen deficiency evaluated for the total oxygen concentration and for the concentration in the linear chains only, respectively.

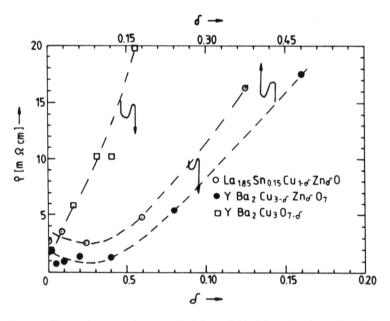

Fig.4   Room temperature resistivity of Cu/Zn-substituted and oxygen deficient ceramic superconductors vs doping concentration.

Interesting details concerning the influence of the substitution can be drawn from the change of the room temperature resistivity vs concentration of substitution as shown in Fig.4. For both ceramics substitution initially leads to a decrease of the resistivity. Only for progressive doping the resistivity increases. The increase is again much stronger for the La-compound as compared to the Y-compound. With respect to oxygen doping no decrease but a rather steep increase of the resistivity is observed. This increase becomes even stronger for total oxygen concentrations above 7 %.

Spectroscopy:

Raman spectra for the Zn-substituted Y-compound were very similar to those of the unsubstituted material except for a strong broadening of the line at 502 cm^{-1}. No reliable Raman spectra for the La-compound could be obtained. In the frequency range between 400 cm^{-1} and 1000 cm^{-1} the IR-absorption spectra showed a strong line at 502 cm^{-1} which shifted upwards and broadened with increasing substitution. For high doping concentrations a new line starts to appear at 683 cm^{-1} as shown in Fig.5. This line is very close to the second IR-absorption line at 590 cm^{-1} which occurs only in

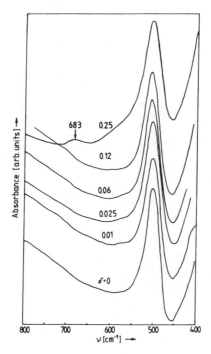

Fig.5　　IR-absorption of $La_{1.85}Sr_{0.15}Cu_{1-\delta}Zn_\delta O_4$ for various Zn-concentrations.

the insulating orthorhombic phase of the La-compounds [8]. In order to study the response of this line on the Zn-substitution we investigated also Zn-substituted LaSr-compounds with a lower Sr concentration.

Fig.6 summarizes the frequency shifts of the IR absorption lines in the two compounds as a function of Zn-concentration. The average lineshift for the $LaSr_{0.15}$-compound is +0.2 cm^{-1} per % substitution. Approximately the same upshift is observed for the 508 cm^{-1} line in $LaSr_{0.05}$ but the second line in this compound exhibits a well expressed downshift. For the Y-compound no change of the IR spectrum with Zn-concentration was observed except for an increase of the main line at 580 cm^{-1}. This is in contrast to the response of the compound to oxygen doping where dramatic changes of the lines have been found [4].

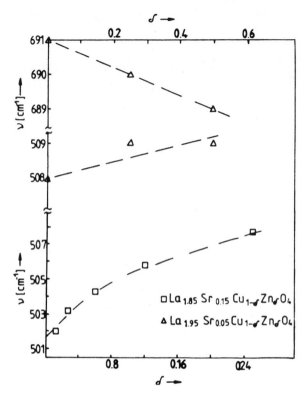

Fig.6 Line shift of the IR-absorption lines in $La_{1.85}Sr_{0.15}Cu_{1-\delta}Zn_\delta O_4$ and $La_{1.85}Sr_{0.05}Cu_{1-\delta}Zn_\delta O_4$ vs Zn-concentration

DISCUSSION

The response of the IR absorption lines to the Zn-substitution can be directly related to the change in the structure of the unit cell. Since the Cu-O distances in the CuO-planes and perpendicular to the CuO-planes increase and decrease with Zn-doping, respectively, (see Fig.1) the observed line shifts identify the two modes as the in plane and the out of plane assymetric stretching vibrations of the Cu-O bonds. The increase of the line width in the Raman and IR vibrational spectra can be interpreted as inhomogeneous broadening by doping induced disorder.

The general appearance of the change of resistivity with Zn-doping is similar for the $LaSrZn$-system, the YZn-system and the YO-system. However, details are very different for the Zn-doping and the O-doping as can be seen eg. from Fig.3. The most evident result from the doping process is the continuous change of the metallic behaviour of the resistivity to a semiconducting behaviour with increasing concentration of substitution. The simultaneous appearance of a metallic temperature dependence of the resistivity for high temperatures and a semiconducting temperature dependence for low temperatures in the intermediate doping range suggests a combined band and hopping mechanism for the transport process. Thus, we suggest that the MOTT-formalism for band conduction and hopping conduction for electrons with localized orbitals applies for these systems. Since the ceramic superconductors are known to be highly crystalline it is rather the electron correlation than a high degree of disorder which localizes the orbitals. For the undoped ceramics an ANDERSON-transition hides the true hopping nature of the transport process and the conductivity appears metallic. The Cu/Zn-substitution produces additional disorder which finally leads to a predominance of the true hopping mechanism. In order to put the interpretation outlined above into a more quantitative form it is useful to consider an energy level diagram as shown in the insert of Fig.2. A two-dimensional electronic system with an energy independent density of states is assumed for simplicity. For the unsubstituted material the mobility edge ϵ_c coincides with the band edge $\epsilon_o = 0$. The FERMI-level ϵ_F is inside the band which renders the system metallic. Increasing disorder from the Zn-substitution shifts the mobility edge upwards until it passes the FERMI-level. In all cases the carriers from the region below and above the mobility edge contribute to the charge transport. For the mathematical description one may start from the KUBO-GREENWOOD formula [5]

$$\sigma(T) = -\int \langle \sigma_\epsilon(0) \rangle \frac{\partial f}{\partial \epsilon} d\epsilon \tag{1}$$

where $\langle \rangle$ means the ensemble average over all atomic positions, f is the FERMI-distribution function and the integration for the contribution from the band states extends from ϵ_c to infinity. $\langle \sigma_\epsilon(0) \rangle$ was estimated by MOTT from the relation

$$\langle \sigma_\epsilon(0) \rangle = (8\pi^3 e^2/\hbar a)(I/U)^2 \tag{2}$$

where a, I, and U are the atomic distance, the overlap integral and the localization potential, respectively. For the ratio I/U a minimum value of $1/60$ is needed to establish an ANDERSON-transition to a metallic state [5]. This results in a minimum metallic conductivity of 350 $\Omega^{-1}cm^{-1}$. Since the conductivity in the ceramics is usually higher and at least at elevated temperatures a contribution from phonons to the disorder is to be expected the ratio may be written in terms of a constant R as

$$I/U = R/60(1 + T/T_o) \tag{3}$$

where T_o determines the temperature for which phonon scattering starts to become effective. The condition for the conservation of the total carrier concentration allows to determine the number of carriers in the hopping region (below ϵ_c) and in the metallic region (above ϵ_c), respectively. For a two-dimensional system this can be done analytically within a nearly free electron and effective mass approximation. From these results and (1) the mobility in the band states is obtained as

$$\mu_{band} = \frac{\langle \sigma_\epsilon(0) \rangle f(\epsilon_c)}{eN(\epsilon_c)k_B T \ln[\exp(-(\epsilon_c - \epsilon_F)/k_B T) + 1]} \tag{4}$$

and the hopping mobility in the states below ϵ_c is obtained from

$$\mu_{hop} = (e\nu_{ph}a^2/6k_B T)\exp(-W/k_B T) \tag{5}$$

where ν_{ph} is a phonon frequency and W is the activation energy for the hopping process. The temperature dependence of the total resistivity is finally determined from

$$\rho(T) = (n_{band}\mu_{band}e + n_{hop}\mu_{hop}e)^{-1} \tag{6}$$

For increasing Zn-substitution increasing values of ϵ_c must be used. The dashed lines in Fig.2 are as evaluated from these relations for a FERMI-level of 104 meV, $R = 2.15$, $T_o = 300$ K and an activation temperature for the hopping process of 70 K. Since the calculation did not yield the exact absolute values for the conductivities a scaling factor S of the order of unity is used for the three curves in Fig.2. (S = 1, $\epsilon_c = 0$ for $\delta = 0$; S = 0.7, $\epsilon_c = \epsilon_F$ for $\delta = 0.06$, and S = 1.8, $\epsilon_c = 2\epsilon_F$ for $\delta = 0.125$).

The good agreement between experimental and calculated results suggests that the combination of band conduction and hopping conduction as described above is a good model for the transport processes in the ceramic superconductors. From the results the crossing of the mobility edge and the FERMI-level can be deduced as the criterium for the quenching of the transition to superconductivity. The strong increase of the room temperature resistivity starts for the same condition.

ACKNOWLEDGEMENT

We acknowledge the preparation of the oxygen deficient Y-compound by S. Pekker from the KFKI in Budapest. One of the autors (HK) acknowledges valuable discussions with V.V. Moshchalkov from the Moscow State University. The work was partially supported by the FFWF in Austria.

REFERENCES

1. J.G. Bednorz and K. A. Müller, Z. Phys. B 64, 189(1986).

2. C.W. Chu, P.H. Hor, R.L.Meng, L. Gao, Z.J. Huang, and Y.Z. Wang, Phys. Rev. Lett. 58, 405(1987)

3. For a summary of the present state of the research field see eg. the Proceedings of the International Conference on High Temperature Superconductors in Interlaken, Switzerland 1988.

4. H. Kuzmany, M. Matus, E. Faulques, S. Pekker, Gy. Hutiray, E. Zoldos, and L. Mihaly, Solid State Communic., 65, 1343(1988)

5. N.F. Mott and E.A. Davis, Electronic Processes in Noncrystalline Materials, Clarendon Press, Oxford 1971

6. J.N. Grin, P.Rogl, K. Hiebl J. Less Common Metals 110, 299(1986)

7. J.D. Jorgensen, H. B. Schüttler, D.G. Hinks, D.W. Capone, H.K. Zhang, M.B. Brodsky, D.J. Scalpino Phys. Rev. Lett. 58, 1024 (1987)

8. M. Stavola, R.J. Cava, and E.A. Rietman, Phys. Rev. Lett. 58, 1571(1987).

RAMAN INVESTIGATIONS ON THE Y-Ba-Cu-O-SYSTEM

P. Knoll and W. Kiefer

Institut für Experimentalphysik der Universität Graz

Universitätsplatz 5, A-8010 Graz, Austria

1. INTRODUCTION

The discovery of ceramic perovskites with superconductivity above 20K[1] and 80K[2] started a new discussion about the superconducting mechanism. Whether the new ceramic superconductors can be described within an ordinary BCS- or Eliashberg-formalism, or more exotic mechanisms as solitons, polarons or excitons are responsible for the superconductivity, is the most interesting question. Raman scattering is based on the electron-phonon interaction, and, therefore, can sensitively probe properties of the pure electronic and phonon states together with their interaction. Because of the long wavelength (compared to interatomic distances) of the photons, this method is restricted to investigations of the centre of the Brillouin-zone, therefore, one can not assume to observe the superconducting mechanism directly. However, important properties as, e.g., the superconducting gap, can in principle be measured[3,4]. In addition, changes in the Raman spectrum due to structural changes or composition of the material in comparison with the change of superconducting properties will give valuable information about the intrinsic properties of the material. This information is needed for an understanding of the superconducting behaviour[5].

Difficulties in the Raman experiments arise from the fact, that metals (even in the superconducting state) have very low scattering intensities because of the screening of the free carriers. This becomes more dramatic, if conducting and non-conducting phases are present at the same time, as is the case for the HTC-sinter-ceramics. Therefore, first we investigate the different phases in the Y-Ba-Cu-O, not with conventional Raman technique but with micro-Raman experiments, where the Raman signal of an area of about only 1 μm^2 is detected. This technique allows to observe the Raman spectra of each phase more seperately even if these phases are distributed over the sample but are homogeneous on a micron scale. With this micro-Raman technique single crystals of $YBa_2Cu_3O_7$ are studied with polarized light in different orientations. From these investigations the spectrum of the superconducting phase is extracted and the symmetry properties of the modes are determined. With these results the changes of the characteristic phonons due to substitution of Ba, Cu and O can be interpreted. All measurements are done with the Dilor OMARS 89 triple spectrometer, multichannel detection and 514,5 nm Ar ion laser excitation. For the micro-Raman experiments an Olympus microscope is attached to the spectrometer. The laser power was carefully set to avoid sample damage (below 1 mW during the micro-Raman investigations and 100 mW line focus for the ordinary Raman experiment).

The crystal structure of $YBa_2Cu_3O_7$ is investigated by several methods[6,7]. The idealized structure is the tetragonal P4/mmm, where ordered O vacancies lower the symmetry to the

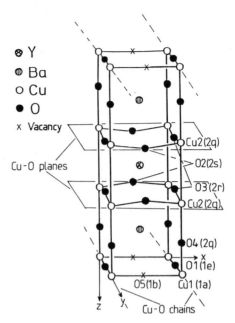

Fig. 1. Crystal structure and Wyckoff sites of the superconducting $YBa_2Cu_3O_7$. Removing the O1 oxygen at the (1e) sites changes the symmetry to a tetragonal one.

orthorhombic Pmmm space group. There is also an alternate suggestion of arranging the O vacancies[8], but in further discussion we will use an elementary cell shown in Fig.1. A factor group analysis gives 15 Raman active modes with symmetry A_g (5 modes), B_{2g} (5 modes) and B_{3g} (5 modes). Further group theoretical treatment shows that only motions at the Wyckoff sites (2t), (2q), (2r) and (2s) can contribute. Therefore, vibrations of the Ba, O4, O3, O2 and Cu2 atoms can be observed in the Raman spectrum.

2. RAMAN SCATTERING FROM DIFFERENT PHASES OF THE Y-Ba-Cu-O SYSTEM

In the Y-Ba-Cu-O system several phases with different compositions exist[9]. Best known are the "green phase" Y_2BaCuO_5 and the 90K superconductor ("black phase") $YBa_2Cu_3O_7$. Depending on the preparation conditions there is always a mixture of different phases. A careful micro-Raman analysis of a 90K superconducting sample showed some amount of CuO and of a "white phase". Sintering at higher temperature will not only produce the green Y_2BaCuO_5 but also the superconducting "black phase" and the "white phase". In Fig. 2 the Raman spectra of the different phases are shown as obtained from the micro-Raman analysis. The superconducting "black phase" is characterized by a very weak signal with one significant mode at 500 cm^{-1} and some small features at 150 cm^{-1}, 330 and 433 cm^{-1}. The semiconducting "green phase" has strong Raman intensities and the phonon frequencies are significantly different from the superconducting phase. The signals of CuO are very weak and at different positions compared with the "black phase". Raman scattering from the "white phase" is rather strong with characteristic lines at 606 cm^{-1}, 575 cm^{-1}, 454 cm^{-1}, 390 cm^{-1} and 315 cm^{-1}. Yet, the detailed composition of this white phase is unknown, but the Raman spectrum does not coincide with any of the known final products of the Y-Ba-Cu-O system[10]. By means of an electron microprobe analysis this white phase was found to mainly consist of Y and Ba with very weak contribution of Cu.

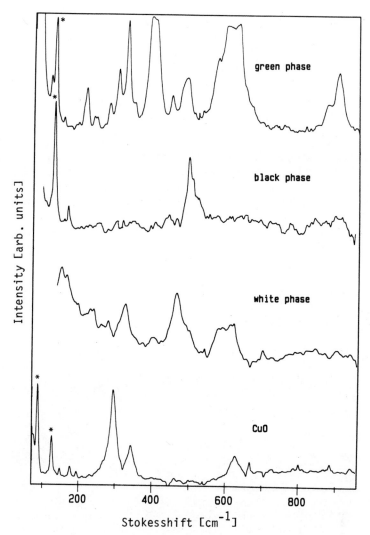

Fig. 2. Raman spectra of different phases in the Y-Ba-Cu-O-system (as indicated) obtained by means of a micro-Raman-analysis (* indicates a laser plasma line).

In addition to the occurance of such different phases, defects and O vacancies can alter the Raman spectrum. This fact may explain why there are so many differences in the Raman spectra reported so far.

3. SINGLE CRYSTAL INVESTIGATIONS

Most detailed information about the solid state properties of a material can be obtained by investigating single crystals. As long as only small crystals of the HTC-compounds are avalaible, Raman scattering is the only tool to study the lattice dynamics in detail. Sofar, only few spectra of single crystals have been published with different results[11,12].

Ordinary sintered HTC ceramics consist of little crystals as shown in Fig. 3. They can come up to a size of some μm. Larger crystals are obtained by growing from a nonstoichiometric melt. Sizes of 1mm and more are reported[13]. The crystals shown in Fig. 4 only have the size of about 50 μm x 50 μm x 5 μm, but are from better quality (no growing zones) as larger crystals, which are also obtained by this method. With crystals of the sintered material of Fig. 3 as well as with the single crystals shown in Fig. 4 micro-Raman experiments were performed. Raman spectra were recorded for different orientations of the polarization of the incident and scattered light. These spectra are shown in Fig. 5. The z-direction was determined by polarization microscopy, x and y direction is not distinguished. Nearly identical results are obtained for both samples. No intensity is observed in crossed polarization, which assigns all observed modes to the A_g symmetry species, in contrast to Ref. 14, where B_{2g} or B_{3g} modes are assumed. The interpretation of the 150 cm^{-1} and 500 cm^{-1} is straightforward: as Ba is the heaviest atom it should be responsible for the lowest frequency, O as the lightest atom corresponds to the 500 cm^{-1} mode; only z motions of these atoms are possible as the modes have A_g symmetry. The two vibrations at 433 cm^{-1} and 330 cm^{-1} should be motions of the O2, O3 and Cu2 atoms. As they have A_g symmetry too, motions of the O sublattice against the Cu sublattice are possible; however, an analysis of the symmetry coordinates is required to get more information about these modes.

During the preparation of the final manuscript, a new Raman work on single crystals was published[15] which is in qualitative agreement with our result. In a recent work[16] the 150 cm^{-1} low frequency mode is assigned to the Cu2 vibration and the Ba motion is thought to be lower in frequency (at 132 cm^{-1}).

Fig. 3.   Microphotograph of single crystals in the sintered $YBa_2Cu_3O_7$ ceramics.

Fig. 4.   Microphotograph of single crystals as grown from a nonstoichiometric melt.

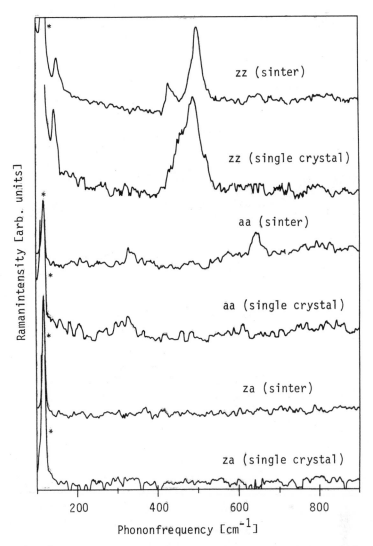

Fig. 5. Polarized Raman spectra from different configurations of single crystals of $YBa_2Cu_3O_7$. The polarization directions of incident and scattered light are indicated with respect to the crystallographic directions, where a indicates either the x or the y direction (* indicates a laser plasma line).

4. SUBSTITUTION OF Ba, Cu AND O

In order to learn more about the superconducting mechanism, all atoms in the elementary cell are systematically replaced. Most work is done at the Y site, which will not affect the superconducting properties very much. Therefore, we concentrated our interest on Ba, Cu and O and replaced them by Ca, Pb and S, respectively[17]. In all cases superconductivity was destroyed in the case of complete substitution. In Fig. 6, Raman spectra of these substituted materials are shown. The substitution of Ca had less influence. There are no dramatic changes in the Raman spectrum; the O4 vibration is only shifted to higher frequencies. But it is not a continuous shift with increasing amount of Ca; at lower Ca concentrations both, the characteristic mode of the $YBa_2Cu_3O_{7-x}$ material and the mode of $YCa_2Cu_3O_{7-x}$ appear at

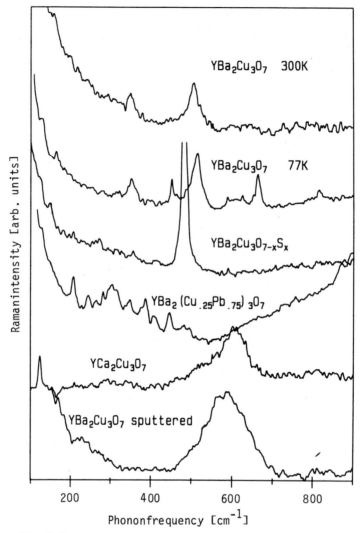

Fig. 6. Raman spectra of various treated samples (as indicated).

the same time in the spectrum, suggesting that there are two phases, one with Ba and one with Ca. These results are in agreement with X-ray diffraction studies. Also the conductivity measurements can be explained by assuming two separated phases, where only one is superconducting. Substitution with Pb changes the Raman spectrum dramatically. The superconductivity is destroyed and a new crystal structure can be assumed. The most interesting behaviour was observed by replacing the O at the (1e) site (Cu-O chains) with S. This was done by heat treatment of a $YBa_2Cu_3O_7$ sample, and exposure of the obtained $YBa_2Cu_3O_6$ samples to gaseous sulfur at temperatures below 700 K to avoid sample decomposition.[17] No superconductivity occurs, the metallic conductivity increases, the phonon spectrum remains nearly unchanged, only the O4 phonon frequency is shifted to lower values by a small amount, which can be explained by partial substitution of O4 by the heavier S. But the Raman intensity of this mode is increased by more than one order of magnitude. This indicates a dramatic change of the electronic band structure and an increase of the electron-phonon interaction. This is in agreement with calculations of the electronic band structure[18], which show a strong contribution of the Cu-O chains to the density of states at the Fermi-level. The results of the S substituted samples support the idea, that the O1 at site (1e) has a great influence on the superconductivity, but more with respect to the electronic band structure than to the phonon frequencies.

Fig. 6 also shows the temperature dependence of the superconducting $YBa_2Cu_3O_7$, which indicates a softening of the phonons with increasing temperature, probably an instability of the lattice at higher temperatures. Sputtering superconducting Y-compounds without annealing will give very broad Raman features, as they are characteristic of strongly disturbed crystalline or amorphous material.

ACKNOWLEDGEMENTS

We thank G. Leising, Technical University, Graz, Austria, for the preparation of the samples as well as for characterizing the samples by means of raster electron microscopy and electron microprobe analysis. One of us (PK) also gratefully acknowledges discussions with him.

REFERENCES

1. J. G. Bednorz and K. A. Müller, Z. Phys. B64: 189 (1986).
2. M. K. Wu, J. R. Ashburn, C. J. Torng, P. H. Hor, R. L. Meng, L. Gao, Y. J. Huang, Y. Q. Wang, and C. W. Chu, Phys. Rev. Lett. 58: 908 (1987).
3. A. A. Abrikosov and L. A. Fal'kovskii, Soviet Phys.-JETP 13: 179 (1961).
4. K. B. Lyons, S. H. Liou, M. Hong, H. S. Chen, J. Kwo, and T.J.Negran, Phys. Rev. B36: 5592 (1987).
5. see e.g. B. Batlogg et al., Phys. Rev. Lett. 58: 2333 (1987).
6. e.g. H. You, R. K. McMullan, J. D. Axe, D. E. Cox, J. Z. Liu, G. W. Crabtree, and D. J. Lam, Sol. Stat. Commun. 64: 739 (1987).
7. e.g. T. Siegrist, S. Sunshine, D. W. Murphy, R. J. Cava, and S. M. Zahurak, Phys. Rev. Lett. 35: 7137 (1987).
8. A. Reller, J. G. Bednorz, and K. A. Müller, Z. Phys. B67: 285 (1987).
9. D. G. Hinks, L. Soderholm, D. W. Capone II, J. D. Jorgensen, I. K. Schuller, and J. D. Grace, Appl. Phys. Lett. 50: 1688 (1987).
10. H. Rosen, E. M. Engler, T. C. Strand, V. Y. Lee, and D. Bethune, Phys. Rev. B36: 726 (1987).
11. R. J. Hemley and H. K. Mao, Phys. Rev. Lett. 58: 2340 (1987).
12. A. Yamanaka, F. Minami, K. Watanabe, K. Inoue, S. Takekawa, and N. Iyi, Jpn. J. Appl. Phys. 26: L1404 (1987).
13. S. Takekawa and N. Iyi, Jpn. J. Appl. Phys. 26: L851 (1987).
14. M. Cardona, L. Genzel, R. Liu, A. Wittlin, Hj. Mattausch, F. Garcia-Alvarado, and E. Garcia-Gonzalez, Sol.Stat.Commun. 64: 727 (1987).
15. P. M. Krol, M. Stavola, W. Weber, L. F. Schneemeyer, S. V. Waszczak, S. M. Zahurak, and S. G. Kosinski, Phys. Rev. B36: 8325 (1987).
16. R. Liu, C. Thomsen, W. Kress, M. Cardona, B. Gegenheimer, F. W. deWette, S. Prade, A. D. Kulkarni, and U. Schröder, preprint.
17. G. Leising, this volume.
18. E. Ortli, Ph. Lambin, J. L. Bredas, J. P. Vigneron, E. G. Derouane, A. A. Lucas, and J. M. Andre, Sol. Stat. Commun. 64, 313 (1987).

DIFFERENCE OF THERMOELECTRIC POWER IN SINGLE AND MULTIPLE

PHASE Y-Ba-Cu-O SUPERCONDUCTORS

T.Porjesz, I.Kirschner, G.Kovacs, T.Karman, and G.Zsolt

Department for Low Temperature Physics
Roland Eötvös University, Budapest, Hungary

INTRODUCTION

Since the discovery of superconductivity above 30 K by Bednorz and Müller[1] and above 90 K by Wu et al.[2], an ever increasing effort was made to investigate the physical properties of high-T_c ceramics. Most of the experiments report the resistivity drop and the diamagnetism as fundamental characteristics. The conclusions drawn from these effects do not, however, characterize a true physical property because of the effect of contacts and/or the superposition of simultaneous phenomena. The thermoelectric power (TEP) measurement seems to be one of the most sensitive tools to investigate the transport properties of high-T_c superconductors.

EXPERIMENTAL

The measurements were carried out on three types of samples. The nominal composition was $Y_1Ba_2Cu_3O_7$ for two kinds of samples (S1, S2) and $Y_1Ba_2Cu_7O_{11}$ (S3) for the third one. The samples were prepared by the now familiar solid state reaction[3,4] but the first of the equivalent compositions - S1 - received a heat treatment 25% shorter than the others. According to magnetization measurements, S2 samples consist of at least 75%, while the S1 and S3 type samples consist of 60 - 70% superconducting phase (or phases). The resistivity measurements have been made by the conventional four-probe method.

The thermoelectric power (TEP) measurements were carried out in an Oxford Instruments Continuous Flow Cryostat. A scheme of the experimental arrangement is shown in Fig. 1. As electrode materials, we have tried bronze, copper and nickel without any significant change in the measured TEP. The temperature difference has been measured by a differential thermocouple being in good thermal contact with the sample but electrically insulated. The actual temperature of the sample was controlled by a high precision temperature controller[5].

The TEP measurement was tested by measuring the thermoelectric power of high purity p- and n-type silicon single crystals and copper rods.

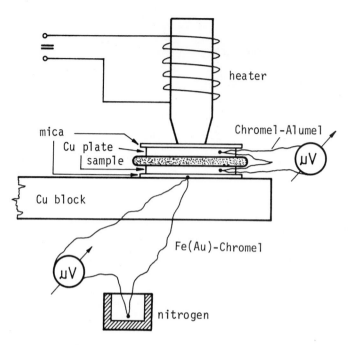

Fig. 1. A schematic diagram for the TEP measurement

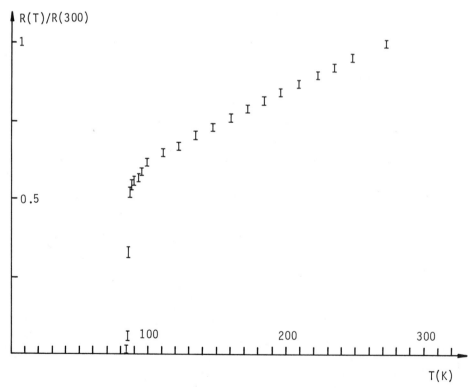

Fig. 2. The resistivity vs. temperature curve characterizing all of the
samples

RESULTS

The resistivity vs. temperature curve characterizing all of the samples is shown in Fig. 2. There was only a slight difference between them. The sharp transition - $\Delta T_c = T(R_{90\%}) - T(R_{10\%}) = 3$ K - demonstrates the good quality of the samples.

The temperature dependence of the TEP is plotted in Fig. 3.

It can be seen from the curves of Fig. 3 that the sample S2 is of n-type while the samples S1 having the same nominal composition and S3 are of p-type above the critical temperature, but surprisingly become of n-type approaching T_c.

The temperature dependences of the absolute thermopowers at higher temperatures are essentially the same and similar to other published results[6,7].

DISCUSSION

According to our and other[8,9] results, samples of nominally equal composition behave strikingly different[10]. For exampel curves of quite different character were measured for materials of the same composition[11,12].

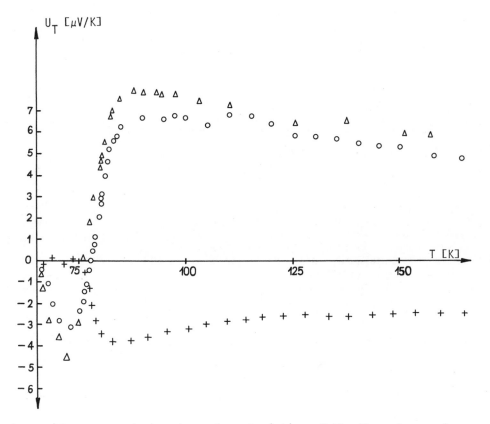

Fig. 3. Thermopower-temperature characteristics of the three types of samples (o) S1, (+) S2 and (Δ) S3

We are not going to deal with results of such substantial contradiction or with non-reproducible[13] results.

However, it is worth studying results, which show a quite unexpected behaviour and differ from the results familiar in the physics of metals or semiconductors[14-17].

These results can be partly characterized according to the sign of the thermopower and the variation of its absolute value in a rather wide range (2 - 25 $\mu V/K$) with respect to the maximum near the critical temperature.

It is interesting to note that even the sign of the thermopower changes in some cases as can be seen in Fig. 3.

These effects can neither be explained by a diffusion current of electrons or holes nor by a phonon drag effect originating from a strong electron-phonon interaction.

It must be assumed, that in these processes both types of conduction play a role with almost equal number of electrons and holes, as can also be concluded from Hall-measurements[18].

The occuring multi-band transport can be sensitive not only to impurities, but also to the preparation techniques and indicates the presence of more than one superconductive phase[19].

The high sensitivity of this method to impurities and preparation techniques could be of a great importance for further investigations. It could help to establish a methodology for carrying out more exact measurements or, if the experimental errors due to extraneous thermopower or different work functions can not be completely eliminated, to make the results at least comparable.

REFERENCES

1. J.G.Bednorz and K.A.Müller, Possible High-T_c Superconductivity in the Ba-La-Cu-O System, Z.Phys. B64, 189 (1986)
2. M.K.Wu, J.R.Ashburn, C.J.Torng, P.H.Hor, R.L.Meng, L.Gao, Z.J.Huang, Y.Q.Wang and C.W.Chu, Superconductivity at 93 K in a New Mixed-Phase Y-Ba-Cu-O Compound System, Phys.Rev.Lett. 58, 908 (1987)
3. I.Kirschner, J.Bankuti, M.Gal and T.Torkos, Superconductivity in La-Ba-Cu-O metallic oxide compounds above 50 K, Phys.Rev. B36, 2313 (1987)
4. I.Kirschner, J.Bankuti, M.Gal, K.Torkos, K.G.Solymos and G.Horvath, Superconductivity in a New YBaCuO Compound at 105 K, Europhys.Lett. 4, 371 (1987)
5. I.Kirschner, G.Kiss, T.Porjesz, R.Remenyi and P.Zentai, High precision temperature controller, Cryogenics 14, 559 (1974)
6. J.Marcus, C.Escribe-Filippini, C.Schlenker, R.Buder, J.Devenyi and P.L.Reydet, Magnetic and transport properties of some La-Ba-Cu-O superconducting oxides, Solid State Comm. 63, 129 (1987)
7. V.Bayot, F.Delannay, C.Dewitte, J.P.Erauw, X.Gonze, J.P.Issi, A.Jonas, M.Kinany-Alaoui, M.Lambricht, J.P.Michenaud, J.P.Minet and L.Piraux, Strong electron-phonon coupling from thermal conductivity measurements in a $YBa_2Cu_3O_7$-type superconducting compound, Solid State Comm. 63, 983 (1987)
8. R.Yaozhong, H.Xuelong, Z.Yong, Q.Yitai, C.Zuyao, W.Ruiping, and Z.Qirui, Transport properties in single phase superconductors $Ba_2YCu_3O_9$, Solid State Comm. 64, 467 (1987)
9. C.Uher and A.B.Kaiser, Thermal transport properties of $YBa_2Cu_3O_7$ superconductors, Phys.Rev. B36, 5680 (1987)
10. G.J.Hyland, Thermopower of high-T_c materials; Inconsistencies and anormalies, Jap.J.Appl.Phys. 26, L1895 (1987)
11. Y.Maeno, Y.Aoki, H.Kamimura, J.Sakurai and T.Fujita, Transport properties and specific heat of $(La_{1-x}Ba_x)_2CuO_{4-y}$, Jap.J.Appl.Phys. 26, L402 (1987)

12. S.Uchida, H.Takagi, H.Ishii, H.Eisaki, T.Yabe, S.Tajima and S.Tanaka, Transport properties of $(La_{1-x}A_x)_2CuO_4$, Jap.J.Appl.Phys. **26**, 440 (1987)

13. J.T.Chen, C.J.McEwan, L.E.Wenger and E.M.Logothetis, Determination of charge carriers in superconducting La-Ba-Cu-O by thermoelectric measurements, Phys.Rev. **B35**, 7124 (1987)

14. U.Gottwick, R.Held, G.Sparn, F.Steglich, H.Rietschel, D.Ewert, B.Renker, W.Bauhofer, S.von Molnar, M.Wilhelm and H.E.Hoenig, Transport properties of $YBa_2Cu_3O_7$: Resistivity, thermal conductivity, thermopower and Hall effect, Europhys.Lett. **4**, 1183 (1987)

15. S.W.Cheong, S.E.Brown, Z.Fisk, R.S.Kwok, J.D.Thompson, E.Zirngiebl, G.Bruner, D.E.Peterson, G.L.Wells, R.B.Schwarz and J.R.Cooper, Normal-state properties of $ABa_2Cu_3O_7$ compounds: Electron-electron correlations, Phys.Rev. **B36**, 3913 (1987)

16. N.Mitra, J.Trefny, M.Young and B.Yarar, Thermoelectic power of Y-Ba-Cu-O, Phys.Rev. **B36**, 5581 (1987)

17. D.Hong-min, L.Li, W.Xie-mei, L.Shu-yuan and Z.Dian-lin, Hall effect of the high-T_c superconducting Y-Ba-Cu-O compound, Solid State Comm. **64**, 489 (1987)

18. Z.G.Khim, S.C.Lee, J.H.Lee, B.J.Suh, Y.W.Park, C.Park, I.S.Yu and J.C.Park, Superconductivity in single-phase $Y_1Ba_2Cu_3O_9$ and thermoelectric power measurement, Phys.Rev. **B36**, 2305 (1987)

19. S.C.Lee, J.H.Lee, B.J.Suh, S.H.Moon, C.J.Lim and Z.H.Khim, Thermoelectric power and superconducting properties of $Y_1Ba_2Cu_3O_7$ and $R_1Ba_2Cu_3O_7$, preprint

20. T.Porjesz, T.Karman, I.Kirschner, G.Kovacs, G.Zsolt and H.Beyer, Anomalous dependence of the thermoelectric power on the temperature in Y-Ba-Cu-O high T_c superconducting samples, ICTP Reports 87/278 (1987)

APATITE AS A PROMISING SOURCE OF RARE EARTHS, STRONTIUM

AND BARIUM FOR HIGH TEMPERATURE SUPERCONDUCTORS

J.I.Skorovarov, V.V.Shatalov, V.D.Kosynkin, and V.A.Semenov

All-Union Research Institute of Chemical Technology, State
Committee on Utilization of Atomic Energy, Moscow, USSR

The discovery of superconductivity in the oxide system La-Ba-Cu[1] and the subsequent research on ceramic oxides, Y-Ba-Cu and other metal oxides[2-4], has attracted the attention of both physicists and chemists interested in the initial components of high temperature superconductor (HTS) systems. Taking into account the wide scope for the application of HTS in technology, chemists have had to consider short- and long-term perspectives for the adequate production of the initial materials, such as compounds of rare earth elements, strontium, barium and copper for metalloceramics. The question concerns guaranteed raw material sources for HTS production. One such source is Kola apatite $Ca_5F(PO_4)_3$, which is processed into large amounts of phosphorus fertilizers for domestic use and export[5,6].

Averaged data for the composition of apatite concentrate are presented in Table 1[7].

It should be noted that strontium compounds are practically absent in phosphorite concentrates, and the content of rare earth elements does not exceed 0.1%, which is 10 times less than in apatite.

Reprocessing of apatite concentrate into fertilizers is carried out with mineral acids - sulphuric, nitric and phosphoric. In addition, thermal methods are used[8].

It is well known that most apatites are presently being processed into common superphosphate using sulphuric acid. In fact, a great amount of rare earths is transferred into fertilizer. Apatite decomposition using sulphuric acid, which is accompanied by the recovery of phosphoric acid extract, makes it possible to extract up to 40%[9] of the rare earth elements.

Using the present flow sheets, an analysis of the distribution of valuable components during the process of apatite reprocessing has proved

Table 1. Composition of Apatite Concentrate

Compo-nent	P_2O_5	CaO	F	SrO	MgO	REO*	Al_2O_3	Fe_2O_3	TiO_2	K_2O	Na_2O	SiO_2
Con-tent,%	39.4	52.0	3.17	2.7	0.15	1.1	0.5	0.45	0.15	0.2	0.9	1.0

*REO - rare earth oxides

that the wasteless method of nitric acid decomposition, utilizing fluorine and calcium as well, is the most expedient, while recovering rare earths and strontium as byproducts[6].

The flow sheet of complex apatite reprocessing for nitrophosphoric fertilizers comprises the following major steps:
- acidic breaking-down of apatite; transfering calcium, fluorine, rare earths and phosphorus into the solution (strontium is retained in the solid phase with the insoluble residue);
- stripping and purifying of strontium;
- crystallization and separation of calcium nitrate;
- extraction of rare earth elements from the nitrophosphoric acid solution (NPS) as phosphate concentrate;
- purification of rare earth elements and their separation into single compounds;
- manufacture of fertilizers.

The decomposition of apatite using nitric acid transfers strontium compounds almost completely into nitrate salt, which then salts out into the solid phase, together with the insoluble apatite residue (strontium concentrate).

The strontium content in the concentrate is 60% of that in the apatite[10,11]. The strontium concentrate contains 30-40% strontium nitrate and can be processed into various compounds, with strontium carbonate being the most useful. The present strontium purification flow sheets are based on the differentiation of the compounds' solubility[12]. The impurities in the strontium concentrate can be divided into three groups. The first group contains minerals which are insoluble in water and diluted nitrate acid. They have an insoluble apatite residue (ilmenite, titanomagnezite, etc.) which can be removed by a water-leaching strontium concentrate.

The second group contains impurities that are soluble in nitric acids - salts of iron and aluminium, fluorides, phosphates. The removal of such impurities is based on their transition to an insoluble state by ammonia. Strontium remains in the solution.

The third group contains calcium salts, the separation of which is based on the solubility difference between strontium and calcium nitrates in water and nitric acid solutions.

It should be noted that the above methods of strontium purification do not facilitate the separation of barium. Thus, the production of strontium carbonate from apatite reprocessing yields strontium carbonate with 3 - 3.5% barium, when a nitric acid flow sheet is used. The high purity strontium compound required for high temperature superconductors can only be obtained by the additional separation of barium.

At present, separation of strontium from barium is carried out by methods which introduce a sulphate-containing reagent into the solution, by precipitation and by the separation of barium sulphate[13-15]. The barium content in the strontium product can be reduced to 0.5%. For a greater degree of purification one can apply crystallization methods for nitrate purification, obtaining strontium nitrate with 0.01% Ba or less, and less than 0.008% Ca[16].

The sulphate method for separating strontium from barium gives a concentrate containing 15% Ba and 30% Sr. This product could be used for the production of barium compounds.

According to the complex nitric acid scheme for apatite processing and after a preliminary separation of calcium and fluorine it seems quite simple to precipitate rare earth elements (REE) into a 10 - 30% phosphate concentrate from nitrophosphoric acid, which considerably raises the efficiency of the REE phosphate filtration unit[11]. Recovery of REE as residue does not exceed 85%.

Removing phosphorus from the phosphate REE concentrate and obtaining pure accumulated REE compounds is one of the major steps in the REE recovery process.

The most conventional chemical purification methods, such as the oxalate method, and the processing of phosphate concentrate using a sodium hydroxide solution, have a number of serious shortcomings, the main one being the inability to obtain a REE product that is pure enough (over 99% REE)[17]. Solvent extraction, which is used to recover and purify REE from nitrophosphate solutions, raises the degree of REE purification, increases the output, and in comparison to chemical methods considerably decreases the reagents cost. The neutral phosphor-organic compound of the TBP type is quite effective for purification-extraction[18].

It is known that the cerium REE group makes up about 90% of the REE apatite, with cerium contents of up to 50% in the total rare earths oxides[6]. The concentration of the yttrium earths is about 7%. Apatite, as a rare earth material, is interesting as a source containing a raised amount of europium (up to 1% in the total RE).

The universal practice of dividing REE into groups and then into single elements widely uses extraction methods in nitrate systems with neutral phosphor-organic compounds[19-21]. In addition, alcylphosphoric acids[22-24], tertiary amines and alcohols[26] are used.

The extraction methods for manufacture of light rare earth elements (lanthanum, cerium, neodymium) are well known. Yttrium is of singular importance with regard to high temperature superconductors. Therefore, more attention should be given to the production methods of its compounds.

The values of yttrium distribution coefficients in nearly all known systems are within the D values of the lanthanide series. Therefore, most methods of separation and thorough purification of yttrium are based on a consequential combination of two or more systems with various D_Y values in the REE series[27-29].

In the USSR an effective method for the extraction of yttrium and its purification from all lanthanide impurities has been developed. The method could be used for treating concentrates with a multiform composition[30-32]. Yttrium is simultaneously cleared from all lanthanides in a single extraction system using naphta sulphoxide (NSO) with molecular mass 120-244, diluted by kerosene, as an extragent. Extraction is carried out in the presence of a salting-out agent (lithium, magnesium, calcium or aluminium chlorides) with the REE concentration in the raffinate: 0.7 - 2.2 mol/dm^3, and that in the salting-out agent: 1.1 - 6 mol/dm^3, or in the presence of rhodanide ions with a molar ratio in the raffinate REE: rhodanide-ions - 3.5 - 230:1. The method is especially effective for the thorough purification of yttrium (99.995 - 99.99999%).

CONCLUSIONS

1. Taking the composition of the apatite concentrate into account and the amount that is available to be treated, apatite can be considered a viable complex raw material source for the production of HTS initial materials.

2. Breaking down the apatite concentrate using nitric acid facilitates the most complete recovery of HTS initial components (rare earth elements, strontium, barium). The total extraction of strontium in the final product reaches 50% of its contents in apatite, and 75% for the rare earth elements (in the total rare earth concentrate).

REFERENCES

1. I.B.Vitman, High temperature superconductivity in ceramic metal oxides, Uspeki Fiz.Nauk. 152, 675 (1987)
2. A.I.Golovashkin, High temperature superconductive ceramics, ref.1, p.553
3. V.L.Ginzburg, D.A.Kirzhnitz, High temperature superconductivity, ref.1, p.575
4. R.B.Goldfarb, A.F.Clark, A.I.Braginski and A.J.Panson, Cryogenics 27, 475 (1987)
5. T.P.Unanjantz, Coll. research on chemistry and technology of fertilizers, pesticides and salts, Publ.H.Nauka, 1966, p.295
6. Complex nitric acid treatment of phosphates, Eds. A.L.Gol'dinov and B.A.Kopylov, L.,Himija, 1982
7. G.A.Golovanov, Flotation of Polish apatite-containing ores, M., Himija, 1970
8. M.E.Posin, Technology of mineral salts, Part II, L., Himija, 1970
9. A.I.Mikhailichenko, E.B.Mikhlin, Y.B.Patrikeev, Rare earth metals, metallurgija, 1987, 232
10. I.F.Muljarchuk et al., in "Chemical Technology", Republican inter-departm. sci.-tech.coll., 1978, No. 1, p.3-4
11. A.L.Gol'dinov et al., Him.prom., 1980, No.1, p.26, No.4, p.225
12. M.A.Kolenkova, G.V.Pasukhin, Coll.Proceedings of Inst. of non-ferr. metals, 1963, XXXV, p.154-159
13. Author sertif. No.1117284 (USSR) Purification method for strontium-containing solutions from barium, N.V.Baryshnikov et al., publ. BI 1984, No. 37
14. S.B.Bukin, A.F.Denisov, M.A.Kolenkova, Purification of concentrated strontium solutions from barium, Moscow Inst.of steels and alloys, M., 1982, p. 12
15. I.S.Frumina, E.S.Kruchkova, S.P.Mushtanova, Analytic chemistry of calcium, M., Nauka, 1974, p.160
16. L.A.Sadokhina, T.V.Zimina, I.F.Poletaev, JPCh 12, 2679 (1986)
17. Methods for separation of rare earth metals, Eds. V.E.Plustshev and L.I.Komissarova, M., IL, 1961, 361
18. S.D.Moiseev, V.A.Semenov, G.L.Shelekhova, Him.prom., No.8, 1981, p.472-475
19. D.F.Peppard, J.Phys.Chem. 57, 294 (1953)
20. D.J.Scargill, Inorg.Nucl.Chem. 4, 304 (1957)
21. D.F.Peppard, J.Inorg.Chem. 4, 326 (1957)
22. E.B.Mikhlin, G.V.Korpusov, JNCh. 10, 2787, 1517 (1965)
23. S.Kolarik, K.Pankova, JNCh. 28, 2325 (1966)
24. C.G.Warren, J.F.Suttle, J.Inorg.Nucl.Chem. 12, 337 (1960)
25. D.I.Baner, US Bureau of Mines R.I. 6809 (1966)
26. E.A.Belousov, S.D.Khakhaev, JNCh. 9, 2659 (1964)
27. C.G.Brown, L.G.Sherrington, J.Chem.Tech.and Biotech. 29, 193 (1979)
28. Pat. 3658486 (USA) 1972
29. Pat. 2750 (Norway) 1971; Pat. 180097 (USA) 1971
30. Auth.sertif. 472566 (USSR) A.I.Mikhailichenko, L.A.Abramov, A.I.Drygin, publ. in BI 1981, No. 47, p.308
31. Auth.sertif. 526182 (USSR) A.I.Mikhailichenko, L.A.Abramov, A.I.Drygin, publ. in Bi 1981, No. 47, p.308
32. Non-ferrous metals, 1979, No. 9, p.129

SECTION 3: THEORY-MECHANISMS

BOSON EXCHANGE MECHANISMS AND THE HIGH T_C OXIDES

J. P. Carbotte

Department of Physics
McMaster University
Hamilton, Ontario, Canada   L8S 4M1

ABSTRACT

We have studied the superconducting properties of an Eliashberg superconductor for arbitrary strength and shape of the kernels. In as much as the basic equations still apply plasmon, exciton, spin fluctuation or some other boson exchange mechanism are included. Comparison with experiment is attempted. Bounds on properties which are applicable in all cases are established.

I - INTRODUCTION

The superconductivity of all conventional systems can be well understood[1] on the basis of the Eliashberg gap equations with electron-phonon spectral density $\alpha^2 F(\omega)$ and Coulomb pseudopotential μ^*. The properties of such systems differ significantly from universal BCS laws[1-3] and it is these differences that carry the information about mechanism.

As yet, the mechanism responsible for the superconductivity of the high T_c oxides is not known although a very large number of suggestions have been put forward.[4] These theories can perhaps be classified into two categories. One uses a weak and the other a strong correlation description of the underlying normal state. In the weak correlation approach,[5,6] band structure theory is assumed to be valid and the Eliashberg equations are retained although it is not necessarily assumed

that the electron-boson exchange spectral density is due to phonons[5,8].
It could arise from excitons, plasmons, spin fluctuations, etc. For the
strong correlation case the starting point is very different. A
localized description, based on a Hubbard[9] Hamiltonian, is used. The
stoichiometric compound La_2 Cu O_4 is a Hubbard-Mott insulator and is
magnetic[10], features which are not incorporated into a band
description. Perhaps the most studied of the models in this class is the
RVB [11,13] (resonating valence bond) but even in this case there does
not yet exist a clear picture of the nature of the superconducting state.
It may have no direct relationship to conventional Cooper pairing.

In this lecture I will be concerned only with weak correlation
models and start from the Eliashberg equations with arbitrary kernels
describing any boson exchange. Even though such an approach is justified
only for phonon exchange, we take the view here that the Eliashberg
equations are more generally applicable if we interpret the kernels to be
effective phenomenological quantities. It may be hard to relate these to
microscopic theory, but this is of no concern to us here since we intend
to model them.

The lecture has two more sections and a conclusion. In section II,
we consider the case of La Sr Cu O for which Weber[14] has calculated an
electron-phonon spectral density. We use this function to make definite
predictions for some superconducting properties and attempt a comparison
with experimental data with a view at obtaining information on mechanism.
In section III, we consider Y Ba Cu O and the very strong coupling limit.
In the course of the discussion, we introduced the idea of upper and
lower bounds on certain dimensionless properties. These bounds need to
be satisfied by all Eliashberg superconductors and constrain severely the
values that can be obtained for these properties in weak correlation
theories.

II - CASE FOR $LA_{1.85}$ $SR_{0.15}$ CU O_4 AS A PHONON SUPERCONDUCTOR

Once the kernels in the Eliashberg equations have been measured by
McMillan-Rowell[15] inversion of tunneling data, thermodynamic[2] and
other properties[3] follow. In Fig. 1, we show results of Eliashberg
calculations for the gap (Δ_0) to critical temperature (T_c) ratio
$2\Delta_0|K_B T_c$. Many of the solid points have been identified in the figure,
for the others, refer to the paper of Mitrovic et al.[3] The horizontal
axis involves the strong coupling parameter T_c/ω_{ln} with ω_{ln} an
appropriate logarithmic[16] weighting of the boson spectral density

$\alpha^2 F(\omega)$ which favours low frequencies. It is quite clear from the figure that most systems fall close to a single trend curve (dashed curve) with some significant fluctuations in a few cases. These fluctuations reflect, in a complicated way, some of the details of the shape and strength of $\alpha^2 F(\omega)$ which are not captured by the single parameter $\omega_{\ell n}$[16]. These are secondary effects, however, and $2\Delta_0|K_B T_c$ can be used to obtain information on the size of $\omega_{\ell n}$ and so on mechanism.

Marsiglio and Carbotte[2] have considered other properties besides $2\Delta_0|K_B T_c$ for a large number of superconductors and again find that the same parameter $T_c/\omega_{\ell n}$ is sufficient to establish qualitative behaviour. Relationships between properties are, therefore, implied which constrain the possible values for the dimensionless thermodynamic ratios.

The electron-phonon spectral density has been calculated for La Sr Cu O by W. Weber[16] who finds that he can get lattice stability and a T_c value of approximately 40K. The value obtained for the mass enhancement λ was 2.6 and $\omega_{\ell n}$ is 14.0 meV so that $T_c/\omega_{\ell n}$ = 0.23. These values fall at the edge of the conventional superconductors plotted on Fig. 1. From this point of view, there is nothing unusual. Also, an important isotope effect has been measured.[17,18] In view of these facts it seems

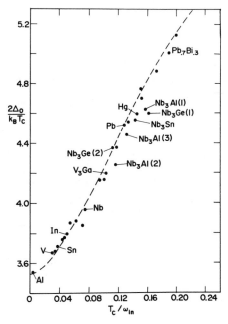

Fig. 1 The gap to critical temperature ratio $(2\Delta_0|K_B T_c)$ as a function of $T_c/\omega_{\ell n}$ for many conventional superconductors.

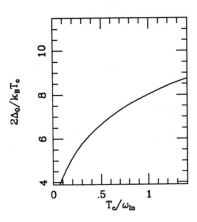

Fig. 2 The dimensionless ratio $2\Delta_0|K_B T_c$ as a function of strong coupling parameter $T_c/\omega_{\ell n}$.

Table 1. The gap to critical temperature ratio $2\Delta_0|K_B T_c$ and the jump in the specific heat at T_c, $\Delta C(T_c)$, for La Sr Cu O.

$$2\Delta_0|K_B T_c$$

FIR: 2.9-4.5, 2.4, 1.6-2.7, 2.5, 0.7-2.7, 2.6, 2 (c axis)

Tunneling: 4.5, 3.0-6.0, 4.1-4.8, 4.5-9.0, 3.5-4.0, 5.2-9.1, <4.5, 5.0-8.7, 4.5-5.8, 8.0-18.0, 4.7, 7.0±2.0

NMR: 1.3±0.2, 5.0

$$\frac{\Delta C(T_c)}{T_c} \; [mJ/mole\ Cu\ atom - K^2]$$

9.9, 11.0, 7.6±1.8, 20.0±5.0, 16.8, 8.8, 22.0-26.0, 10.0±2.0, 6.5

worthwhile to compare results of theoretical calculations based on Eliashberg theory and Weber's[16] spectrum with experiment. The ratio of the isotropic gap edge to T_c is predicted to be 5.3[19]. In table 1 we summarized results obtained from a survey of the experimental literature. This tabulation with references is part of the thesis of F. Marsiglio to which the reader is referred for details. It is clear that there is a great deal of variation with experiment. The reason for this is not clear. The two main types of experiment that have traditionally yielded the gap are, far infrared absorption (FIR) and tunneling. Both are difficult. It is clear, from the table, that FIR measurements give small values of the gap, around and even lower than the BCS value of 3.54, while tunneling generally gives larger values with some as large as 9. A further complicating factor is that the gap is likely to be very anisotropic in the oxides while our theoretical estimate is for an isotropic system. In addition, it is not at all clear that a well defined gap value exists. For example, all infrared optical measurements give some absorption inside the gap. In spite of all these uncertainties, it can still be tentatively concluded that our theoretical estimate is somewhat high, indicating that some additional higher frequency boson exchange may be present. Further experiments of greater reliability are required, however, before we can be more definitive.

A second quantity of interest is the normalized specific heat jump at T_c given by $\Delta C(T_c)/\gamma T_c$ where γ is the Sommerfeld constant. A direct normal state specific heat determination of γ is not possible so that indirect methods need to be used. Because of the uncertainties involved in the indirectly derived values of γ, we will use a different approach here. Noting from table 1 that there is some agreement as to the value of the jump at T_c ($\Delta C(T_c)$), we fix the value of $\Delta C(T_c)/T_c$ to 17 mJ/mole K^2 and use our calculated $\Delta C(T_c)/\gamma T_c = 2.8$ to deduce a γ of 6 mJ/mole K^2. This value is close to several quoted values. The Sommerfeld γ is related to the single spin electronic density of states $N(0)$ at the Fermi energy. We get for $N(0)$ a value of 0.36 states/ev Cu atom which compares poorly with a value of 0.90 quoted by Freeman et al.[26] from band structure calculations. This may mean that we are using a value of λ which is too large and would be evidence against an electron-phonon mechanism. It could also mean that the band structure estimates for $N(0)$ are not reliable because the oxides are highly correlated systems.

In summary, comparison with experiment of predictions of Eliashberg theory based on Weber's[14] electron-phonon spectral density are inconclusive at the present time because of uncertainties in experimental data. Some of the data favours a small gap value and argues for some non-phonon contribution to the Eliashberg kernels as does the very small value that we have obtained for the single spin electronic density of state. While we cannot be categoric, it is important to realize that the work presented here could, in principle, allow us to come to a definitive conclusion as to mechanism, if the data available was more precise.

III - THE CASE OF Y BA CU O

The important observation that there is almost no isotope effect in Y_1 Ba_2 Cu_3 $O_{y-\delta}$[21,22] argues against a phonon mechanism. Lacking information on the electron-phonon spectral density, it is not unreasonable to assume that the characteristic phonon energy $\omega_{\ell n}$ in Y Ba Cu O is not so different from that for La Sr Cu O. As a first approximation, we take it to be 14.0 mev with a certain amount of softening or stiffening introduced through a constant scaling factor δ. This leads to a model spectral density of the form[23] $\alpha^2_{mod}F(\omega) = B \alpha^2 F_{La Sr Cu O}(\delta\omega)$ with B a second scaling factor. While, as we have already acknowledged, Y Ba Cu O is not likely to be a pure phonon superconductor, it is interesting to predict what some of its properties would be if it was, and see if experimental results agree or disagree with such predictions.

In the results to be described below, δ is chosen to get a definite, but arbitrary, value for the strong coupling parameter $T_c/\omega_{\ell n}$ and B is then adjusted so that T_c = 96K. In Fig. 2, we show our results for $2\Delta_0|K_B T_c$[23] as a function of $T_c/\omega_{\ell n}$. If no stiffening or softening is accounted for, Y Ba Cu O would have a $T_c/\omega_{ln} = 0.6$ which is well beyond the usual strong coupling regime for which $T_c/\omega_{ln} \lesssim 0.25$. (We refer to this new regime as the very strong coupling limit). It is clear from Fig. 2 that the predicted value for $2\Delta_0|K_B T_c$ is about 7.0. In table 2, we summarize the many experiments with which we are familiar. References can be found in the thesis of F. Marsiglio. Clearly, there is no consensus as to gap value. In general, the FIR values are small while tunneling sometimes gives higher results. Despite the large variability from experiment to experiment, we can, nevertheless, conclude unambiguously that, on the whole, the data does not support the large value predicted. Instead, a combined phonon--exciton mechanism, with a larger effective value of ω_{ln} is indicated with attendant smaller $2\Delta_0|K_B T_c$. It is interesting to note that Carbotte et al.[24] have shown that an upper limit exists on $2\Delta_0|K_B T_c$ for an Eliashberg superconductor independent of origin, size, and shape of the electron-boson exchange spectral density. Our best numerical estimate at this time is around 13.0, a value which is somewhat greater than our initial estimate[24]. The limiting value arises for a delta function spectrum with Einstein oscillator ω_{ln} going to zero. The physical limit is likely to be reached before this mathematical limit. For example, to get with $\mu^* = 0$, a dimensionless ratio of 9.0, we need a λ value of more than 20, while to get 10.0, we need $\lambda = 50.0$. This limit should be useful in analyses of experiments, since it puts a severe upper bound on $2\Delta_0|K_B T_c$ which, if violated, would imply that Eliashberg theory does not apply.

Table 2. The gap to critical temperature ratio $2\Delta_0|K_B T_c$ and the jump in the specific heat at T_c, $\Delta C(T_c)$, for Y Ba Cu O

$$2\Delta_0|K_B T_c$$

FIR: 3.2±0.3, 3.5±0.3, 2.0, 3.5, 1.6-3.4, 2.3-3.5, 2.5-4.2, 3.3, 3.2, 3.5, ~8.0, 3.8

Tunneling: 3.7-5.6, 11.0, 4.5-6.0, 3.9-4.8, 4.8, 3.81-4.52, 10.0, 13.0, 3.9, 4.8, 3.2±0.4, 7.0-13.0, 3.9

NMR: 7.5, 1.3±0.2

Other: 5.0±0.2, 3.4±1.5, 4.8±0.5

$$\frac{\Delta C(T_c)}{T_c} \ [mJ/mole\ Cu\ atom\ -\ K^2]$$

18.0, 15.5, 13.0, 16.0, 7.0, 11.0±2.0, 23.0±5.0, 13.0, 13.0, 11.3

In Fig. 3, we show results for the normalized specific heat jump at T_c and the dimensionless quantity $T_c^2 \gamma / H_c^2(0)$ where $H_c(0)$ is the zero temperature thermodynamic critical magnetic field.[23] In BCS theory, these would have values 1.43 and .168, respectively, and would be material independent. We see that, as $T_c/\omega_{\ell n}$ increases, $\Delta C(T_c)/\gamma T_c$ first increases, as is observed for the conventional strong coupling materials, but then peaks after which it starts dropping. Use of a different spectral density does not change the qualitative behaviour obtained. We note that in the very strong coupling limit $\Delta C(T_c)/\gamma T_c$ can fall substantially below 1.43. A similar situation holds for $\gamma T_c^2/H_c^2(0)$ which first shows a shallow minimum before it starts rising above .168 at higher values of $T_c/\omega_{\ell n}$ and is 0.32 for $T_c/\omega_{\ell n} = 1.4$. These results are quite unexpected and could be used to characterize the very strong coupling regime.

Experimental results on $\Delta C(T_c)/T_c$ are summarized in table 2. Although there is again uncertainty about γ, most experiments give a value of the normalized jump near or less than BCS. This value is consistent both with a value of $T_c/\omega_{\ell n} \equiv 0.6$ or with $T_c/\omega_{\ell n}$ close to zero. Since the gap ratio for $T_c/\omega_{\ell n} = 0.6$ is too large we would need to conclude that $T_c/\omega_{\ell n}$ is small. This is a strong argument for a non-phonon mechanism.

The maximum in the normalized specific heat jump that we observe in Fig. 3 is suggestive. While the value and position of this maximum does

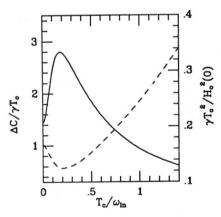

Fig. 3 The normalized specific heat jump $\Delta C(T_c) / \gamma T_c$ and dimensionless ratio $T_c^2 \gamma / H_c^2(0)$ as a function of strong coupling parameter $T_c/\omega_{\ell n}$.

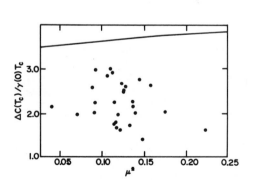

Fig. 4 Maximum normalized specific heat jump (solid line) and values (solid dots) for many real superconductors.

depend somewhat on the shape of the electron-boson spectral density $a^2F(\omega)$, Blezius and Carbotte[25] were able to establish an absolute maximum for $\Delta C(T_c)/\gamma T_c$ which applies to all spectral densities for a given value of Coulomb pseudopotential μ^*. This bound places a severe constraint on possible values of the normalized jump.

In Fig. 4, we show the value of the maximum $\Delta C(T_c)/\gamma T_c$ as a function of μ^*. Also, shown as black dots on the same figure are the results for many conventional superconductors. For an identification of the actual materials represented, the reader is referred to the original paper. While some points fall close to the theoretical maximum, others do not, because their spectral densities are far from optimum.

For the second upper critical magnetic field[26] a strong coupling index η_{Hc2} is defined by $H_{c2}(T, t^+) \equiv \eta_{Hc2}(T, t^+) H_{c2}^{BCS}(T, t^+)$ where H_{c2} is the actual upper critical field and H_{c2}^{BCS} is its BCS limit. Here T is the temperature and t^+ is a normal impurity scattering parameter related to the electron lifetime τ through $t^+ = 1/2\pi\tau$. Just as in the case of the specific heat jump, we can prove that there is a theoretical maximum for the strong coupling correction $\eta_{Hc2}(T_c, t^+)$[27] for a given t^+ and μ^* which must be satisfied by all Eliashberg superconductors. Our results for this maximum are shown in Fig. 5 as a function of μ^* for $t^+ = 50.0$ mev. Also shown are results for some real materials based on full

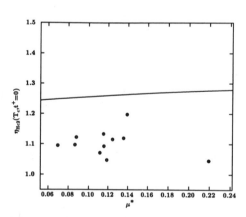

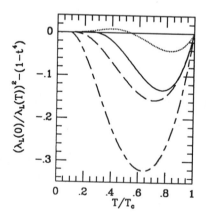

Fig. 5 The maximum strong coupling correction to the upper critical field at T_c (solid line) and values for many real superconductors (solid dots).

Fig. 6 The temperature dependence of the London limit penetration depth $[\lambda_L(0)|\lambda_L(T)]^2 - (1-t^4)$ with $t = T/T_c$ for various values of $T_c/\omega_{\ell n}$.

Eliashberg calculations. It is comforting that they all fall below the solid maximum curve. One falls close, indicating that the solid curve represents a serious constraint on possible experimental values.

The temperature dependence of the London[28] penetration depth $(\lambda_L(T))$ in Y Ba Cu O has been measured in muon spin relaxation experiments.[29]   In Fig. 6, we show our theoretical results[28] for $[\lambda_L(T)/\lambda_L(0)]^2 - (1-t^4)$ as a function of reduced temperature $t = T/T_c$ for four suitably chosen values of T_c/ω_{ln}, namely 0.074 (solid curve), 0.2296 (dotted curve), 0.661 (dashed curve) and 1.175 (long-short dashed curve). For reference, the BCS curve, which is not shown, would fall below the dashed curve and have a minimum value of about -0.21.[28] We see that as the coupling strength is increased, the minimum in the curve moves up towards smaller negative values. For $T_c/\omega_{ln} = 0.296$, it has developed a positive region at low temperature followed by a very shallow minimum around $t = 0.8$. This is typical strong coupling behaviour. As the coupling is increased further, however, the trend is reversed and the curve becomes more BCS like. In the very strong coupling regime, the absolute value of the minimum can be very much larger than BCS.[30] This characteristic temperature variation is distinctive to large values of T_c/ω_{ln} and is not confirmed experimentally [29].

V - CONCLUSIONS

We have calculated the superconducting properties of an Eliashberg superconductor as a function of the characteristic strong coupling parameter T_c/ω_{ln}. In principle, a measurement of a dimensionless superconducting ratio would allow us to deduce a value for T_c/ω_{ln} and so get information on mechanism from a knowledge of the size of the characteristic boson exchange energy ω_{ln}. Also, measurements of several ratios would put severe consistency checks as to the applicability of Eliashberg theory. A comparison of our theoretical result with experiment is inconclusive because of large variabilities in existing data on such quantities as the gap to critical temperature ratio.

In the course of the work bounds in several superconducting dimensionless ratios were established which put severe restrictions on the kind of properties that can be exhibited by Eliashberg superconductors independent of origin, size or shape of electron-boson exchange spectral density. Quantities considered are the normalized jump in the specific heat, the gap to critical temperature ratio, and the slope of the second upper critical field at T_c.

This work was done in collaboration with R. Akis, J. Blezius, F. Marsiglio, E. Schachinger, and M. Schossmann.

REFERENCES

1) J.M. Daams and J.P. Carbotte, Jour. Low Temp. Phys. 43, 263 (1981).

2) F. Marsiglio and J.P. Carbotte, Phys. Rev. B33, 6141 (1986).

3) B. Mitrovic, H.G. Zarate, and J.P. Carbotte, Phys. Rev. B29, 184 (1984).

4) Many papers in Novel Superconductivity, edited by S.A. Wolf and V.Z. Kresin (Plenum Press, New York, 1987).

5) J. Ruvalds, Phys. Rev. B35, 8869 (1987).

6) V. Kresin, Phys. Rev. B35, 8716 (1987).

7) C.M. Varma, S. Schmitt-Rink, and E. Abrahams, Sol. State Comm. 62, 681 (1987).

8) D. Allender, J. Bray, and J. Bardeen, Phys. Rev. B7, 1020 (1973).

9) V.J. Emery, Phys. Rev. Lett. 59, 2794 (1987).

10) D. Vaknin, S.K. Sinba, D.E. Moncton, D.C. Johnston, J.M. Newsam, C.R. Safinya, and H.E. King, Jr., Phys. Rev. Lett. 58, 2802 (1987)..

11) P.W. Anderson, Science 235, 1196 (1987).

12) S. Kivelson, D. Rokhar, and J. Sethna, Phys. Rev. B35, 8865 (1987).

13) P.W. Anderson, G. Baskaran, Z. Zou, and T. Hsu, Phys. Rev. Lett. 58, 2790 (1987).

14) W. Weber, Phys. Rev. Lett. 58, 1371 (1987).

15) W.L. McMillan and J.M. Rowell in, Superconductivity, edited by R.D. Parks (Dekker, New York, 1969) Vol. 1. p. 562.

16) P.B. Allen and R.C. Dynes, Phys. Rev. B12, 905 (1975).

17) B. Batlogg et al., Phys. Rev. Let. 58, 2333 (1987).

18) L.C. Bourne et al., Phys. Rev. Lett. 58, 2337 (1987).

19) M. Schossmann, F. Marsiglio, and J.P. Carbotte, Phys. Rev. B36, 3627 (1987).

20) A.J. Freeman, J. Yu, and C.L. Fu, Phys. Rev. B36, 7111 (1987).

21) B. Batlogg et al., Phys. Rev. Lett. 59, 912 (1987).

22) T.A. Faltens et al., Phys. Rev. Lett. 59, 915 (1987).

23) F. Marsiglio, R. Akis, and J.P. Carbotte, Phys. Rev. B36, 5245 (1987).

24) J.P. Carbotte, F. Marsiglio, and B. Mitrovic, Phys. Rev. B33, 6135 (1986).

25) J. Blezius and J.P. Carbotte, Phys. Rev. B36, 3622 (1987).

26) F. Marsiglio and J.P. Carbotte, Phys. Rev. B36, 3633 (1987).

27) R. Akis, F. Marsiglio, E. Schachinger, and J.P. Carbotte, Phys. Rev. B (sub).

28) J. Blezius and J.P. Carbotte, Phys. Rev. <u>B33</u>, 3509 (1986).

29) D.R. Harshman et al., Phys. Rev. <u>B36</u>, 2386 (1987).

30) J. Blezius, R. Akis, F. Marsiglio, and J.P. Carbotte (to be published).

MECHANISMS OF HIGH-T$_c$ SUPERCONDUCTIVITY

Werner Weber

Kernforschungszentrum Karlsruhe
Institut für Nukleare Festkörperphysik
P.O.B. 3640, D-7500 Karlsruhe, FRG

ABSTRACT

A survey is given on various models proposed for the electronic pairing in the new cuprate superconductors. This survey includes the results of the one-electron energy band calculations, as well as studies on electron-phonon interaction. The limits of the validity of energy band theory are discussed, and it is argued that - due to strong correlation effects of the Cu d electrons, the electronic structure is severely modified. This leads to the consideration of magnetic and excitonic momodels for the pairing, which try to take these correlation effects into account.

I. INTRODUCTION

Since the discovery of high-Tc superconductivity in various copper oxides [1-2] a huge quantity of theoretical papers has addressed the problem of the electronic ground state in these materials and of the coupling mechanism for superconductivity. Historically, most researchers first used the conventional concepts which have been very successful for the "old" high-T$_c$ materials like the 4d and 5d transition metal compounds of A15 type Nb3Sn, V3Si etc. or the refractory materials like NbN. These concepts include one-electron band theory within the framework of density functional theory in local density approximation (LDA), the assumption of electron-phonon interaction as well as the validity of BCS theory for superconductivity.

Meanwhile there exists overwhelming experimental evidence which invalidates most of the basic assumptions underlying the conventional approach. In particular, it has been found that strong correlation effects exist in the shell of Cu d valence electrons.

These correlation effects have first been postulated in Anderson's resonance valence bond (RVB) model [3] where a "magnetic" coupling mechanism for the pairing has been suggested. A variety of other "magnetic" models has since been proposed[4]. Yet the interaction by spin excitations may not be the only new coupling mechanism which is possible in the limit of large electronic correlations. In addition, there are also charge excitation mechanims possible.

In this review, a brief survey of the theoretical development is given. In Section 2 the approaches using conventional concepts are surveyed, such as results of energy band theory and of calculations of the electron-phonon interaction. Section 3 is devoted to the discussion of the new approaches leading to either "magnetic" or "excitonic" coupling mechanisms. The paper ends with concluding remarks in Section IV.

II. THE OLD CONCEPTS

1. Energy Bands

All bandstructure results within LDA theory[5,6] exhibit very broad bands formed by σ-bonded Cu d orbitals of e_g symmetry; i.e., $d(x^2-y^2)$ and $d(3z^2-r^2)$ orbitals, and oxygen p orbitals. Very large (pd) energy transfer (hopping) matrix elements ($\approx$-1.5 eV) are obtained, as well as an almost perfect hybridization between these orbitals; i.e., for the orbital energies $E_d \approx E_p$ is found. In these bands, the direct hopping matrix elements of neighbour O p orbitals are $\approx$0.5 eV.

Typical energy band results for the La_2CuO_4 system are shown in Fig. 1. The broad σ-bonded p-d bands (approx. 8 eV from the bottom of the bonding part to the top of the anti-bonding part) are almost completely filled. For pure La_2CuO_4 the Fermi energy is situated in the middle of the anti-bonding part of these bands, leaving only one hole per unit cell. Due to the two-dimensional nature of the electronic structure there exist two singularities at half-filling. These are i) a large peak in density of states and ii) a $2k_F$ nesting anomaly, as the Fermi surface approximately forms a square prism. As a consequence a very strong Peierls instability of the crystal structure has been expected.

The instability should lead to a doubling of the unit cell, to the existence of a gap over most (or all) parts of the Fermi surface and thus to the possibility of insulating behaviour. (Band theory predicts all me-

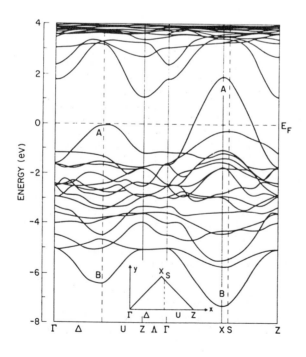

Figure 1
Energy bands for La_2CuO_4 (from Ref. 5). Note the broad band complex denoted by A and B consisting of strongly hybridized Cu $d(x^2-y^2)$ and oxygen p(x), p(y) planar orbitals. The Fermi energy is situated in the anti-bonding part (A) of these bands.

terials with an <u>odd</u> number of electrons per unit cell to be metallic, while those with an <u>even</u> number may be nonmetals).

2. Electron-Phonon Interaction

Utilizing the results of band theory, calculations of the electron-phonon interaction have been performed, employing a tight binding method with a realistic orbital basis[7]. The results of these calculations included the renormalization of the phonon dispersion curves due to the interaction of the phonons with the conduction band electrons, the Fermi surface averages of the squares of the electron-phonon matrix elements, the Eliashberg function $\alpha^2(\omega)F(\omega)$ for strong coupling theory of superconductivity and determination of T_C by numerical solution of the Eliashberg equations.

The study confirmed the results of the energy band calculations that a strong tendency towards a Peierls instability exists for pure La_2CuO_4 (see Fig. 2). The lattice vibration involved was predicted to be alternating breathing-type distortions of the planar squares of oxygen around the Cu ions in the CuO_2 planes of La_2CuO_4. Pure La_2CuO_4 was expected to be a Peierls-gap nonmetal in analogy to $BaBiO_3$[7].

With doping of divalent Sr or Ba for La, the tendency towards the instability was found to be suppressed. For dopant concentrations of

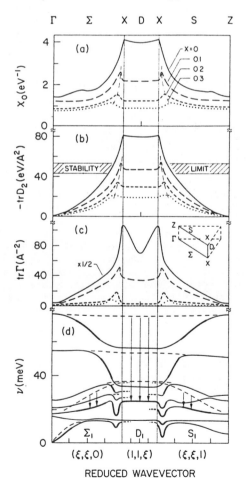

Figure 2
Results of the electron-phonon calculations, as a function of wave vector q. (a) Bare susceptibility χ_o for four different dopant concentrations x (x=0 corresponds to La2CuO4). (b) Corresponding curves for the phonon renormalization D2. Also indicated is the stability limit due to finite bare-phonon stiffness. (c) Linewidth with its very rapid increase for small x (note the scale change). (d) Effect of the phonon renormalization on the dispersion curves (full lines, with D2; dashed lines, without D2). Results are for a calculation with x=0.15 and a specific set of bare-phonon force constants. Only the fully symmetric Σ_1 and S1 branches and some of the D1 phonons are shown. The dispersion of the breathing-type branch is indicated by the strong solid lines (from Ref. 7).

REDUCED WAVEVECTOR

0.1-0.15 per unit cell, the "breathing" phonons should be barely stable, but would now exhibit very strong electron-phonon coupling. This coupling was sufficiently large to yield values of T_c in the range 30-40 K.

At this early stage of development - before the discovery of the 90 K superconductor $YBa_2Cu_3O_7$ - the explanation of the superconductivity in La_2CuO_4 by conventional concepts appeared to be rather convincing. It should be noted, however, that an analoguous calculation for $YBa_2Cu_3O_7$ has later produced theoretical values of the electron-phonon coupling much too small to yield T_c values near 90 K [8].

3. Fundamental Problems of One-Electron Theory

Yet much before these results were obtained, it became obvious that various experimental results were in contradiction with the predictions of band theory and the electron-phonon calculations. For example, La_2CuO_4 was indeed found to be a nonmetal - yet it also turned out to be an anti-ferromagnet with a large magnetic moment of 0.6 μ_B per Cu [9]. No Peierls-type distortions have so far been found by crystallographic studies. In addition, the electronic gap appears to be very large (3 eV). Furthermore, other materials like La_2NiO_4 or YBa_2CuO_6 [10] are also magnetic non-metals, while band theory predicts them to be metals [11]. In summary, there appear to exist fundamental problems for the validity of band theory in this class of materials.

III. NEW APPROACHES

1. Electronic Correlations

The observation that cuprates can exhibit nonmetallic behaviour is reminiscent of the oxides and halides of most 3d transition metals. P.W. Anderson was the first to point out the similarity of the cuprates to these latter compounds and to propose a new approach to deal with superconductivity in the cuprates [3].

Oxides of 3d transition metals like NiO or FeO are all insulators, irrespective whether the number of valence electrons per unit cell e/uc is even or odd. According to Mott and van Vleck, the reason for this behaviour is the large energy necessary to change the specific valence of an ion with incomplete 3d shells, i.e., being in a certain valence configuration d^n with n<10. A charge fluctuation between adjacent 3d ions of the type 2. $d^n \rightarrow d^{n+1} + d^{n-1}$ requires the energy U which is large compared to any gain in kinetic energy t due to hopping of the electrons through the lattice. The suppression of charge fluctuations leads to a strong correlation of the electronic motion, and for integer numbers of e/uc, the electrons are localized completely, leading to a gap in the excitation spectrum. In addition, magnetic order is observed, in general anti-ferromagnetism.

The simplest model to describe such a situation is the Hubbard model. It assumes a single non-degenerate orbital per atom, a hopping matrix element t to nearest neighbours and a repulsive energy U when a site is doubly occupied by spin↑ and spin↓ electrons. The Hamiltonian is thus given by

$$H = \sum_{ij}' t(a_i^+ a_j + c.c) + U \sum_i n_{i\sigma} n_{i-\sigma}$$

where the prime indicates nearest neighbour sums. For the half-filled band e/uc=1, the limit t/U<<1 yields a ground state with preferential

156

antiferromagnetic order of neighbour spins. This spin arrangement allows virtual hopping to neighbour sites resulting in an energy gain of order t^2/U. In other words, neighbour electron spins interact with an exchange constant $J=-t^2/U$.

In the case of 3d compounds with ligands exhibiting closed electronic shells like O^{--}, there should be no direct exchange coupling possible. For these materials, Anderson had introduced the "superexchange" mechanism, with $J_{sx} \approx -t'^2/U$ and $t'=t^2/(E_d-E_p)$. Here t is the hopping matrix element between a 3d and ligand orbital, while E_d and E_p are the respective orbital energies. In this way, the more complicated case of an 3d oxid or halide can also be mapped onto the one-orbital Hubbard model.

2. "Magnetic" Coupling Models

For the case $e/uc \neq 1$, the Hubbard model always yields metallic behaviour - although strongly modified compared to simple metals, as the electronic motion has to be highly correlated. Even small amounts of doping lead to the suppression of the long range anti-ferromagnetic order. However, no comprehensive theory of the Hubbard model does exist for the case $e/uc \neq 1$.

Anderson [3] has proposed that the essential physics of the cuprates can be projected onto the $e/uc \neq 1$ Hubbard model in a square planar lattice - representing the Cu sites of the CuO_2 planes. Without doping, there are Cu^{++} ions, corresponding to a d^9 configuration. Due to the two-dimensional symmetry of the lattice, the usual long-range antiferromagnetic order of the Néel state does not exist (or is barely stable). Instead it is postulated that the spins on the square lattice can be described as a distribution of spin singlet pairs - not necessarily only between nearest neighbour sites. However, these pairs are not fixed to certain sites, but may fluctuate on the lattice like resonating bonds on a benzene molecule fluctuate over the ring - thus the name resonating valence bond (RVB) model. With doping, the pairs may break up in a way that a spinless charge carrier (boson-like "holon") is created while the remaining Fermion is a neutral object, as is always attached to a Cu ion (neutral fermion-"spinon").

The "holons" may form pairs via virtual exchange of spin excitations. In contrast to phonon exchange with typical oxygen frequencies of $\approx 40-60$ meV, the spin excitation energies are expected to be of order $J_{sx} \approx 100-200$ meV. This increase in the energy of the intermediate boson thus would allow much larger values of T_c.

Up to now, various other models have been presented proposing "magnetic" coupling mechanisms [12,13]. The school of "magnetic" mechanisms has been supported by important experimental results, in particular by the finding of very strong magnetic fluctuations in pure La_2CuO_4 - even at much larger temperatures than the anti-ferromagnetic ordering temperature T_N [14,15]. A common feature of the "magnetic" models is their tendency towards p or d wave pairing, due to the repulsive nature of the underlying interaction.

3. The Role of Oxygen

With doping, the charge carriers may either be situated on the Cu or the O sites or on both of them. If they were found on Cu, a significant fraction of Cu^{+++} ions had to be expected. Yet the latter possibility is not confirmed by spectroscopic investigations - instead there are a lot of data suggesting that the charge carriers are predominantly situated on the oxygen ions. The most direct evidence for this behaviour has been found from oxygen core level spectra using electron energy loss spectroscopy [16,17]. These results do not eliminate the possibility of magnetic coupling, yet it raises the question whether a single orbital Hubbard model is still sufficient to describe the essential physics of the cuprates [12,18].

Another important problem is the position of the oxygen p band relative to the Cu d^9 configuration. In the case of the undoped cuprates La_2CuO_4 and $YBa_2Cu_3O_6$ this difference describes the energy of a "charge transfer" excitation of the form $d^9 O^{--} \rightarrow d^{10} O^-$. This energy E_{CT} is independent of U and this provides an additional material parameter to be described for the 3d compounds [19].

For CuO, a rather small value of $E_{CT} \approx 1.5$ eV has been observed. Assuming a similar situation for the high-T_c cuprates Varma et al. [20] have suggested that low-lying and rather localized charge transfer excitations should exist. They have estimated an excitation energy of ≈ 0.5 eV. These excitations are postulated to mediate the pairing of the charge carriers. In addition, an isotropic pairing mechanism analogous to the electron-phonon coupling was expected – resulting in s-wave pairing like in conventional BCS theory. Various experiments indeed suggest BCS-type behaviour of the condensate, e.g., the temperature dependence of the London penetration depth [21]. On the other hand, optical data [22,23] indicate a very large value of $E_{CT}(\approx 4$ eV). This result would not support the charge transfer excitation model.

Summarizing the spectroscopic data, the author considers the following electronic structure the most probable one for the cuprates (see Fig. 3): In hole representation, Cu^{++} with d^9 configuration has one hole in the d shell. Due to the large correlation energies involved, a double occupancy of this hole state (corresponding to d^8, i.e. Cu^{+++}) requires a large energy $U(\approx 6-8$ eV). On the other hand, the d^9 state is not sharp, but has a width of order J_{sx} (≈ 0.2 eV) due to superexchange coupling to neighbour states.

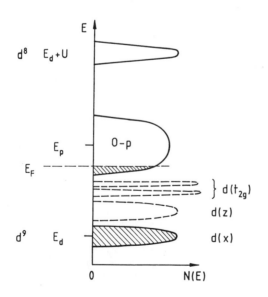

Figure 3
Model of the electronic structure of the cuprates (in hole representation). See text for details (from Ref. 24).

Approximately 4 eV above this "lower" Hubbard d hole band, there is a band of O p holes with an estimated width of 3-4 eV, due to i) direct pp hopping ($t_{pp} \approx 0.5$ eV) and ii) indirect hopping through Cu of order $t_{pd}^2/(E_p-E_d) \approx 0.5$ eV. This O p hole band is empty in the case of the non-metallic cuprates La_2CuO_4 and $YBa_2Cu_3O_6$ (corresponding to a O^{--} situation). With doping of either Sr, Ba or O, this band will be partially occupied. Because of the very small carrier concentration (≈ 0.1 hole/oxygen), correlation effects are unimportant in this band.

4. The d-d Excitation Model

In all magnetic 3d oxides and halides there exists another charge excitation [25] which has so far been ignored in the present discussion. If we consider the open d shell of an isolated 3d ion, i.e., in spherical symmetry, the ground state is multiply degenerate. For Cu^{++} with d^9 there is a five-fold degenerate ground state. This degeneracy is lifted in the crystal field, mostly due to interactions with the nearest neighbour ligands. For 3d oxides, these valence conserving charge excitations lie in the range of 0.5 - 3 eV; i.e., in general below the charge transfer excitation gap [25,26] (see Fig. 3). Evidence for the presence of d-d excitations in the cuprates has been found from optical transmission studies of $YBa_2Cu_3O_7$ films [22,23].

A perfect ligand octahedron leads to a cubic crystal field, with two degenerate orbital states of eg symmetry, viz. $d(x^2-y^2)$ and $d(3z^2-r^2)$ and three degenerate states of t_{2g} symmetry, viz. d(xy), d(yz) and d(zx). As the two e_g orbitals undergo strong σ-bonded interactions with the O p orbitals, while the t_{2g} orbitals have weaker π-type interactions, it is most probable that the ground state orbital is of eg symmetry (this is also in agreement with results from band theory). Consequently, the ground state of d^9 is still doubly degenerate - the prerequisite for the Jahn-Teller effect. Any distortion of the ligand octahedron will lift the degeneracy - with the energy gain being linear in the distortion coordinate. In chemistry, Cu^{++} is well known to exhibit a very large Jahn-Teller splitting, up to 0.5 eV [27].

For the crystal structures of the high-T_c cuprates, it appears at first glance as if the Jahn-Teller case if not given, since the Cu site symmetries are always lower than cubic. However, one should rather argue the opposite way: the Jahn-Teller energy is so large that only crystal structures are formed which incorporate this energy in the arrangements of the atoms. For La_2CuO_4 this is possible by a very large axial elongation of the oxygen octahedron - the ratio of the Cu-O bond lengths b_{axial}/b_{plane} = 1.25 versus 1.1 for the corresponding ratio of La_2NiO_4 (Ni^{++} is not a Jahn-Teller ion in octahedral coordination) [27]. For the Perovskite-defect structure $YBa_2Cu_3O_7$ the oxygen octahedra are even incomplete - and missing ligands also cause the splitting of the e_g levels, probably a bigger one than the octahedral distortion.

As the big splitting of the eg levels is caused by the strong σ-type interactions with the p orbitals of the O^{--} ions, any doping of the O p hole band will lead to a strong interaction between the p holes and the $d(e_g)$ orbitals. This is the basic idea of the author's d-d excitation model for the pairing in the cuprates [24].

It is argued that a O p hole changes the d^9 ground state of a neighbour Cu^{++} ion by admixing the $d(3z^2-r^2)$ orbital to the $d(x^2-y^2)$ one. This change can be described by virtual excitations into the $d(3z^2-r^2)$ state. The tendency to $d(3z^2-r^2)$ admixture can be made plausible by inspection of Fig. 4. It is evident that the Coulomb repulsion of the planar $d(x^2-y^2)$

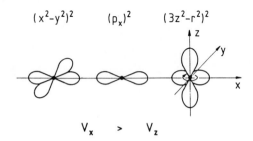

$(x^2-y^2)^2$ $(p_x)^2$ $(3z^2-r^2)^2$

V_x > V_z

<u>Figure 4</u>
Coulomb interaction
between an oxygen p(x)
hole and a Cu d(x^2-y^2)
hole (V_x) or a d($3z^2-r^2$)
hole (V_z). The diffe-
rence in the charge
distributions leads
to $V_x > V_z$ (from Ref. 24).

and p(x) or (p(y)) orbitals is considerably larger than that of the p(x)
with the axial d($3z^2-r^2$) orbital. The difference V of the two Coulomb
repulsion integrals is estimated to be ≈0.3-0.5 eV.

In other words, a p hole moving through the lattice, carries with it
a "polaronic" cloud of d($3z^2-r^2$) excitations, similar as a conduction
electron also carries a cloud of lattice distortions or virtual phonon
excitations. A second O p hole on the O square around the Cu^{++} ion is
attracted to this orbital "polaron" and by enhancing the excitation can
bind with the first hole to form a pair.

In simplest approximation, the pair binding energy is given in analogy to electron-phonon interaction by the excitation energy E_{JT} (instead
of a phonon energy h<ω>) and an exponential term exp(-1/λ) with
$λ ≈ ΔV^2/(E_{JT} \cdot t_{pp})$, so that

$$T_c ≈ E_{JT} \exp(-1/λ).$$

Using $E_{JT} ≈ 5000$ K and $λ ≈ 0.1-0.3$, values for T_c certainly have the
right order of magnitude.

As simple as the basic idea of this coupling mechanism may appear,
as complicated is the theoretical approach. The Hamiltonian is much more
complex than that of the simple Hubbard model - there is a spin and an
orbital degree of freedom on the Cu site, and there is another orbital
on the oxygen sublattice. Presently certain limiting cases are investigated [28].

However, the model allows certain qualitative conclusions. It limits
the occurrence of the coupling mechanism

i) to the presence of a "strong" Jahn-Teller ion
ii) to the presence of the charge carriers on the ligand sublattice.

These are rather limiting restrictions as far as further possible superconductors are concerned. For instance, as Ni^{++} is not a Jahn-Teller
ion, doped La_2NiO_4 should not be found superconducting. Corresponding Cu^{++}
halides are probably excluded, as, in case of doping, the charge carriers
are not necessarily situated on the halide lattice.

In addition, the d^9 charge excitation will always result in a coupling to vibrations of the O octahedra. Consequently, O isotope effects
in the cuprates are therefore expected. The fact that they are found to be
rather different in magnitude, may be related to different values of the
Jahn-Teller excitation energy in doped La_2CuO_4 and in $YBa_2Cu_3O_7$.

IV. CONCLUDING REMARKS

In this survey, the author's view of the essential theoretical problems for the high T_c cuprates has been presented. It has been argued that the conventional – and elsewhere very successful – concepts like one-electron band theory or the usual electron-phonon coupling are not appropriate for these materials, as they lead to conflict with experiment. The problems of the one-electron approximation are caused by strong correlation effects. In particular, valence fluctuations of the Cu d^9 configuration appear to be strongly suppressed. These effects – similar as in other oxides and halides of 3d transition elements – have to be treated using new theoretical approaches. In the limit of large electronic correlation energies, anti-ferromagnetism is present, and one school of models suggests pairing of electrons via the exchange of virtual spin excitations. In addition to the "magnetic" coupling mechanism, there are also pairing mechanisms possible due to exchange of virtual charge excitations – either of charge transfer type or of valence conserving excitations in the d^n shell involving Jahn-Teller coupled orbitals.

The theoretical treatment of these various models is extremely complicated. Yet all of these models have to be investigated in great detail to find further insight into the physics of these fascinating materials. In the author's view this will be the theoretical challenge in solid state physics for the years to come.

REFERENCES

1. J.G. Bednorz and K.A. Müller, Z. Phys. B 64, 189 (1986)
2. M.K. Wu, J.R. Ashburn, C.J. Torng, P.H. Hor, R.L. Meng, L. Gao, Z.J. Huang, Y.O. Wang, and C.W. Chu, Phys. Rev. LEtt. 58, 908 (1987)
3. P.W. Anderson, Science 235, 1196 (1987);
   P.W. Anderson, G. Baskaran, Z. Zou, and T. Hsu, Phys.Rev. Lett. 26, 2790 (1987)
4. see, e.g. Proc. 23rd Yamada Conf. Superconductivity in Highly Correlated Fermion Systems (ed.s M. Tachiki, Y. Muto and S. Maekawa), North Holland, Amsterdam (1987)
5. L.F. Mattheiss, Phys. Rev. Lett. 58, 1028 (1987);
   J. Yu, A.J. Freeman, and J.H. Xu, Phys. Rev. Lett. 58, 1035 (1987)
6. S. Massida, J. Yu, A.J. Freeman, and D.D. Koelling, Phys. Lett. 122, 1987 (1987); J. Yu, S. Massida, A.J. Freeman, and D.D. Koelling, Phys. Lett. 122, 203 (1987)
7. W. Weber, Phys. Rev. Lett. 58, 1371 (1987)
8. W. Weber and L.F. Mattheiss, Physica 148B, 271 (1987)
9. D. Vaknin, S.K. Sinha, D.E. Moncton, D.C. Johnston, J.M. Newsam, C.R. Safinya, and H.E. King, Jr., Phys. Rev. Lett. 58, 2802 (1987), and references therein
10. J.M. Tranquada, D.E. Cox, W. Kunnmann, H. Moudden, G. Shirane, M. Suenaga, P. Zolliker, D. Vaknin, S.K. Sinha, M.S. Alvarez, A.J. Jacobson, and D.C. Johnston, Phys. Rev. Lett. 60, 156 (1988)
11. F. Herman, R.V. Kasowski, and W.Y. Hsu, Phys. Rev. B, in print.
12. V. Emery, Phys. Rev. Lett. 58, 2794 (1987)
13. C. Gros, R. Joynt, T.M. Rice, Z. Phys. B 68, 425 (1987)
14. G. Shirane, Y. Endoh, R.J. Birgeneau, M.A. Kastner, Y. Hidaka, M. Oda, M. Suzuki, and T. Murakami, Phys.Rev. Lett. 59, 1613 (1987)
15. G.A. Kourouklis, A. Jayaraman, W. Weber, J.P. Remeika, G.P. Espinosa, A.S. Cooper, and R.G. Maines,Sr,Phys.Rev. B 36, 7218 (1987)
16. N. Nücker, J. Fink, J.C. Fuggle, P.J. Durham, and W.M. Temmerman, Phys.Rev. B, submitted

17. J. Fink, N. Nücker, J.C. Fuggle, P.J. Durham, and W.M. Temmerman, this volume
18. F.C. Zhang and T.M. Rice, preprint
19. J. Zaanen, G.A. Sawatzky, and J.W. Allen, Phys. Rev. Lett. 55, 418 (1985)
20. C.M. Varma, S. Schmitt-Rink, and E. Abrahams, Solid State Commun. 62, 681 (1987)
21. D.R. Harshman, G. Aeppli, E.J. Ansaldo, B. Batlogg, J.H. Brewer, J.F. Carolan, R.J. Cava, M. Celio, A.C.D. Chaklader, W.N. Hardy, S.R. Kreitzman, G.M. Luke, D.R. Noakes, and M. Senba, Phys. Rev. B 36, 2386 (1987)
22. H.P. Geserich, G. Scheiber, J. Geerk, H.C. Li, G. Linker, W. Assmus, and W. Weber, Europhys. Lett. (submitted)
23. H.P. Geserich, G. Scheiber, J. Geerk, H.C. Li, W. Weber, H. Romberg, N. Nücker, and J. Fink, this volume
24. W. Weber, Z. Phys. B, to be published
25. see, e.g. J.W. Allen, in "Magnetic Oxides", (ed. D.J. Craik), Wiley, London (1975) p. 349
26. R. Newman and R.M. Chrenko, Phys. Rev. 114, 1507 (1959)
27. D. Reinen, Comments Inorg. Chem. 2, 227 (1983)
28. A.L. Shelankov, X. Zotos, and W. Weber, in preparation

ANISOTROPY EFFECTS IN THE SYSTEM La-Sr-Cu-O [*]

W.Pint, M.Prohammer and E.Schachinger

Institut für Theoretische Physik
Technische Universität Graz
A-8010 Graz, Austria

ABSTRACT

Recent experimental and theoretical studies on the upper critical magnetic field, the Hall effect and the thermopower indicate that the system La-Sr-Cu-O is highly anisotropic. This theoretical study attempts to describe the thermodynamic data consistently within the framework of an anisotropic strong coupling theory. The results of the theoretical study agree well with the data within experimental accuracy.

1. INTRODUCTION

It is well established by experiment [1,2] and theory [3] that the high-T_C superconductor $La_{1.85}Sr_{0.15}CuO_4$ (LSCO) is a highly anisotropic system. Anisotropy, on the other hand, affects greatly the superconducting properties by increasing the critical temperature T_C [4], decreasing the gap edge Δ_0 [5], the specific heat jump at T_C ΔC [6] and the thermodynamic critical field $H_C(T)$ [7]. Thus, anisotropy has to be included in any attempt to analyze experimental data by theoretical models. There is also some evidence from tunneling spectroscopy that this superconductor might be a conventional strong coupling system [8].

There were two attempts to analyze LSCO experimental thermodynamics data using strong coupling Eliashberg theory. In both cases anisotropy was neglected. Ashauer et al. [9] studied various models for the electron-phonon interaction spectral function $\alpha^2F(\omega)$ based on theoretical computations by Weber[10] and on the measured phonon density of states[11] mainly to prove that a conventional theory is not necessarily unreasonable in describing high-T_C superconductivity. Schossmann et al. [12] engaged in a more quantitative analysis of experimental data using Weber's $\alpha^2F(\omega)$ to describe the electron-phonon interaction. They reported a rather large mass enhancement factor λ of 2.6 and found pronounced inconsistencies between theoretical predictions and experimental findings. They emphasized that these discrepancies were large enough to indicate a possible breakdown of conventional theory of superconductivity for the high-T_C's.

We believe that these discrepancies can largely, if not entirely, be

* Research supported in part by Forschungsfonds of the Styrian Government, Austria.

compensated for by taking into account the anisotropic features of the system. Thus, we organize this paper as follows: Section 2 gives a summary of the theoretical anisotropy model used for our calculations, Section 3 describes the results of these calculations and, finally, in Section 4 some conclusions are drawn.

2. THEORY

In a general anisotropic Eliashberg theory [6], formulated on the imaginary axis, the renormalized Matsubara frequencies $\tilde{\omega}_k(n)$, the Matsubara gaps $\Delta_k(n)$ and all material parameters are expanded into series using a special orthonormal system of functions, the Fermi Surface Harmonics (FSH)[13]. The simplest possible model consists of two spherical Fermi surface regions which allows us to keep only the zeroth order terms in the FSH series expansions. Thus the anisotropy of the electron-phonon interaction can be described by the separable model [4]:

$$\alpha^2F(\omega)_{\underline{k},\underline{k}'} = (1+a_{\underline{k}})<\alpha^2F(\omega)>(1+a_{\underline{k}'})$$ (1)

where $\underline{k}$ and $\underline{k}'$ present the incoming and outgoing quasiparticle momentum vectors for the electron-phonon scattering process, $a_{\underline{k}}$ describes the deviation of the real Fermi surface in direction $\underline{k}$ from an isotropic one of the same volume and $<\alpha^2F(\omega)>$ is the isotropic (Fermi surface average) electron-phonon interaction spectral function found by tunneling experiments or band structure calculations.

If the two FS regions have the same weight, the mean square anisotropy $<a^2>$ of the electron-phonon interaction is the only additional material parameter which enters the theory. (It was shown by Niel et al. [7] that this model gives accurate results for anisotropic indium.)

Usually $<a^2>$ is determined from T_c degradation with increasing impurity concentration. This data is not available for LSCO and we have to treat $<a^2>$ as an additional parameter which can be used to establish agreement between theory and experiment.

3. RESULTS

We can choose from two sets of data for $\alpha^2F(\omega)$. Both are displayed in Fig. 1 for a mass enhancement factor $\lambda=2.6$. The full line presents Weber's data calculated from first principle band structure analysis (S1)[10]. The dashed line, on the other hand, corresponds to an $\alpha^2F(\omega)$ calculated from Renker et al.'s [11] phonon density of states by simple rescaling (S2). If we neglect the very strong coupling of electrons at frequencies around 10 meV in spectrum S1, the two spectra are in good agreement.

To understand the role both spectra play in a strong coupling theory we calculate, in a first step, the Coulomb pseudopotential μ^* which describes the electron depairing by the repulsive Coulomb interaction using isotropic Eliashberg theory. For a T_C of 36 K spectrum S1 gives a μ^* of 0.123, while spectrum S2 gives a μ^* of 0.568. The cutoff frequency was chosen to be 498 meV and 576 meV respectively.) This result is not surprising if one calculates the Allen-Dynes frequency ω_{log} [14]. Spectrum S1 gives 13.9 meV and S2 gives 19.2 meV. From the T_C equation by Leavens and Carbotte [15] it becomes immediately transparent that spectrum S2 is much more effective and therefore needs a stronger Coulomb repulsion to give the same T_C as spectrum S1. We can tnerefore reduce the λ of spectrum S2 considerably as "reasonable" values of μ^* are usually assumed to be in the range of 0.05÷0.3 and we do not want to discuss possible effects of disorder [16].

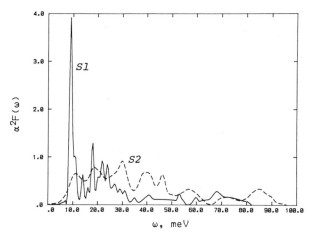

Fig. 1. Model electron-phonon interaction spectral function $\alpha^2F(\omega)$ for
La$_{1.85}$Sr$_{0.15}$CuO$_4$ with a mass enhancement factor $\lambda=2.6$. The full
line corresponds to results reported by Weber [10], while the dashed
line represents rescaled phonon density of states measured by
Renker et al.[11].

To describe anisotropy we have to choose a value for $\langle a^2 \rangle$. As T_c is
also a function of the impurity content of the specific sample we have to
decide on a value for $t^+=1/(2\pi\tau_{tr})$, where τ_{tr} is the transport relaxation
time. A good estimate for τ_{tr} can be found from the residual resistivity
ρ_0 of the sample using Drude theory:

$$\tau_{tr} = 4\pi/(\rho_0\Omega_p^2) \quad . \tag{2}$$

Here Ω_p is the Drude plasma frequency. Good samples have a residual
resistivity of 100÷300 $\mu\Omega$cm [17,18] . This gives together with the measured
Ω_p of 0.83 eV [19] values of t^+ in the range of 1.5÷4.5 meV. This is not at
all in the dirty limit ($t^+>15$ meV). Thus, we can fully characterize our
theoretical model samples by the values of $T_c,\lambda,\langle a^2 \rangle,t^+,$ μ^* and the spectrum
model. (μ^* is calculated from anisotropic linear Eliashberg equations [20].)

The thermodynamics are calculated from the solutions of the full nonlinear anisotropic Eliashberg equations and require two more material
parameters. Firstly, the number of atoms per unit volume N_A. It is easily
calculated from the volume of the unit cell $\Omega=1.88\cdot10^{-28}$ m^3 [21] which contains 14 atoms and we find $N_A=7.423\cdot10^{28}$ atoms/m^3. Secondly, we need either
the Sommerfeld specific heat constant γ, or the quasiparticle density of
states at the Fermi energy $N(0)$. The former is quoted in numerous experimental papers but is usually calculated from the specific heat jump ΔC at
T_c using the BCS relation $\Delta C/(\gamma T_c)=1.43$. This relation does not hold in
the case of LSCO and we prefer to use $N(0)$ instead. It was calculated by
Allen et al.[3] from band structure analysis to give $N(0)=1.045$ states/eV,
Cu-atom or $1.108\cdot10^{28}$ states/eV,m^3. This value is significantly smaller
than values reported by experiment [17,21] (where the $1+\lambda$ renormalization
was not performed). It is, on the other hand, about 60% greater
than the value found by Schossmann et al.[12] from calculations using isotropic Eliashberg theory.

Finally, the gap edge Δ_0 is calculated by analytically continuing
the Fermi surface average of the Matsubara gaps $\bar{\Delta}_k(n)$ using Padé approxi-
mants [22] . A typical result for the quasiparticle density of states
$N(\omega)/N(0)$ is presented in Fig. 2. It shows two peaks which represent the

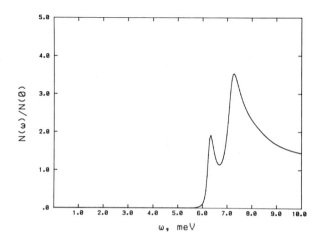

Fig. 2. Quasiparticle density of states at T=0 K calculated by analytically
continuing the Fermi surface average Matsubara gaps by Padé approx-
imants for a S2-type spectrum and $<a^2>$ =0.1. The two peaks are the
result of the anisotropy model chosen for our calculations. (See
also Section 2.)

two square root singularities in the density of states as a result of our
anisotropy model. The smearing comes from the fact that the analytic contin-
uation was performed for a rather high temperature, namely 1.8 K.

Table 1. Comparison of theoretical model calculations to experimental data
for the system $La_{1.85}Sr_{0.15}CuO_4$. S1 indicates Weber-like $\alpha^2F(\omega)$
spectra, while S2 indicates spectra which are proportional to the
phonon density of states. (T_C is in K, t^+ in meV, $\Delta C/T_C$ in mJ/g-at/K^2,
$\mu_0H_C(0)$ in mT, $dH_C(T)/dt$ at T_C in mT/K, Δ_0 in meV, γ in mJ/g-at/K^2
and R is the rms gap anisotropy in %.)
^{a}gap in c-direction, Ref. 26.

Spectrum	S1	S1	S2	S2	S1	S2	exp
T_C	36	36	36	36	36	36	35÷39
μ^*	0.108	0.081	0.143	0.177	0.132	0.117	
t^+	3.0	3.0	3.0	3.0	3.0	3.0	
λ	2.4	2.2	1.7	1.7·	2.6	1.7	
$<a^2>$	0.1	0.2	0.1	0.2	0.0	0.0	
$\Delta C/T_C$	3.13	2.76	2.22	2.07	3.53	2.38	2÷2.4
$\Delta C/\gamma(0)T_C$	2.62	2.46	2.34	2.08	2.79	2.50	
$\mu_0H_C(0)$	388	371	331	325	404	336	
$-dH_C/dT\|T_C$	22	21	18.5	17.9	23.4	19.2	14÷20
Δ_0	7.4	6.9	6.30	5.5	8.25	7.2	3.1÷9.8
$2\Delta_0/k_BT_C$	4.67	4.26	4.06	3.55	5.32	4.64	2^a÷6.3
$\gamma(0)$	1.2	1.12	0.95	0.95	1.27	0.95	
R	11	16	13	20	0.0	0.0	

Table 1 presents the results of our calculations for four anisotropic
and two isotropic theoretical model samples and includes experimental
data for easy comparison. For the model samples with S1-type $\alpha^2F(\omega)$ spectrum
we use the fact that λ can be reduced with increasing anisotropy by keeping
μ^* close to the reasonable value of 0.1 in order to achieve better agreement with experiment. The second model sample with a λ of 2.2 and an
$<a^2>$ of 0.2 gives a $\Delta C/T_C$ of 2.76 mJ/g-at/K^2 which is already close to

the range of 2 $\pm$ 30% mJ/g-at/K² found by Junod et al.[23]. The slope
$-dH_c/dT$ at T_c also agrees nicely with the value of 20 mT/K quoted by Renker
et al.[11] but disagrees with the value of 14 mT/K reported by Finnemore et
al.[21]. The value for $2\Delta_0/(k_BT_c)$ of 4.26 is also well in the experimental
range and agrees to the average value of 4.5 quoted by Kirtley et al.[24]
found by tunneling measurements.

It is no surprise that the situation improves substantially when we
investigate theoretical model samples on the basis of S2-type $\alpha^2F(\omega)$
spectra. It is possible to reduce λ significantly and its value of 1.7 is
comparable to that of the A-15 superconductor Nb_3Sn. The rather strong
coupling of the electrons to high frequencies ($\omega > 50$ meV) leads to an
additional decrease in the thermodynamic data as one can expect from specific heat functional derivatives [25]. A comparison of the results for the two
anisotropic S2-type spectrum model samples reveals how $<a^2>$ can be used
to "tune" the theoretical results almost perfectly into the experimental
data.

The two isotropic models are included in Table 1 to demonstrate that
anisotropy is an essential feature of our analysis. The S1-type model
sample is well off the experimental results while the S2-type model sample
has thermodynamic data which are already at the upper end of the experimental range. Including anisotropy improves substantially the agreement
between theory and experiment.

4. CONCLUSION

We calculated thermodynamic data and the gap edge Δ_0 for a
$La_{1.85}Sr_{0.15}CuO_4$ system using anisotropic Eliashberg theory. It is possible
to find good agreement between theory and experiment assuming a rather
substantial rms anisotropy of ≈ 45 % for the electron-phonon interaction
and keeping in mind the still very substantial uncertainties in the experimental data. The results of our analysis also suggest that the $\alpha^2F(\omega)$
spectrum in this material will have less weight at frequencies around
10 meV than was suggested by Weber [10]. In contrast, the coupling to higher
frequencies becomes more pronounced. Furthermore, our calculations confirmed the value for the quasiparticle density of states at the Fermi
energy reported by Allen et al.[3].

The calculated values for $\Delta C/\gamma(o)T_c$ are in the range 2.3 - 2.8 and
it is not possible to use the BCS value of 1.43 for $\Delta C/\gamma(o)T_c$ to calculate
$\gamma(o)$ from the specific heat jump.

Despite the fact of the still unresolved controversy on the partial
isotope effect for oxygen (theoretical value for $\beta_0=0.3$ and the experiment
shows a β_0 of o.16 [27]) we believe that this study strongly supports the
idea that the high-T_c superconductor LSCO is a conventional superconductor
with very pronounced anisotropic features.

REFERENCES

1. Y.Hidaka, Y.Enomoto, M.Suzuki, M.Oda and T.Murakani, Jpn. J. Appl.
    Phys. 26:L377 (1987).
2. S.Shamoto, M.Onoda, M.Sato and S.Hosoya, Jpn. J. Appl. Phys.
    26:1131, Suppl. 26-3 (1987).
3. P.B.Allen, W.E.Pickett and H.Krakauer, Phys. Rev. B 36:3926 (1987).
4. D.Markovitz and L.P.Kadanoff, Phys. Rev. 131:563 (1963).
5. P.G.Tomlinson and J.P.Carbotte, Phys. Rev. B 13:4738 (1976).
6. H.G.Zarate and J.P.Carbotte, J. Low. Temp. Phys. 55:67 (1984).

7. L.Niel, N.Giesinger, H.W.Weber and E.Schachinger, Phys. Rev. B 32:2976 (1985).
8. J.Moreland, A.F.Clark, L.F.Goodrich, H.C.Ku and R.N.Shelton, Phys. Rev. B 35:8711 (1987).
9. B.Ashauer, W.Lee and J.Rammer, Z. Phys. B 67:147 (1987).
10. W.Weber, Phys. Rev. Lett. 58:1371 (1987).
11. B.Renker, F.Gompf, E.Gehring, N.Nücker, B.Ewert, W.Reichart and H.Rietschel, Z. Phys. B 67:15 (1987).
12. M.Schossmann, F.Marsiglio and J.P.Carbotte, Phys. Rev. B 36:3627 (1987).
13. W.H.Butler and P.B.Allen, in: "Superconductivity in d- and f-Band Metals", D.H.Douglas, ed., Plenum, New York (1976), p.73.
14. P.B.Allen and R.C.Dynes, Phys. Rev. B 12:905 (1975).
15. C.R.Leavens and J.P.Carbotte, J. Low Temp. Phys. 14:501 (1974).
16. E.Schachinger and J.P.Carbotte, Solid State Commun. 53:997 (1985).
17. S. Uchida, H.Takogi, K.Kishio, K.Kitazawa, K.Fucki and S.Tanaka, Jpn. J. Appl. Phys. 26:L443 (1987).
18. J.M.Tarascon, Science 235:1373 (1987).
19. S.Tajima, S.Uchida, S.Tanaka, S.Kaube, K.Kitazawa and K.Fucki, Jpn. J. Appl. Phys. 26:L432 (1987).
20. J.M.Daams, E.Schachinger and J.P.Carbotte, J. Low Temp. Phys. 42:69 (1981).
21. D.K.Finnemore,R.N.Shelton, J.R.Clem, R.W.McCallum, H.C.Ku, R.E. McCarley, S.C.Chen, P.Klavins and V.Kogan, Phys. Rev. B 35:5319 (1987).
22. R.C.Leavens and D.S.Ritchie, Solid State Commun. 52:137 (1985).
23. A.Junod, A.Bezinge, D.Cattani, J.Cors, M.Decroux, O.Fischer, P.Genoud, L.Hoffmann, J.-L.Jorda, J.Muller and E.Walker, Jpn. J. Appl. Phys. 26:1119, Suppl. 26-3 (1987).
24. J.R.Kirtley, C.C.Tsuei, S.I.Park, C.C.Chi, J.Rozen and M.W. Shafer, Phys. Rev. B 35:7216 (1987).
25. F.Marsiglio, J.P.Carbotte and E.Schachinger, J. Low Temp. Phys. 65:305 (1986).
26. Z.Schlesinger, R.T.Collins, M.W.Shafer and E.M.Engler, Phys. Rev. B 36:5275 (1987).
27. B.Batlogg, G.Kourouklis, W.Weber, R.J.Cava, A.Jayaraman, A.E. White, K.T.Short, L.W.Rupp and E.A.Rietman, Phys. Rev. Lett. 59:912 (1987).

INFLUENCE OF PRESSURE AND SUBSTITUTION OF COPPER BY TRANSITION METALS ON HIGH-T$_c$ SUPERCONDUCTORS

Jacek Kasperczyk

Institute of Physics , Pedagogical University
Al.Zawadzkiego 13/15, PL-42201 Czestochowa (Poland)

INTRODUCTION

The discovery of high-temperature superconductors of the LaBaCuO and YBaCuO type [1,2] promoted searching for new microscopic mechanisms of superconductivity. A variety of models was proposed to explain the anomalous properties of this new family of superconductors. Some of them neglect totally the classical (BCS) pairing caused by the electron- phonon interaction and introduce a novel mechanism, e.g. exchange of excitons [3] or the RVB state [4]. Other models consider electron-phonon interaction associated with different effects, e.g. an enhancement of the phonon mode due to spin density wave [5] or develop the idea of vibronic modes [6].

It seems that pure electron-phonon coupling is not sufficient. Although the pairing mechanism is not quite clear at the moment a concensus exists that the general structure of BCS theory should be preserved. Therefore, in the following we will not specify the nature of superconducting coupling. We propose to consider the problem of pairing between fermions (holes) by exchange of quanta of a boson field (phonon, exciton or mixed).

SOME EXPERIMENTAL RESULTS

After the discovery of 40K and 90K superconductors, experiments on substitution of some of their components by other elements were made. For example, oxygen was replaced by sulphur or fluorine [7] in the YBaCuO system. The substitution of nonmagnetic yttrium or lanthanum by other rare earths carrying nonzero magnetic moments surprisingly shows nearly no effect on critical temperature [8]. This indicates a rather weak coupling of the localized magnetic moments to the superconducting pairs.

In this paper, we focus our attention on a partial substitution of copper by other transition elements from the first long period of the periodic table of elements. Although a decrease of superconducting critical temperature is observed in general, some anomalous features occur.

First, the lowering of T_c depends on the impurity concentration x, although the critical concentrations are relatively high [9], at least in comparison with classical superconductors. This concentration dependence of T_c is in agreement with predictions of the Abrikosov-Gor'kov theory [10]. On the other hand, for a given concentration of the impurity elements a direct correlation between the lowering of T_c and the value of the localized magnetic moments is not observed as should be expected from the A-G theory. The values of the effective magnetic moments can be easily estimated from measurements of the magnetic susceptibility and the Curie-Weiss law above the superconducting critical temperature in the absence of the Meissner effect. For example, iron and cobalt are carrying magnetic moments of $3.29 \div 3.61 \mu_B$ and $2.87 \div 3.20 \mu_B$, respectively. However, the critical temperature of $YBa_2(Cu_{0.9}Fe_{0.1})_3O_{7-\delta}$ is higher than for $YBa_2(Cu_{0.9}Co_{0.1})_3O_{7-\delta}$.

Pressure studies were also made in the YBaCuO and the LaBaCuO systems [11]. Their critical temperatures depend on external pressure to different extent. For example, the pressure derivative dT_c/dp is equal to about 0.07 K/kbar for $YBa_2Cu_3O_{7-\delta}$ and its stoichiometric modifications [11]. For the $La_{2-x}A_xCuO_4$ (A=Ba,Sr) compounds the value of dT_c/dp is an order of magnitude larger and depends strongly on the concentration x . Quite recently the value of 1.5 K/kbar was reported for the ceramic system of the $(Y_{1.2}Ba_{0.8})_4Cu_4O_{16-\delta}$ type [12]. However, it should be pointed out that the pressure derivatives of the critical temperature in the high-temperature superconductors are higher than in the classical ones.

BASIC EQUATIONS

Let us consider a superconducting system with a concentration x of paramagnetic impurity. In the standard way the system can be described by a Hamiltonian, which consists of the BCS part and the term of s-d interaction between current carriers and localized magnetic moments [10].

The most essential point of the paper consists in including some details of the electronic structure of our system, in contrast to standard models where they are neglected and only $N(\varepsilon_F)$, the value of the density of states (DOS) at the Fermi surface, is used. The DOS was calculated with high precision by Freeman's group [14]. To make some analytical calculations possible, we can roughly approximate the DOS by a root singularity of the form :

$$N(\varepsilon) = N_o + \alpha \mid \varepsilon - \varepsilon_K \mid^{1/2} \tag{1}$$

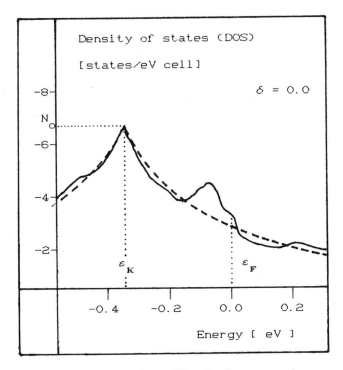

Fig.1. DOS of the $YBa_2Cu_3O_{7-\delta}$ system from [14] — full line, the root approximation (eq.6) — dashed line.

The peak at ε_K near the Fermi energy comes from copper and oxygen contributions. Substitution of copper by other transition metals changes the number of electrons and leads to a shift of the Fermi level if we assume that no essential changes occur with the shape of the DOS. Therefore, the superconducting properties of the system can change as well.

Within the Abrikosov-Gorkov scheme one introduces the Dyson's equation for the electron Green's function determined by the frequency ω, the Bloch single-particle energy ε_k and the fermionic order parameter Δ, mutually renormalized [13]. After some algebra the following set of equations is obtained :

$$\Delta = N_o V \int_o^{\omega_o} M \, \text{th}(\beta\omega/2) \, \text{Re}(\tilde{\Delta}/(\tilde{\omega}^2 - \tilde{\Delta}^2)^{1/2}) \, d\omega \qquad (2)$$

$$\tilde{\omega} = \omega + i \, M \, \Gamma_1 \tilde{\omega} / (\tilde{\omega}^2 - \tilde{\Delta}^2)^{1/2} \qquad (3)$$

$$\tilde{\Delta} = \Delta + i \, M \, \Gamma_2 \tilde{\Delta} / (\tilde{\omega}^2 - \tilde{\Delta}^2)^{1/2} \qquad (4)$$

where :

$$\Gamma_1 = \pi \times N_o (V_1^2 + S(S+1)V_2^2/4) \qquad (5)$$

$$\Gamma_2 = \pi \times N_o (V_1^2 - S(S+1)V_2^2/4) \qquad (6)$$

$$M = 1 + \sum_{\nu=1,-1} \nu((\varepsilon_F^* + \nu(\tilde{\omega}^2 - \tilde{\Delta}^2)^{1/2})^{1/2} \, \alpha/2N_o \qquad (7)$$

$\varepsilon_F^* = \varepsilon_F - \varepsilon_K$, $\beta = (k_B T)^{-1}$, S is the value of impurity spin and ω_o denotes the energy scale of the bosons responsible for the superconducting pairing. V_1 and V_2 are parameters describing potential and magnetic scatterring of carriers, respectively.

CALCULATION OF CRITICAL TEMPERATURE

The above set of equations is still complex and requires a knowledge of the boson spectrum. To simplify this problem, let us consider an Einstein-like spectrum of the form given by $D(\omega) = D_o \delta(\omega - \omega_o)$, without specifying its detailed physical nature. In the limit of temperature close to T_c this leads to the formula :

$$T_c = (\omega_o/2k_B)(\operatorname{arth}((N_oV(1-(8(\varepsilon_v-\varepsilon_K))^{-1/2}$$

$$\sum_{\nu=1,-1} \nu \ (\varepsilon_F^*+\nu\omega_o+((\varepsilon_F^*+\nu\omega_o)^2+\Gamma^2)^{1/2}))^{1/2}))^{-1}))^{-1} \qquad (8)$$

where ε_v is the edge of the valence band $(N(\varepsilon_v)=0)$ and

$$\Gamma = \pi \times N_o \ (V_1^2 + (p_A^2-p_{Cu}^2) \ \tilde{V}_2^2) \qquad (9)$$

for the magnetic moments p_A, obtained by means of the Curie-Weiss law from susceptibility measurements [9] . The parameters V_1, V and $\tilde{V}_2$ are assumed to be independent of the concen-

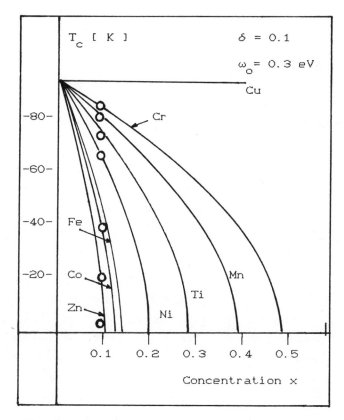

Fig. 2. The calculated critical tempera-
rature as a function of concentration x
for the $YBa_2(Cu_{1-x}A_x)_3 O_{7-\delta}$ compounds,
(A=Cr,Mn,Ti,Ni,Fe,Co,Zn), experimental
values [9] - open circles. It should be
pointed out that the critical concentrations are found to be very high.

tration x within a sufficiently large region. The values of
these parameters are fitted to experimental cricital temperatures for mother compound of the $YBa_2Cu_3O_{7-\delta}$ type and its modifications doped with Cr and Fe. Since we do not detail the
boson field we can now examine our model having in view the
value of ω_o. The best results were obtained when ω_o exceeded
the Debye energy (e.g. $\omega_o = 0.3$ eV, Fig.2), although full
agreement with measurements of T_c can not be achieved because
of model simplifications. It should be also stressed that in
some cases the oxygen change on substitution is not well controlled and the above comparison of experimental and calculated T_c's for a given δ is not valid generally.

In summary, we find some indications that either a non-
phononic mechanism enhances the phonon-mediated coupling or
operates alone.

PRESSURE DERIVATIVE OF T_c

If one takes into account the sensitivity of topology of
the Fermi surface to pressure, the derivative of critical temperature can be rewritten as follows :

$$dT_c/dp = (\ dT_c/d\varepsilon_F^*)\ (\ d\varepsilon_F^*/\ dp\ ) \qquad\qquad (10)$$

Using Eq.(9)the derivative $dT_c/d\varepsilon_F^*$ is easily calculated leading to a rather long expression which has a sharp peak associated with the peak in the DOS, i.e. for $\varepsilon_F^* = 0$. Since the
DOS at the Fermi level in the LaBaCuO system is greater than
in the YBaCuO system, we may expect the Fermi level in LaBaCuO
to be closer to the root singularity at ε_K. Therefore, the value
of pressure derivative of cricital temperature can be greater
for the lanthanum compounds than for the yttrium ones as is
actually observed [15]. This prediction is rather qualitative
because of the general inability of the BCS-like models to explain a pressure dependence of T_c in the case of dT_c/dp greater
than about 0.1 K/kbar. Therefore, application of the real space bipolaron theory of superconductivity [15], in which pressure effects are included in a natural way, is recommended
and leads to better quantitative results [16].

ACKNOWLEDGEMENTS

The author wishes to thank the Organizing Committee of the International Discussion Meeting on High-T$_c$ Superconductors (Mauterndorf, Austria, February 7-11,1988) for a financial support. This work was partially supported by the Institute of Physics - Polish Academy of Sciences (Warsaw) under RPBP 01.09 Programme, contract No. 15.1 .

REFERENCES

[1] J. G. Bednorz and K. A. Müller, Z. Phys. B64: 189 (1986).

[2] M. K. Wu, J. R. Ashburn, C. T. Torng, P. H. Hor, R. L. Meng, L. Gao, Z. J. Huang, Y. Q. Wang and C. W. Chu, Phys. Rev. Lett. 58: 908 (1987).

[3] A. A. Abrikosov, J. of Less. Common Met. 62: 451 (1978).

[4] P. W. Anderson, Science 235: 1196 (1987).

[5] E. W. Fenton, Sol. State Commun. 63: 993 (1987).

[6] J. Klamut, private communication.

[7] S. R. Ovshinsky, Phys. Rev. Lett. 58: 2479 (1987).

[8] P. H. Hor, R. I. Meng, Y. Q. Wang, L. Gao, Z. J. Huang, J. Bechtold, K. Forster and C. W. Chu, Phys. Rev. Lett. 58: 1891 (1987).

[9] G. Xiao, F. H. Streitz, A. Gavrin, Y. W. Du and C. L. Chien, Phys. Rev. B 35: 8782 (1987).

[10] A. A. Abrikosov and D. P. Gor'kov, JETP 39: 1781 (1960), in Russian.

[11] J. E. Schirber, D. S. Ginley, E. L. Venturini and B. Morosin, Phys. Rev. B 35: 8709 (1987).

[12] J. Baszyński, M. Maćkowiak and M. Zdanowska-Frączek, Phys. Lett. A 126: 130 (1987).

[13] M. E. Palistrant and A. T. Trifan, " The Theory of Doped Superconductors under Pressure", Shtintza, Kishiniev (1980), in Russian.

[14] S. Massidda, J. Yu, A. J. Freeman and D. D. Koelling, Phys. Lett. A 122: 198 (1987).

[15] B. K. Chakraverty, J. Phys. Lett. 40: L 99 (1979).

[16] J. Kasperczyk, submitted to ICM'88 (Paris).

TEST OF PHONON MEDIATED PAIRING FROM RAMAN SCATTERING ON HIGH−TEMPERATURE SUPERCONDUCTORS

Karl Svozil

Institut für Theoretische Physik, Technische Universität Wien

Karlsplatz 13, A-1040 Vienna, Austria

Bitnet: E1360DAB at AWIUNI11

and

Rudolf Lassnig

Institut für Experimentalphysik, Universität Innsbruck

Technikerstraße 25, A-6020 Innsbruck, Austria

A strong enhancement of the Raman activity at frequencies close to twice the superconducting gap can be expected for conventional phonon−mediated pairing of the high temperature superconducting systems. For other pairing mechanisms this effect is absent.

1. Introduction

In view of the persisting controversies concerning the origin of the high temperature superconductivity of the Cu oxides it is of great importance to establish criteria for phonon mediated superconductivity [1,2]. In this communication we present the renormalization of the phonon modes due to polarization effects, yielding an enhanced Raman activity in the superconducting phase close to the (real) quasiparticle creation threshold of twice the superconducting gap 2Δ. In the case of nonphonon−mediated superconductivity this effect is absent.

For the high–temperature superconductors there should exist relevant phonons with $2\Delta/\omega \sim 1$, whereas in usual superconducting systems this ratio is much smaller than one. This gives raise to a remarkable phenomenon: when the phonon frequency is of the order of or greater than two gap values, then real quasiparticle – antiquasiparticle production sets in, strongly enhancing the phonon polarization $\Pi(\omega)$ in the vicinity of $\omega \sim 2\Delta$. This renders strong contributions to the phonon renormalization correspond-ing to enhanced Raman activity for frequencies $\Omega = [\omega_0^2 - 2\omega_0\Pi(\Omega)]^{1/2}$, where ω_0 is the bare phonon frequency and Π is the (Coulomb screened) phonon polarization.

The physical interpretation of this effect is quite evident: The phonon couples to the electronic charge density and the resulting phonon–charge density mixture exhibits a low–energy and a high–energy branch. Only the branch above 2Δ is broadened, since the electron–hole excitation is impossible below $\omega \sim 2\Delta$. The whole effect is strongly enhanced at $\omega \sim 2\Delta$ as a consequence of the diverging polarisability.

2. Phonon polarization

In the following we report calculations of the lowest order phonon polarization $P^{(2)}$ for $T \neq 0$, which are applied for the renormalization of the phonon propagator [3–7]. This result is then used to predict a strong enhancement of the Raman activity, here referred to as "Resonant Raman Scattering" (RRS) via the effect discussed above.

The calculation bases upon a model often used in the Eliashberg theory of strong coupling superconductivity [8,9]: An expansion of the full phonon propagator D yields

$$[D(k, i\omega_n)]^{-1} = [D_0(k, i\omega_n)]^{-1} - \Pi(k, i\omega_n) \quad , \tag{1}$$

where $D_0(|\ k\ | \sim \omega_0/c, i\omega_n) = 2\omega_0/[\omega_0^2 - (i\omega_n)^2 + i\delta]$ is the bare phonon propagator and the polarization Π is given in lowest order, including Coulomb screening

$$\Pi = g^2(1 - V_cP)^{-1}P \quad , \tag{2}$$

$$P^{(2)}(q, i\omega_n) = -k_BT \int \frac{d^3k}{(2\pi)^3} \sum_{m=-\infty}^{\infty} \text{Tr}[\tau_3G(k+q, i\omega_n + i\omega_m)\tau_3G(k, i\omega_m)] \tag{3}$$

In (2), V_c stands for the Coulomb potential.

After taking the traces and performing the frequency summation via extension of the contour of the Poisson formula to infinity, one obtains in the dynamic screening range $[|\ q\ | \sim 0]$ after analytic continuation to real frequencies

$$P^{(2)}(\omega) = 4 \int \frac{d^3k}{(2\pi)^3} \frac{\Delta^2}{(\epsilon_k^2 + \Delta^2)^{1/2}(\omega^2 - 4\Delta^2 - 4\epsilon^2)} \tanh[\frac{1}{2}\beta(\epsilon_k^2 + \Delta^2)^{1/2}] \quad , \tag{4}$$

where $\beta = (k_BT)^{-1}$ and $\Delta \sim \Delta_{T=0}[1 - T/T_c]^{1/2}$ from the BCS–theory is assumed. The integral in (4) can be solved by substituting $d^3k/(2\pi)^3 \sim N(0)d\epsilon$, where $N(0)$ is the density of electron states at the Fermi surface. It is given by $[x = \mathrm{asinh}(\epsilon/\Delta)]$

$$P^{(2)}(\omega) = -8N(0)\Delta^2 \int_0^\infty dx \frac{\tanh[\frac{1}{2}\beta\Delta\cosh(x)]}{4\Delta^2[\cosh(x)]^2 - \omega^2} \qquad (5)$$

The dimensionless phonon–electron and Coulomb coupling strength can be parameterized by $\lambda = 2g^2 N(0)/\omega_0$ and $\mu = V_c N(0)$. These variables should however not be confused with the dimensionless coupling parameters of the Eliashberg theory [8,9], where the coupling strength is weighted over the phonon spectrum and the momentum ranges. Only in the case of a strongly peaked (Einstein) spectrum the λ–parameters coincide.

One should be able to observe a strong peak of the Raman activity in the super-conducting phase at frequency shifts close to 2Δ due to the diverging polarizability $P(2\Delta)$, as can be seen from (5). This requires the existence of phonon modes in this range as well as electron (hole)–phonon coupling; both are necessary conditions for the conventional pairing mechanism.

3. Numerical solutions

We have numerically studied the phonon frequency shifts due to renormalization. In Figs. 1 and 2 the computed values for the renormalized phonon peak frequency Ω obtained from $D(\Omega) = 0$, that is

$$\Omega^2 - \omega_0^2 - 2\omega_0\Pi(\Omega) = 0 \qquad (6)$$

are plotted as functions of the dimensionless coupling parameters λ and μ defind in section 2.

At the same time one should be able to observe a continuum contribution due to a broadening of the phonon modes close to 2Δ.

4. Comparison with experiment

Assuming gap values of 50–400 cm^{-1}, corresponding to 6–48 meV, enhancement of the Raman energy shift should occur at about twice that value. Raman spectra are

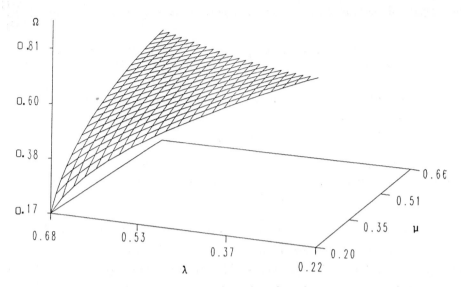

Fig. 1 Renormalized phonon frequency in units of ω_0 for $T = 20K$ and $\Delta = 1.2\omega_0$ as function of λ and μ.

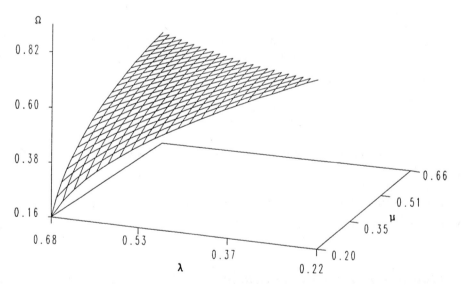

Fig. 2 Renormalized phonon frequency in units of ω_0 for $T = 0K$ and $\Delta = 1.2\omega_0$ as function of λ and μ.

published for transparent materials with the K_2NiF_4 structure, such as Sr_2TiO_4 [10]. Furthermore, there are two Raman studies on $La_{1.85}Sr_{0.15}CuO_{4-\delta}$; (i) in the 50–500 cm^{-1} frequency shift range at temperatures 6–300 K [11]; and (ii) in the 50–900 cm^{-1} frequency shift range at temperatures 44–300 K [12]. Both groups record no significant change in the Raman spectrum of the superconducting phase up to frequency shifts of 500 cm^{-1}. At 666 cm^{-1} a broad peak appears at low temperature, an effect which has been ascribed to a planar breathing mode [12]. Hence, so far no clear signature of a resonant enhancement of the Raman spectrum, even for the first class of materials of the $(La_{1-x}M_x)_2CuO_4$ type has been observed. Whether the absence of this effect can be explained by nonexisting phonon modes at 2Δ is unclear at present.

For the $YBa_2Cu_3O_7$ system experimental results are more promising [13]. By subtracting the normal state Raman intensity from the superconducting state one, a significant enhancement of the Raman activity has been observed in the frequency range of 30–100 meV. It is not unreasonable to interpret this increased Raman activity as an indication for a resonance effect close to twice the superconducting gap and thus as a criterion for phonon mediated superconductivity.

The authors thank H. Weber for bringing to our attention some references cited below, as well as stimulating discussions with E. Gornik.

References

[1] T. M. Rice , Z. Phys. **B67**, 141 (1987)

[2] H. Rietschel, Phys. Bl. **43**, 357 (1987)

[3] J. I. Balkarei and D. I. Khomskii, *Lebedev Institute report* 1967 (unpublished).

[4] C. A. Balseiro and L. M. Falicov, Phys. Rev. Lett. **45**, 662 (1980).

[5] H. G. Schuster, Solid State Commun. **13**, 1559 (1973).

[6] K. Machida, Prog. Theor. Phys. **66**, 41 (1981).

[7] K. Svozil, Phys. Rev. **B31**, 4688 (1985); Phys. Rev. **B35**, 7113 (1987).

[8] J. R. Schrieffer, *Theory of Superconductivity* (Benjamin, New York 1964).

[9] D. Scalapino, in *Superconductivity*, ed. by R. D. Parks (Dekker, New York 1969).

[10] G. Burns, F. H. Dacol and M. W. Schafer, Solid State Commun. **62**, 687 (1987)

[11] S. Blumenroeder, E. Zirngiebl, J. D. Thompson, P. Killough, J. L. Smith and Z. Fisk, Phys. Rev. **B35**, 8840 (1987)

[12] S. Sugai, M. Sato, S. Hosoya, S. Uchida, H. Takagi, K. Kitazawa and S. Tanaka, in *Proc. 18th Int. Conf. on Low Temperature Physics*, Jpn. J. Appl. Phys. **26**, Supplement 26-3, 1003 (1987); see also K. Ohbayashi, N. Ogita, M. Udagawa, Y. Aoki, Y. Maeno and T. Fujita, Jpn. J. Appl. Phys. **26**, L420 (1987)

[13] A. V. Bazhenov, A. V. Gorbunov, N. V. Klassen, S. F. Kondakov, I. V. Kukushkin, V. D. Kulakovskii, O. V. Misochko, V. B. Timofeev, L. I. Chernyshova and B. N. Shepel, in *Novel Superconductivity*, ed. by A. Wolf and V. Kresin (Plenum Press, New York 1987), p. 893, in particular Fig. 4 [the correct page order is 893,895,894,896]

ARE ONE-DIMENSIONAL STRUCTURAL FEATURES ESSENTIAL FOR

SUPERCONDUCTIVITY AT 90K ?

Arndt Simon, Horst Borrmann, Wolfgang Bauhofer
Reinhard K. Kremer, and Hansjuergen Mattausch

Max-Planck-Institut fuer Festkoerperforschung

D-7000 Stuttgart 80, F.R.G.

The crystal structures of the novel high-T_c oxocuprates[1,2,3,4] have certain features in common, but also exhibit essential differences. $La_{1.8}M_{0.2}CuO_4$(M=Ba,Sr; $T_c \approx$30 to 40K) contains copper in an average oxidation state +2.2. The coordination octahedron around Cu is tetragonally elongated (d(Cu-O)=190 pm (4x) and 240 pm (2x)). Taking the drastic differences of the Cu-O distances into account, the structure contains CuO_2 layers formed from corner-sharing CuO_4 squares according to $_\infty^2[CuO_{4/2}]$ which are sandwiched between rigid slabs of composition $_\infty^2[(La,M)_2O_2]$. Obviously, the layers are only weakly coupled to the surrounding slabs.

In the structure of $YBa_2Cu_3O_{7-x}$ ($T_c \approx$ 90 K for x→0) the same kind of CuO_2 layers pairwise surround the Y atoms. The CuO_4 squares are orthorhombically distorted and the layers are buckled. Between the slabs of composition $_\infty^2[YCu_2O_4]$ structural units Ba_2CuO_{3-x} are inserted which contain parallel ribbons $_\infty^1[CuO_2O_{2((1-x)/2)}]$. Obviously these ribbons are necessary for 90 K superconductivity. They consist of linear CuO_2 units (d(Cu-O) = 185 pm) which are bridged by oxygen atoms (O1) resulting in square planar Cu coordination with a rhomboidal distortion of the squares (d(Cu-O1) = 194 pm). The ribbons are again weakly coupled to the slabs via O-Cu contacts near 230 pm. The oxygen deficiency in the structure is restricted to the O1 atoms and determines the oxidation state of Cu in the ribbons. For x=0 these Cu atoms are essentially trivalent. Removing the loosely bound O1 atoms leads to a lowering of T_c. For x=0.5 corresponding to Cu^{2+} the phase becomes semiconducting. Finally, for x=1.0 an array of isolated linear CuO_2 units with Cu^+ is left. This oxygen removal can be performed in a reversible reaction.

Our hypothesis for the electron-pairing mechanism in superconducting $YBa_2Cu_3O_{7-x}$ is based on a dynamical simulation of the topochemical redox process in terms of a phonon mediated valence fluctuation of Cu between the oxidation states +1 and +3. For that purpose a lattice vibration is needed which changes the environment of the Cu atoms towards the characteristic coordinations for Cu^+ and Cu^{3+}.

All established structures of oxocuprates contain Cu^{3+} (isoelectronic with Ni^{2+}) in a square (or rectangular) planar coordination and Cu^+ in a linear coordination. It is important that the Cu-O distances in both coordinations are essentially the same (d(Cu-O) ≈ 184 pm) and slightly shorter than in oxocuprates(II) (d(Cu-O)$\geqslant$189 pm). Simple valence bond theory takes account of the bonding in terms of a dsp^2 hybrid for the (low-spin) d^8 system of Cu^{3+} and in terms of a sp hybrid for the d^{10} system of Cu^+.

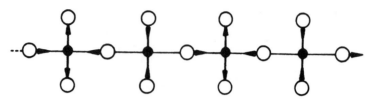

<u>Fig. 1</u> CuO_3 ribbon in $YBa_2Cu_3O_7$; the indicated lattice vibration leads to the tendency for d^{10} and d^8 configuration of adjacent Cu atoms.

The fully occupied CuO_3 ribbon in $YBa_2Cu_3O_7$ chemically contains Cu in an oxidation state near +3. Yet, the CuO_2 dumbbell characteristic for Cu^+ is preformed structurally. The pairwise increase of d(Cu-O1) around one Cu site enhances the tendency towards a d^{10} configuration simultaneously favoring a d^8 configuration for the adjacent Cu atom in the ribbon (Fig.1). The indicated zone edge mode should result in a pairwise attraction of electrons via strong electron-phonon coupling. The valence instability of copper is assisted by the configurational instability of the oxygen dianion.[5,6]

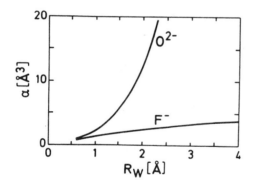

Fig. 2
Polarizabilities α
of O^{2-} and F^- as a
function of the
Watson radius R_w.

It has been pointed out by Bilz[7] that the polarizability of the O^{2-} ion in oxides changes dramatically with the anion-cation distance (see Fig. 2; $\Delta\alpha\sim\Delta d^{12}$ in case of pronounced anisotropy) which in case of the CuO_3 ribbon leads to a dynamic delocalization of an oxygen 2p-electron to the more distant Cu atom. Simultaneously this Cu atom has the tendency to localize two electrons in the d^{10} configuration. In the case of the CuO_2 layer in $La_{1.8}M_{0.2}CuO_4$ the resulting electronic resonant state can be associated with a mode that leads to a quadrupolar distortion of the Cu environment, changing C.N.=4 to 2+2. The CuO_3 ribbon in $YBa_2Cu_3O_7$ meets the structural requirements for the discussed valence fluctuation of Cu more efficiently.

From the above the onset of superconductivity around 90 K seems intimately related to the presence of the one-dimensional CuO_{3-x} ribbons in $YBa_2Cu_3O_{7-x}$ and related "123" systems, where yttrium is replaced by

other rare earth metals. A fragmentation of the ribbons by increased O
vacancy concentration only slightly decreases T_c. Around x=0.3 a rather
sharp drop to $T_c \approx 60$ K is observed. Finally, for x=0.5 semiconducting
behavior is observed due to the only presence of Cu in the oxidation
state +2. What is special about the composition $YBa_2Cu_3O_{6.7}$? It is
interesting to note that the smallest fragment of the ribbon which still
contains one Cu coordinated by 4 O atoms (a necessary preposition for
the proposed valence fluctuation) has the composition $Cu_3O_8=CuO_{2.67}$.
Our experimental investigations are mainly concerned with the importance
of the CuO_{3-x} ribbons for 90 K superconductivity.

1.) Mitzi et al. show that samples of $La_{3-x}Ba_{3+x}Cu_6O_{14+\delta}$ ("336")
exhibit an onset of superconductivity at approx. 45 K (x=0), 55 K
(x=0.25), 60 K (x=0.5), 85 K (x=0.75) and 90 K (x=1.0).[8] The compositions of the single-phased samples (a) $La_{1.38}Ba_{1.63}Cu_3O_{7+\delta}/2$ and (b)
$La_{1.13}Ba_{1.88}Cu_3O_{7+\delta}/2$ are rather close to the composition of the corresponding "123" phase, in particular for the sample (b). Yet, from the
X-ray powder diagrams it is concluded that essential differences exist
for the structure types of "336" and "123". Whereas "123" contains CuO_2
twin layers and CuO_{3-x} ribbons, "336" is said to contain only layers
linked via additional oxygen atoms following the earlier suggestion of
Er-Rakho et al.[9]

In the X-ray diagrams very weak reflections have been interpreted
in terms of a superstructure calling for a unit cell twice as large as
that of the "123" structure, achieved by a special rearrangement of
oxygen atoms around the Cu atoms near barium in the "123" structure.
In "123" all Cu atoms in the z=0 plane are equivalent in which case the
small unit cell is sufficient to describe the structure. In the "336"
structure the same Cu atoms are made non-equivalent by placing four and
six oxygen atoms (square planar and octahedral coordination, respectively) around adjacent Cu atoms. Thus, the doubled unit cell is created.

It is evident that all weak reflections assigned to the "336"
structure model by Mitzi et al. actually belong to the diagram of the
"123" structure in just one exception. The small peak near $2\theta \approx 16°$
indexed as 100 should not occur with the "123" diagram. Suppose the
diagrams were taken with crystal-monochromatized radiation, then the
$\lambda/2$-reflections of the strongest 113/200 lines occur in exactly the
observed position of 100. So there is no experimental evidence for a new
"336" structure type. In fact, recent neutron diffraction data of
$La_3Ba_3Cu_6O_{14.39}$ confirm the isomorphism with the tetragonal variant of
$YBa_2Cu_3O_{7-x}$.

2.) Nguyen et al.[10] describe a series of compounds with the general
composition $La_{2-x}Sr_{1+x}Cu_2O_{6-x}/2$, for $0 \leqslant x \leqslant 0.14$. The structure (Fig.3)
results from an "intergrowth" of slabs of composition $^2_\infty[(La,Sr)_2O_2]$, as
they occur in $La_{1.8}M_{0.2}CuO_4$, and twin layers $^2_\infty[(La,Sr)Cu_2O_4]$ which are
one structural feature of $YBa_2Cu_3O_7$. The geometrical details of the twin
layers, i.e. slight displacement of the Cu atoms out of the oxygen
planes and the distances d(Cu-O) = 193 pm are similar to those in
$YBa_2Cu_3O_7$. As in $La_{2-x}Sr_{1+x}Cu_2O_{6-x}/2$ no CuO_3 ribbons but only twin
layers exist, the compounds can be taken to test whether the twin layers
as such lead to 90 K superconductivity.

We have prepared $La_{2-x}Sr_{1+x}Cu_2O_{6-x}/2$ phases within the range $0.2 \leqslant$
$x \leqslant 0.5$. Stoichiometric amounts of La_2O_3 (4N, preheated at 600°C, 10^{-5}
mbar), $SrCO_3$ (p.a. quality) and CuO (p.a.) were reacted according to
Nguyen. Finally the samples were kept at 500°C in 1 bar O_2 for 3h.

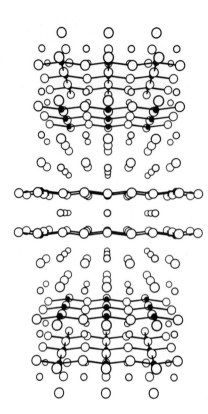

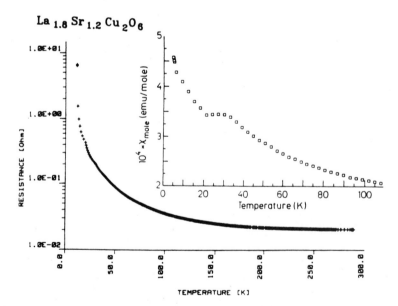

Fig. 4 Electrical resistivity and magnetic susceptibility of
$La_{1.8}Sr_{1.2}Cu_2O_6$ as a function of temperature.

The temperature dependencies of the electrical resistivity and magnetic susceptibility of $La_{1.8}Sr_{1.2}Cu_2O_6$ are shown in Fig. 4. With decreasing temperature the resistance increases monotonically. Only at approx. 30 and 20 K small discontinuities are apparent. From the magnetization measurement the effect at 30 K might arise from a superconductivity minority phase, e.g. $La_{2-x}Sr_xCuO_4$, at an impurity level of approximately 100 ppm. The measurements on $La_{1.8}Sr_{1.2}Cu_2O_6$ as well as other samples reveal no transition to bulk superconductivity down to 5K.

3.) One could think of a compound that contains only Cu in CuO_3 ribbons as an ideal candidate for high temperature superconductivity. According to Teske and Mueller-Buschbaum[11] the compounds A_2CuO_3 (A=Ca, Sr) generally meet this requirement (Fig. 5). Unfortunately, though, the structures do not meet the requirements in detail. The Cu-O distances within the CuO_2 dumbbells are 196 pm and along the chains 189 (Ca) and 196 pm (Sr). The distances in the dumbbell are much too large for the linear coordination of Cu^+. Furthermore, these large distances indicate a strong bonding of the O atoms to the adjacent alkaline earth ions. The ribbons are not loosely coupled to a rigid frame work as in "123". In spite of these facts we prepared[12] $Ca_{2-x}Na_xCuO_3$ (x < 0.3) which according to modified Guinier photographs[12] are single phase (with significantly broadened line shapes). No anomalies of the electrical resistivity were found down to 4K.

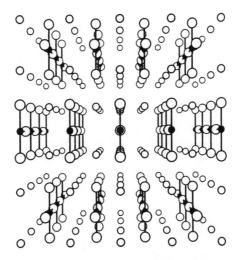

Fig. 5 A_2CuO_3 (A= Ca, Sr). The Cu-O dumbbells are outlined.

4.) In "123" the Y can be substituted by several RE with no significant change of T_c. Thus, $HoBa_2Cu_3O_7$ obtained by reacting Ho_2O_3, $BaCO_3$ and CuO at 950°C in O_2, slowly cooling to 300°C (15°/h) after 4h shows a transition to the superconducting state near 91 K. Single pellets out of sets of identically treated pellets repeatedly exhibited unusual electrical properties. Fig. 6 is an example. The resistivity curve of the freshly prepared sample has a small depression near 130 K and, more obvious, becomes superconducting at a lowered T_c = 85 K. After 3 days at room temperature in demoisterized air the same sample shows no anomaly down to approx. 93 K when it becomes superconducting.

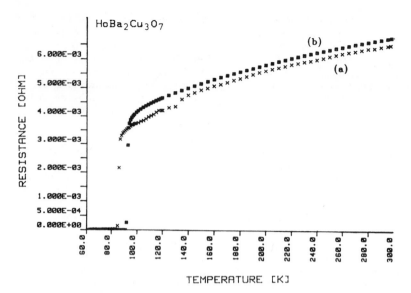

Fig. 6   Electrical resistivity of $HoBa_2Cu_3O_7$:
a) freshly prepared sample   b) the same sample after 3 days

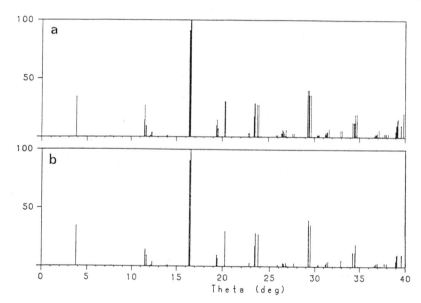

Fig. 7   X-ray Guinier powder pattern of $HoBa_2Cu_3O_7$:
a) freshly prepared sample
b) the same sample after 3 days

Anomalies of that kind, even superconductivity at temperatures well
above 100 K have been reported by several groups.
We are not aware of any structural observations associated with these
effects. In our cases the Guinier diagrams of the fresh and 3 days old
samples differed significantly as shown in Fig. 7. Whereas the old
sample has the well known diagram of the "123" structure (a= 388.7,
b=382.4, c= 1168.0 pm) the diagram of the fresh sample looks more complicated. It can be decomposed into the diagrams of two slightly
different orthorhombic cells (a= 381.5, b= 387.8, c= 1164.0 pm and
a=379.7, b= 386.2 and c=1158.3 pm).

References

1. Bednorz, J.G., and Mueller, K.A., Possible High-T_c Superconductivity in La-Ba-Cu-O System, <u>Z.Phys. B</u> 64:189 (1986).

2. Cava, R.J., Dover, R.B., Batlogg, B., and Rietman, E.A., Bulk Superconductivity at 36K in $La_{1.8}Sr_{0.2}CuO_4$, <u>Phys.Rev.Lett.</u> 58:408 (1987).

3. Wu, M.K., Ashburn, J.R., Torug, C.J., Hor, P.H., Meng, R.L., Gao, L., Huang, Z.J., Wang, Y.Q., and Chu, C.W., Superconductivity at 93K in a New Mixed Phase Yb-Ba-Cu-O Compound System at Ambient Pressure, <u>Phys.Rev.Lett.</u> 58:908 (1987).

4. Beno, M.A., Soderholm, L., Capone II, D.W., Hinks, D.G., Jorgenson, J.D., Schuller, I.K., Segre, C.U., Zhang, K., and Grace, J.D., Structure of the Single Phase High Temperature Superconductor $YBa_2Cu_3O_7$, <u>Appl.Phys.Lett.</u> 51:57 (1987).

5. Simon, A., Superconductivity - A Chemical Phenomenon?, <u>Angew.Chem.Int.Ed.Engl.</u> 26:579, (1987).

6. Callaway, J., Kanhere, D.G., and Mistra, P.K., Polarization-induced Pairing in High Temperature Superconductivity, <u>Phys.Rev. B</u> 36:7141 (1987).

7. Bussmann, A., Bilz, H., Roenspiess, R., and Schwarz, K., Oxygen Polarizability in Ferroelectric Phase Transitions, <u>Ferroelectrics</u> 25:343 (1980).

8. Mitzi, D.B., Sun, J.Z., Webb, D.J.,Beasley, M.R., Geballe, T.H., and Kapitulnik, A., Adriatico Research Conference on High Temperature Superconductors, Trieste (1987).

9. Er-Rakho, L., Michel, C. Provost, J., and Raveau, B., A Series of Oxygen-defect Perovskites Containing Copper(II) and Copper(III): The oxides $La_{3-x}Ln_xBa_x[Cu_{5-y}^{II}Cu_{1+2y}^{III}]O_{14+y}$, <u>J.Solid State Chem.</u> 37:151 (1981).

10. Nguyen, N., Er-Rakho, L., Michel, C., Choisnet, J., and Raveau, B., Intergrowth of layers of "defect perovskites" and layers of the type sodium chloride. The oxides $La_{2-x}A_{1+x}Cu_2O_{6-x/2}$ (A=Ca,Sr), <u>Mat.Res.Bull.</u> 15:891 (1980).

11. Teske, Chr.L., and Mueller-Buschbaum, Hk., Zur Kenntnis von Ca_2CuO_3 und $SrCuO_2$, <u>Z.anorg.allg.Chem.</u> 379:234 (1970).

12. Simon, A., Eine Methode zur Untersuchung extrem luftempfindlicher Substanzen mit der Guinier-Methode, <u>J.Appl.Crystallogr.</u> 3:11 (1970).

THEORY OF OXYGEN ORDERING AND ABSORPTION IN

1,2,3 HIGH-T$_c$ SUPERCONDUCTORS

J.P.A. Westerveld[*], D.M.R. Lo Cascio[*], H. Bakker[*]
and D.O. Welch[**]

* Natuurkundig Laboratorium der Universiteit van
  Amsterdam, NL-1018 XE The Netherlands
**Brookhaven National Laboratory, Upton, NY 11973, USA

The oxygen content and the oxygen ordering in 1,2,3 high T$_c$ compounds plays an important role in the superconducting properties of these materials. Experiments [1,2] show that the absorption and desorption of oxygen mainly occurs in the CuO planes, which are located at maximum distance from the trivalent ion, and that the ordering of the oxygen atoms in linear Cu-O-Cu-O-chains in these planes leads to the orthorhombic structure at lower temperatures. Full occupancy of the oxygen positions in these chains along the direction of the b axis leads to 7 oxygen atoms per formula unit. Complete order means that all oxygen positions in the perpendicular direction (along the a axis) are vacant. Disordering in combination with oxygen desorption causes the orthorhombic to tetragonal phase transition at high temperatures. In the tetragonal phase with about 6.5 oxygen atoms per formula unit the oxygen atoms are distributed over the 'b' and 'a' positions in equal amounts. This causes the a and b lattice parameters to be equal and leads to the tetragonal phase.

In a previous paper [3] we presented a thermodynamic model which describes the oxygen content and the atomic ordering as a function of temperature and oxygen pressure. A negative site energy e was assigned to oxygen atoms in the CuO planes and a positive repulsive energy v between neighbouring oxygen atoms was postulated. This repulsive interaction leads to atomic ordering at lower temperatures. In fact, the problem is an Ising problem, which was solved in pair approximation. This approximation is a somewhat better approximation than a mean field or Bragg-Williams approximation. Three equations were obtained between seven quantities, namely the occupancy of one of the sublattices, the total oxygen content, the number of oxygen-oxygen nearest-neighbour pairs, temperature, outside oxygen pressure and both the negative site energy e and the positive repulsive energy v. If, at a certain oxygen pressure (for example 1 atm.), both the orthorhombic-to-tetragonal transition temperature and the oxygen content are known, the number of unknown parameters reduces to three, namely e and v and the number of oxygen-oxygen pairs, so that these quantities can be obtained by solving the equations. From experimental data, obtained by neutron diffraction, values of e and v can be evaluated. Data by Jorgensen et al. [1] were upscaled in such a way that full occupancy of the 'b' sublattice at lower temperatures was obtained. The values derived from

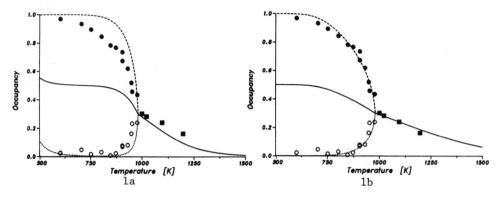

Fig.1a-b Occupancy of the oxygen sublattices by oxygen atoms as a
function of temperature at an oxygen pressure of 1 atm.,
experiment and model. Data points from [1]; solid line: average
occupancy; dashed line: occupancy of the 'oxygen' sublattice;
dotted line: occupancy of the 'vacancy' sublattice. a: concentration independence of the site energy e; b: concentration
dependence of e.

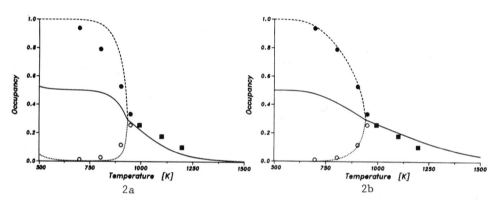

Fig.2a-b As in figure 1, but now with an oxygen pressure of 0.2 atm.

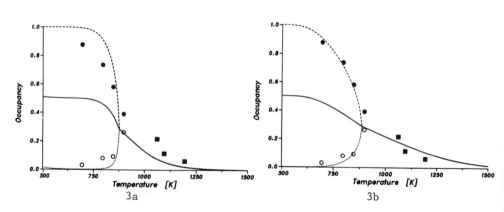

Fig.3a-b As in figure 1, but now with an oxygen pressure of 0.02 atm.

their data at 1 atm. of oxygen pressure are then e = -3.68 eV and v = 0.185 eV. A similar experiment by La Placa et al. [2] yields e = -3.62 eV and v = 0.218 eV. The agreement between both sets of values is satisfactory. Moreover, Freitas and Plaskett [4] performed a different type of experiment. They carried out measurements of the electrical resistivity as a function of temperature under various oxygen pressures. In this experiment the orthorhombic-to-tetragonal transition temperature is manifest in a change of slope of the measured curves. Evaluation of these results gives the value e = -3.68 eV in good agreement with the e values from the neutron diffraction experiments.

Using the above values for the energy parameters, the sublattice occupancies can be derived at any temperature and oxygen pressure. Results are given in the figures 1a, 2a and 3a and are compared with the experimental results from [1]. The agreement between prediction and experiment is fairly good.

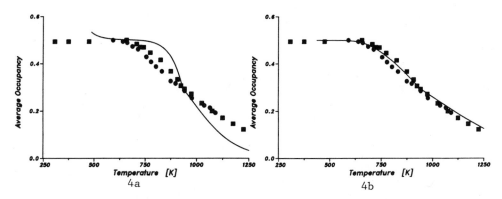

Fig.4a-b Average oxygen occupancy of the oxygen in the CuO planes as a function of temperature at an oxygen pressure of 0.2 atm., experiment and model. Squares from [5], circles from [6]. a: concentration independence of the site energy e; b: concentation dependence of e.

However, from fig.4a it is observed that the agreement between the model predictions and results of TGA measurements of the total oxygen content as a function of temperature [5,6] is less satisfying. Apparently the model predicts a change of slope at the transition temperature which is much too abrupt. A similar conclusion was drawn by Salomons et al. [7] on the basis of their measurement of the pressure dependence of the oxygen content at various temperatures. They were able to modify the model by introducing, analogous to metal hydrides, a concentration dependence of the site energy e. It turned out that a simple linear dependence of e on the total oxygen concentration c yielded a much better agreement between theory and their experiments. Accepting this extension of the model, we reanalyzed the neutron diffraction data from [1] and found in this way, e = -4.01+1.10*c. Results using this modification are given in the figures 1b, 2b and 3b. The agreement between theory and experiment turns out to be almost ideal.

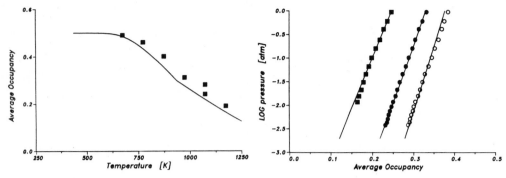

Fig.5 Average oxygen occupancy of the oxygen in the CuO planes as a function of temperature at an oxygen pressure of 0.2 atm.. Model with concentration dependence of the site energy. Data points from [8].

Fig.6 Oxygen Pressure - Oxygen Con-Concentration Isotherms. Data points from [7]. Lines: model with concentration dependence of the site energy. Squares: 1081 K; filled circles: 926 K open circles: 838 K.

Moreover, the TGA measurements [5,6] are now represented much better by the model predictions (see fig.4b). Besides, in fig.5 measurements of the same quantity by Verwey [8] are shown and compared with our model. These data points are somewhat higher than those from [5,6]. Since we used in this analysis energy parameters somewhat different from Salomons et al. [7], we also checked the consequences of the use of these new parameters for the fit to the Salomons et al. data. From fig.6 it is observed that this fit remains excellent.

In conclusion, we can state that a simple three parameter model is able to represent the measured oxygen content and oxygen ordering quite satisfactorily.

Acknowledgement

We thank the Dutch Foundation for Fundamental Research on Matter (FOM) for financial support. Part of this work was performed under the auspices of the U.S. Department of Energy, Division of Materials Science Office of Basic Energy Sciences under Contract Number DE-AC02-76CH00016.

REFERENCES

1. J.D.Jorgensen, M.A.Beno, D.G.Hinks, L.Soderholm, K.J.Volin
   R.L.Hitterman, J.D.Grace, Ivan K.Schuller, C.U.Segre, K.Zhang and
   M.S.Kleefisch, Phys.Rev. B 36, 3608 (1987).
2. S.La Placa, R.Boehme and D.E.Cox, personal communication.
3. H.Bakker, D.O.Welch and O.W.Lazareth,Jr,
   Solid State Commun. 64, 237 (1987).
4. P.R.Freitas and T.S.Plaskett, Phys.Rev.B 36, 5723 (1987).
5. P.K.Gallagher, H.M. O Bryan, S.A.Sunshine and D.W.Murphy
   Mat.Res.Bull., 22, 995 (1987).
6. P.Strobel, J.J.Caponi, C.Chaillout, M.Marezio
   and J.L.Tholence, Nature 327, 306 (1987).
7. E.Salomons, N.Koeman, R.Brouwer, D.G. de Groot and R.Griessen
   Sol.State Commun. to be published.
8. H.Verwey and W.H.M.Bruggink
   submitted to J.Phys.Chem.Sol.

Cu d-d ORBITAL TRANSITIONS AND CHARGE-TRANSFER EXCITATIONS IN HIGH-T$_c$ SUPERCONDUCTORS

H.P. Geserich[1], G. Scheiber[1], J. Geerk[2], H.C. Li[2]
W. Weber[2], H. Romberg[2], N. Nücker[2], J. Fink[2], and
B. Gegenheimer[3]

[1]Inst. f. Angew. Physik, Universität Karlsruhe
 D-7500 Karlsruhe, FRG
[2]Kernforschungszentrum Karlsruhe, Institut für Nukleare
 Festkörperphysik, P.O.B. 3640, D-7500 Karlsruhe, FRG
[3]Max-Planck-Institut für Festkörperforschung
 D-7000 Stuttgart 80, FRG

INTRODUCTION

Numerous theoretical models have been proposed as a possible explanation for the high superconducting transition temperatures T_c of the superconducting cuprates. Electron-phonon coupling, magnetic-exchange mechanisms, or coupling by low-lying electronic excitations such as plasmons or charge-transfer excitations have been considered as explanations for the high-T_c's. In order to obtain information on the nature of low energy electronic excitations, we have investigated such excitations by optical reflectivity and absorption measurements and by high-energy electron energy-loss spectroscopy (EELS) in transmission. The optical data have been published already previously.[1] The EELS measurements have been extended to 50 eV in order to obtain the dielectric constants of single-crystalline $YBa_2Cu_3O_{6.8}$. Thus information on the stronger bonded electronic states could be derived. The data are interpreted by a large charge-transfer gap of about 4 eV. In this gap, weak absorption maxima appear which are analyzed in terms of valence conserving d-d transitions of Cu^{++}.

EXPERIMENTAL

For the optical experiments, the $YBa_2Cu_3O_{7-y}$ films were prepared by a two-step method by sputtering from a single composite target onto a randomly oriented sapphire substrate at a temperature near 750°C and then in a second step, these films were intercalated with oxygen at 1 atm and 430°C. In this case a crystalline material with a strong texture due to grains grown with the c-axis perpendicular to the film plane was obtained.[2] Single crystals were grown from the melt with an excess of CuO. For EELS studies, thin films with a thickness of about 1000 Å were cut from the single crystals by an ultra-microtome and mounted on standard electron-microscope grids. The structure of the single crystals was controlled by electron diffraction. Since we know that the oxygen content is extremely sensitive to radiation damage by the e beam which transforms O_7-compounds into O_6-compounds, the O content was controlled by measuring O 1s core excitation edges.[3] EELS spectra were taken with a 170 keV spectrometer.[4] The energy and the momentum transfer resolution was chosen to be 0.18 eV and 0.04 $Å^{-1}$, respectively.

RESULTS

In Fig. 1 we show the reflectance spectrum of a typical $YBa_2Cu_3O_{6.8}$ film. For comparison, the spectrum of a superconducting single crystal, obtained for $\vec{E} \perp \vec{c}$, and thin film reflectance data of Bozovic et al.[5] are shown. Nearly the same spectral position of the plasma edge is found for the two materials, indicating the same value of the carrier concentration which is related to the 0 content. In the visible and UV region, the reflectance values of the films decrease to values below those of the single crystal. We believe that this effect is due to the topology of the film surface.

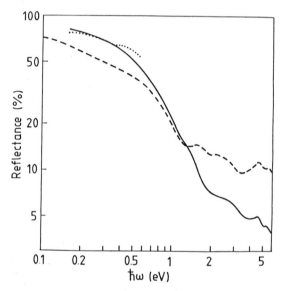

Fig. 1.   Reflectance spectra of $YBa_2Cu_3O_{6.8}$. Thin film, d = 4300 Å (full line); single crystal, $\vec{E} \perp c$ (dashed line); thin film, d = 1μ from Ref. 5 (dotted line).

In Fig. 2, we show reciprocal transmittance spectra of $YBa_2Cu_3O_7$ films for various film thicknesses. There are three absorption maxima at 0.6, 1.4 and 3 eV and a rise in absorption appears at about 4 eV. It is remarkable that the structures in the spectra are almost independent of the film thickness. This indicates that they are not caused by a contamination of the film surface.

We now come to the optical constants as derived from the energy-loss spectra. In Fig. 3 we show a typical loss function $Im(-1/\varepsilon)$ for a single crystal of $YBa_2Cu_3O_{6.8}$ with q = 0.1 $Å^{-1}$ perpendicular to the c axis. It is derived from the loss spectra by removing contributions from the direct beam at zero energy and contributions due to multiple losses. Finally, the spectrum is corrected for the angular weighting. Since at present, only single-crystalline films are available which are considerably smaller than the beam diameter (about 0.5 mm) the contributions due to the direct

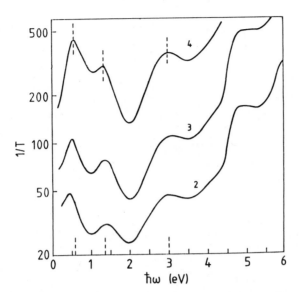

Fig. 2. Reciprocal transmittance of $YBa_2Cu_3O_7$ films with different thicknesses, d = 2900 Å (2); d = 3400 Å (3), d = 4300 Å (4).

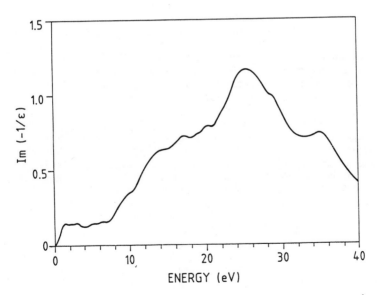

Fig. 3. Energy-loss function $Im(-1/\varepsilon)$ for a $YBa_2Cu_3O_{6.8}$ single crystal for momentum transfer q = 0.1 Å^{-1} perpendicular to the c axis.

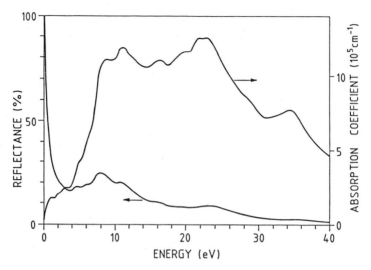

Fig. 4.   Reflectance and absorption coefficient for $YBa_2Cu_3O_{6.8}$ as derived from the loss function shown in Fig. 3.

beam were rather strong and therefore the correction resulted in a con‑
siderable uncertainty of the loss function below $\sim$ 2 eV. From the loss
function the dielectric functions ε_1 and ε_2, the reflectance, and the
absorption coefficient were derived by a Kramers-Kronig transformation.
The latter two functions are shown in Fig. 4. The reflectance is not far
from the data shown in Fig. 1 and other data in the literature thus giving
some confidence on the procedure for the correction of contributions due
to the direct beam. The absorption coefficient μ shows at low energy
similar features as the reciprocal transmittance shown in Fig. 2. There
are peaks or shoulders at $\sim$1 eV, $\sim$2 eV and $\sim$3 eV close to the peaks in
the reciprocal transmittance, but less pronounced. Moreover, there is
again a stronger rise at about 4 eV. The assignment of the peaks above
4 eV is at present speculative. Mattheiss and Hamann [6] give values of 10
and 16 eV for the binding energy of the Ba 5p and the O 2s states, re‑
spectively. In the absorption coefficient a doublet at 8.8 and 11 eV
(probably the Ba $5p_{3/2}$ and $5p_{1/2}$ doublet) and a peak at 16 eV is observed.
The features between 20 and 22 eV, and that at 34 eV may be explained by
Ba 5p $\rightarrow$ 5d, 4f and Y 4p $\rightarrow$ 4d excitations, respectively. It is remarkable
that the maximum in the loss function at 25.5 eV is very close to the
calculated free electron value at 26 eV when for the calculations of the
density of the electrons, the valence band electrons and low-lying core
levels up to the Y 4p and Ba 5s states are taken into account. At the
end of the description of our EELS results it should be mentioned that
the present data are at variance with the EELS data published by Yuan
et al. [8] and by Chen et al. [9]. Reasons for this may be a different O
content of the samples or radiation damage by the electron beam.

DISCUSSION

Probably the most important point of the present data for the low-
energy absorption coefficient (optical data and EELS data) is the close
similarity to data on NiO. In this compound the strong rise in absorption
above $\sim$3 eV was interpreted as a charge-transfer transition $d^nO^{--} \rightarrow d^{n+1}O^-$.

The weak low-energy (E < 3 eV) absorption features were assigned to valence conserving transitions within the respective d^n configuration, which are not allowed for dipole selection rules. Although there is a close similarity of the $YBa_2Cu_3O_7$ and the NiO data, there are also some characteristic differences. Firstly, the absorption coefficient in $YBa_2Cu_3O_7$ is below the charge-transfer gap at about 4 eV of the order of 10^5 cm^{-1} while that of NiO is between 10^2 and 10^3 cm^{-1}. Secondly, the absorption peaks for NiO are much narrower than those for $YBa_2Cu_3O_7$. Finally, the charge-transfer is by 1 eV higher than in NiO, while that of CuO appears to be much lower [3] near 1.5 eV. The increase in absorption and the broadening of the peaks may be caused by the presence of charge carriers in $YBa_2Cu_3O_7$, which may interact with the d electrons thus leading to more dipole allowed broadened d-d excitations. The increase of the charge-transfer gap in the cuprates is probably caused by the presence of additional, very electro-positive cations.

The two Cu sites in $YBa_2Cu_3O_7$ have approximately tetragonal symmetries and therefore, the d states are split in two e_g states ($d(x^2-y^2)$ and $d(3z^2-r^2)$) and in two t_{2g} states ($d(y,z),d(z,x)$, degenerate and $d(x,y)$). There are many spectroscopic data [3,10,11] which strongly suggest that Cu is close to a $3d^9$ configuration, i.e., the highest $3d(x^2-y^2)$ state is half filled. One thus would expect three d-d transitions, and the lowest one would be the $3d(3z^2-r^2)$ to $3d(x^2-y^2)$ excitation within the e_g orbitals.

The present data and the interpretation have important consequences for the validity of models based on a pairing mechanism due to electronic excitations. A large charge-transfer gap of 4 eV rules out a model based on low-lying charge transfer excitations.[12] The existence of low-lying d-d transitions between e_g states, on the other hand, is a further explicit indication that charge fluctuations of the type $2d^9 \rightarrow d^8+d^{10}$ are strongly suppressed due to the large electronic correlation energy. This is the basis for all models which propose a pairing by magnetic excitations. On the other hand, the low-lying d-d excitations may also lead to a pairing of the conduction holes in the high-T_c cuprates (assumed to be O 2p holes), according to a model which was proposed by one of the authors.[13,14]

REFERENCES

1. H.P. Geserich, G. Scheiber, J. Geerk, H.C. Li, G. Linker, W. Assmus, and W. Weber, Cu^{++} d-d Transitions in the Optical Spectra of Superconducting $YBa_2Cu_3O_7$-δ Films, Europhysics Lett., submitted
2. H.C. Li, G. Linker, F. Ratzel, R. Smithey, and J. Geerk, Preparation of Y-Ba-Cu-O Superconducting Thin Films by Magnetron Sputtering, Appl. Phys. Lett. in press
3. N. Nücker, J. Fink, J.C. Fuggle, P.J. Durham, and W.M. Temmerman, Evidence for Holes on Oxygen in the High-T_c Superconductors $La_{2-x}Sr_xCuO_4$ and $YBa_2Cu_3O_7$, Phys. Rev. B in press
4. J. Fink, Electron Energy-Loss Spectroscopy Applied to Solids, Z. Phys. B, 61:463 (1985)
5. I. Bozovic, D. Kirillov, A. Kapitulnik, K. Char, M.R. Hahn, M.R. Beasley, T.H. Geballe, Y.H. Kim, and A.J. Heeger, Optical Measurements on Oriented Thin $YBa_2Cu_3O_7$-δ Films: Lack of Evidence for Excitonic Superconductivity, Phys. Rev. Lett., 59:2219 (1987)
6. L.F. Mattheiss and D.R. Hamann, Electronic Structure of the High-T_c Superconductor $YBa_2Cu_3O_{6.9}$, Solid State Commun., 63:395 (1987)
7. R. Newman and R.M. Chrenko, Optical Properties of Nickel Oxide, Phys. Rev., 114:1507 (1959)

8. J. Yuan, L.M. Brown, and W.Y. Liang, Electron Energy-Loss Spectroscopy of the High-Temperature Superconductor $YBa_2Cu_3O_{7-x}$, J. Phys. C: Solid State Phys., 21:517 (1988)

9. C.H. Chen, L.F. Schneemeyer, S.H. Liou, M. Hong, J. Kwo, H.S. Chen, and J.V. Waszczak, Electronic Excitations of $YBa_2Cu_3O_{7-x}$ Superconductor: A Study by Transmission Electron Energy-Loss Spectroscopy with an Electron Microprobe, Phys. Rev. B, submitted

10. A. Fujimori, E. Takyama-Muromuchi, Y. Uchida, and B. Okai, Spectroscopic evidence for strongly correlated electronic states in La-Sr-Cu and Y-Ba-Cu oxides, Phys. Rev. B, 35:8814 (1987)

11. A. Bianconi, A. Congin Castellano, M. DeSantis, P. Rudolf, P. Lagarde, A.M. Flank, and A. Marcelli, $L_{2,3}$ XANES of the High-T_c Superconductor $YBa_2Cu_3O_{\sim7}$ with Variable Oxygen Content, Solid State Commun., 63:1009 (1987)

12. C.M. Varma, S. Schmitt-Rink, and E. Abrahams, Charge Transfer Excitations and Superconductivity in "Ionic" Metals, Solid State Commun., 62:681 (1987)

13. W. Weber, A. Cu d-d Excitation Model for the Pairing in the High-T_c Cuprates, Z. Phys. B, submitted

14. W. Weber, Mechanism of High-T_c Superconductivity, in this volume

ELECTRONIC STRUCTURE OF La$_{2-x}$Sr$_x$CuO$_4$ AND YBa$_2$Cu$_3$O$_{7-y}$

J. Fink[1], N. Nücker[1], H. Romberg[1], J.C. Fuggle[2]
P.J.W. Weijs[2], R. Schoorl[2], P.J. Durham[3]
W.M. Temmerman[3], and B. Gegenheimer[4]

[1]Kernforschungszentrum Karlsruhe, Institut für Nukleare
Festkörperphysik, P.O.B. 3640, D-7500 Karlsruhe, FRG
[2]Department of Molecular Spectroscopy, Research Institute
for Materials, University of Nijmegen, Toernooiveld
6525 ED Nijmegen, The Netherlands
[3]Daresbury Laboratory, Daresbury, Warrington WA4 4AD, UK
[4]Max-Planck-Institut für Festkörperforschung
D-7000 Stuttgart 80, FRG

INTRODUCTION

Since the discovery of the new superconducting cuprates, numerous
models for the mechanism for the high superconducting transition temperatures have been developed.[1-3] At present there is no clear evidence for
one or another model. Therefore, it is extremely important to elaborate
the electronic structure of this new class of materials and to determine
the relevant parameters by experiments in order to guide theory in the
choice of a model. In particular, there is a strong discussion whether
band-structure calculations using the local density approximation (LDA)
yield reasonable information on the electronic structure or whether the
electron-electron interaction of the 3d electrons on the Cu sites is important for the high-T$_c$ superconductors similar to the oxides, chalcogenides and halides of the late 3d transition metals. Having accepted the
strong e-e correlation, a further question is, what is the influence of
the correlation effects on the electronic structure and what is the nature
of the charge carriers in the new superconductors. This contribution
reviews recent high-energy spectroscopic investigations on the electronic
structure of the superconducting cuprates.[4-6] Recent results on single
crystals have been included. X-ray induced photoemission (XPS), Auger-
electron spectroscopy (AES), bremsstrahlung-isochromat spectroscopy (BIS)
and electron energy-loss spectroscopy (EELS) were used.

EXPERIMENTAL

Bulk samples were prepared starting from the oxides, which were mixed
together, followed by pressing and sintering twice, as described in Ref. 4.
Most of the samples measured were determined to be single phase by X-ray
diffraction. Superconductivity was verified by measurements of the resistivity and AC$_\circ$susceptibility. For high-energy EELS in transmission, films
about 1000 Å thick were cut from bulk samples and from single crystals by

an ultra-microtome and mounted on standard electron-microscope grids. The superconductivity and the structure of the films were established by measurements with a SQUID magnetometer and by electron diffraction, respectively. XPS, AES and BIS were performed by an instrument with a large solid angle X-ray monochromator and an energy resolution of 0.8, 0.81 and 1.0 eV, respectively. The EELS measurements were performed using a 170 keV spectrometer.[7] The energy resolution was chosen to be 0.4 eV. We emphasize that EELS in transmission measures bulk properties and is not surface sensitive like other spectroscopies, such as photoelectron spectroscopy or X-ray absorption spectroscopy (XAS) in the partial yield mode.

RESULTS AND DISCUSSION

In Fig. 1 we show typical experimental XPS and BIS spectra at 1486.7 eV for La_2CuO_4. For comparison we show calculated XPS and BIS spectra which were obtained within the self-consistent field local density-functional (LDF) approximation using the method of linearized muffin-tin orbitals within the atomic-sphere approximation.[8] We emphasize that the calculated curve is not just a density-of-states (DOS) curve weighted by the atomic cross section. Rather it is a weighted sum over the angular momentum components of the local density of states on each atom. The weight factor is a squared electron-photon matrix element. While the method has been used with success for a number of alloys and compounds, here there are strong deviations between theory and experiment. The calculated XPS peak is much too narrow and about 2 eV below E_F while the measured spectrum has a maximum at 4 eV below E_F. Since, according to the calculations, the XPS spectra are dominated by Cu 3d states and the cross section for O 2p states is about five times smaller, the comparison between theory and experiment indicates that the bare Cu 3d energy E_B is shifted by the mean e-e correlation energy to the LDA bandstructure value E_{LDA} to higher energy: $E_{LDA} = E_B + U_{eff}<n>$. Depending on the occupation number $<n>$, values for U_{eff} of several eV can be derived. This indicates a rather large e-e interaction U_{eff} when compared with the total bandwidth. This is in line with the fact that in all oxides, chalcogenides and halides of the late 3d transition metals, electron-electron correlations are generally accepted to be important for the electronic properties.[9] A similar high value $U_{eff} > 4$-5 eV has been derived from our AES measurements [5] and from resonance photoemission spectroscopy [10] ($U_{eff} \sim 6$ eV).

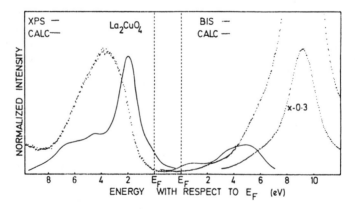

Fig. 1. Experimental and calculated XPS and BIS spectra at 1486.7 eV for La_2CuO_4.

It is interesting to note that due to the correlation induced shift to lower energy, the Cu 3d density of states at E_F is strongly reduced compared to the calculated one. Upon doping with Sr, there is within error limits no change of the XPS spectrum.[4] There is no indication of a Fermi edge which would be expected for the superconducting material. From these experimental results, we can conclude that there is almost no change in the electronic structure of the Cu 3d electrons when transforming the semiconductor La_2CuO_4 to a superconductor by doping with Sr acceptors. This is a further indication that the DOS at E_F has only a small amount of Cu 3d character. Similar results on the high-T_c cuprates were obtained in all other high-energy spectroscopy investigations which are sensitive to the Cu 3d states. Thus, e.g., almost no changes are found in Cu 2p near-edge X-ray absorption spectra [13] upon variation of x and y.

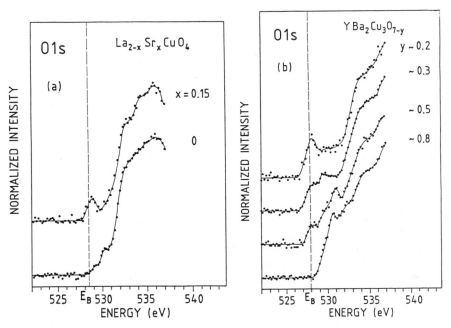

Fig. 2.   Oxygen 1s absorption edges of (a) $La_{2-x}Sr_xCuO_4$ and (b) $YBa_2Cu_3O_{7-y}$ measured by electron energy-loss spectros-copy. The binding energy of the O 1s level, as determined by X-ray induced photoemission, is shown by the broken line. In the framework of an interpretation of the spectra by the density of unoccupied states, this line would correspond to the Fermi energy.

On the other hand, strong changes are observed in measurements which are sensitive to the O 2p states. Such measurements are excitations of O 1s electrons into the local unoccupied part of the DOS at the O atoms having p symmetry. In Figs. 2(a) and (b) we show the O 1s absorption edges of $La_{2-x}Sr_xCuO_4$ and $YBa_2Cu_3O_{7-y}$ for various values of x and y. The binding energy of the O 1s level as determined by XPS measurements, is shown by

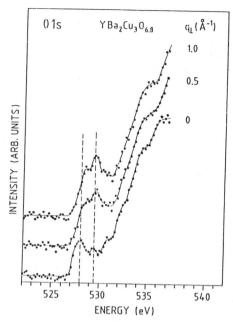

Fig. 3. Oxygen 1s absorption edges of a single crystal of
YBa$_2$Cu$_3$O$_{6.8}$ as a function of momentum transfer q .
For q$_\perp$ = 0 : $\vec{q}$ = q$_\parallel$ is about parallel to the c
axis. For q$_\perp$ = 1 Å^{-1} : $\vec{q}$ is almost in the a,b
plane.

the broken line at E$_B$. In the framework of an interpretation of the spectra
by the density of unoccupied states, this line would correspond to the
Fermi energy E$_F$. At 3–5 eV above threshold, in all samples there is a
strong increase of spectral weight. For La$_{2-x}$Sr$_x$CuO$_4$, this intensity is
due to La 5d and 4f states hybridized with O p states, for YBa$_2$Cu$_3$O$_{7-y}$,
the rise in intensity is related to Ba 5d and 4f and Y 4d states. Of central interest are the states near threshold. The spectrum of the undoped
La$_2$CuO$_4$ shows no intensity near threshold which is in agreement with the
semiconducting behaviour of this material. The doped La$_{2-x}$Sr$_x$CuO$_4$ with
x = 0.15 shows a peak near threshold which we know from previous studies [4]
to have intensity roughly proportional to the Sr concentration up to x =
0.25. Since this peak does not appear in the XPS spectra upon doping, we
believe that this peak has dominantly O 2p and almost no Cu 3d character,
i.e., it is caused by holes in an O 2p band (see below). Similar observations were obtained for the YBa$_2$Cu$_3$O$_{7-y}$ system. For y ∿ 0.8 no intensity at
threshold is observed in agreement with transport measurements which have
revealed semiconducting properties for y > ∿0.5. With decreasing y, a peak
at E$_B$ increases which again indicates a peaked DOS at E$_F$ with dominantly
O 2p character. In Fig. 3 we show recent angular dependent O 1s spectra
for a YBa$_2$Cu$_3$O$_{6.8}$ single crystal. For q$_\perp$ = 0 ($\vec{q}$=q$_\parallel$ = 0.35 Å^{-1} ∥ $\vec{c}$), dominantly O 2p$_z$ states are reached with lobes parallel to the c axis while
for q$_\perp$ = 1 Å^{-1} ($\vec{q}$=q$_\perp$ + q$_\parallel$, q$_\parallel$ = 0.35 Å^{-1}) O 2p$_{x,y}$ states are probed. The
measurements indicate that both O 2p$_z$ states (probably from O4 atoms in
the BaO plane) and O 2p$_{x,y}$ states (probably from O2 and O3 atoms in the
two-dimensional CuO$_2$ planes and the O1 atoms in the one-dimensional CuO
chains) are close to E$_F$. The most likely explanation for the energy
difference of about 1.5 eV between the peak of the p$_z$ states at 528 eV
and that of the p$_{x,y}$ states at 529.5 eV may be a different binding energy
of the 1s levels of the O4 atoms and those of the O1, O2 and O3 atoms.
Another explanation would be that the O 2p holes at the O1, O2 and O3
atoms are at higher energy above E$_F$ than those on O4 atoms. Independent

of the two explanations the measurements probably indicate that holes are created on all the four O sites. This is in agreement with the band-structure calculations (see below).

In Fig. 4 we show for comparison calculations of the O 1s absorption edges based on the same approximation as described for the calculations of the XPS and BIS spectra. At threshold, the calculations predict for the semiconducting compounds (x=0 and y=1) a finite spectral weight due to holes in the two-dimensional $Cu\ 3d_{x^2-y^2}$–O $2p_{xy}$ band typical of a metal. The comparison between calculated and experimental (no DOS at E_F) spectra is a further very explicite indication that there is a breakdown of the LDF approximation due to e-e correlations. For the superconducting compounds (x=0.15, in the calculation x=0.5, and y=0) there are differences between calculations and experiment but less pronounced. The details are discussed elsewhere.[6]

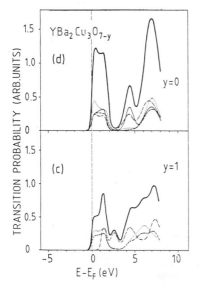 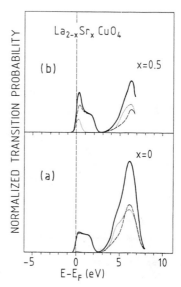

Fig. 4. Calculated oxygen 1s absorption edges in the framework of a density-functional formalism for $La_{2-x}Sr_xCuO_4$ ((a): x=0, and (b): x=0.5) and for $YBa_2Cu_3O_{7-y}$ ((c): y=1, (d): y=0). For $La_{2-x}Sr_xCuO_4$: Total: (full curve), O atoms in the CuO_2 planes (broken line), O atoms in the LaO planes (dotted curve). For $YBa_2Cu_3O_{7-y}$: Total (thick full curve), O1 (thin full curve), O2 (broken curve), O3 (dotted curve) and O4 (chain curve). The O sites are explained in the text.

The results discussed in this contribution do not only indicate a breakdown of LDA bandstructure calculations for the high-T_c superconductors but also lead to a model of the electronic structure [14] which was previously developed for other transition metal compounds [9] and which is illustrated in Fig. 5. In the LDA band-structure calculation for La_2CuO_4 a half filled $Cu\ 3d_{x^2-y^2}$–O $2p_{x,y}$ band results (see Fig. 5(a)). There are about 9.5 3d electrons on the Cu atoms, i.e., on every second Cu site there are two $Cu\ 3d_{x^2-y^2}$ electrons. Due to the strong Cu 3d on-site correlation, there is a large splitting of the Cu 3d sites into a lower and a higher Mott-Hubbard band. The strong e-e interaction pushes the second electron on each second Cu atom to the O atoms which leads to a filled O 2p band between the lower and the upper 3d band. An insulating state is reached for pure La_2CuO_4 as shown in Fig. 5b. Upon doping by Sr or O, holes are created in the O 2p band, the DOS of which can be seen directly in the O 1s edges. As shown in Fig. 5, in a first approximation the Cu 3d

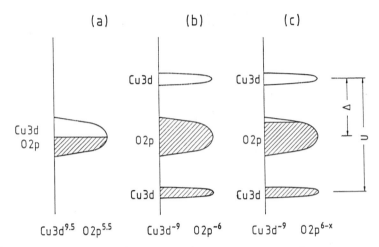

Fig. 5. Model for the electronic structure of high-T_c cuprates. (a) band-structure calculation in the local density approximation. (b) model with strong 3d-3d interaction at the Cu sites for the semiconductors. (c) the same as (b) for the superconductors.

states are not changed upon doping. Charge fluctuations of the type $2d^9 = d^8 + d^{10}$ are strongly suppressed in this model by the large correlation energy U_{eff}. The formation of $3d^8$ states (Cu^{3+}) is strongly prohibited in this model which is experimentally supported by our XPS measurements on the Cu 2p level.[4]

REFERENCES

1. T.M. Rice, A Theorist's View of High-T_c Superconductivity, Z. Phys. B, 67:141 (1987).
2. see, e.g., Proc. 18th Int. Conf. Low Temp. Physics, Kyoto (1987), Jap. J. of Appl. Phys. 26, Suppl. 26-3 (1987).
3. see, e.g., Proc. of Yamada '87, Conf. Superconductivity in Highly Correlated Fermion Systems, Sendai (1987), North-Holland, Amsterdam (1987).
4. N. Nücker, J. Fink, B. Renker, D. Ewert, C. Politis, P.J.W. Weijs, and J.C. Fuggle, Experimental Electronic Structure Studies of $La_{2-x}Sr_xCuO_4$, Z. Phys. B, 67:9 (1987).
5. J.C. Fuggle, P.J.W. Weijs, R. Schoorl, G.A. Sawatzky, J. Fink, N. Nücker, P.J. Durham, and W.M. Temmerman, Valence Bands and Electron Correlation in the High-T_c Superconductors, Phys. Rev. B, 37:123 (1988).
6. N. Nücker, J. Fink, J.C. Fuggle, P.J. Durham, and W.M. Temmerman, Evidence for Holes on Oxygen in the High-T_c Superconductors $La_{2-x}Sr_xCuO_4$ and $YBa_2Cu_3O_{7-y}$, Phys. Rev. B, in print.

7. J. Fink, Electron Energy-Loss Spectroscopy Applied to Solids, Z. Phys. B, 61:463 (1985).
8. W.M. Temmerman, G.M. Stocks, P.J. Durham, and P.A. Sterne, Electronic structure of La-Cu and Y-Ba-Cu oxides: ground-state properties and photoemission spectra, J. Phys. F: Met. Phys., 17:L135 (1987).
9. J. Zaanen, G.A. Sawatzky, and J.W. Allen, Band Gaps and Electronic Structure of Transition-Metal Compounds, Phys. Rev. Lett., 55:418 (1985).
10. R.L. Kurtz, R.L. Stockbauer, D. Mueller, A. Shik, L.E. Toth, M. Osofsky, and St.A. Wolf, Resonant photoemission study of super-conducting Y-Ba-Cu-O, Phys. Rev. B, 35:8818 (1987).
11. Zhi-xun Shen, J.W. Allen, J.J. Yeh, J.-S. Kang, W. Ellis, W. Spicer, I. Lindau, M.B. Maple, Y.D. Dalichaouch, M.S. Torikachvili, J.Z. Sun, and T.H. Geballe, Anderson Hamiltonian Description of the Experimental Electronic Structure and Magnetic Interaction of Copper Oxide Superconductors, Phys. Rev. B, 36:8414 (1987).
12. J.C. Fuggle et al., to be published.
13. A. Bianconi, A. Congiu Castellano, M. DeSantis, P. Rudolf, P. Lagarde, A.M. Flank, and A. Marcelli, $L_{2,3}$ XANES of the High-T_c Superconductor $YBa_2Cu_3O_{\sim7}$ with Variable Oxygen Content, Solid State Commun., 63:1009 (1987).
14. V.J. Emery, Theory of High-T_c Superconductivity in Oxides, Phys. Rev. Lett., 58:2794 (1987).

SUSCEPTIBILITY AND MÖSSBAUER STUDIES

OF ORTHORHOMBIC AND TETRAGONAL EuBa$_2$(Cu$_{1-x}$^{57}Fe$_x$)$_3$O$_{7-\delta}$

E. Ikonen[1], J. Hietaniemi[1], K. Härkönen[2,4], M. Karppinen[2], T. Katila[1], J. Lindén[1], L. Niinistö[2], H. Sipola[3,4], I. Tittonen[1] and K. Ullakko[3]

[1] *Department of Technical Physics,* [2] *Laboratory of Inorganic and Analytical Chemistry,* [3] *Laboratory of Engineering Materials, Helsinki University of Technology, SF-02150 Espoo*
[4] *Technical Research Centre of Finland, Reactor Laboratory, SF-02150 Espoo, Finland*

INTRODUCTION

Oxygen stoichiometry plays an important role in the new high-temperature superconductors of the type YBa$_2$Cu$_3$O$_{7-\delta}$. Variation of the parameter δ from 0 to 1 changes the crystal structure from orthorhombic to tetragonal. Superconductivity has been reported only for values $\delta \leq 0.5$. With these 1-2-3 compounds, substitution of copper atoms by transition metal elements has been studied extensively. Doping with iron atoms allows to probe the oxygen symmetry around the lattice sites of copper by ^{57}Fe Mössbauer spectroscopy. Several research groups have reported Mössbauer studies of iron-doped 1-2-3 compounds.[1-6] The rare-earth site has been investigated for example by the ^{151}Eu Mössbauer resonance.[1,7]

We have synthesized orthorhombic and tetragonal EuBa$_2$(Cu$_{1-x}$^{57}Fe$_x$)$_3$O$_{7-\delta}$ samples, where the parameter δ varies from 0.1 to 1.0. Instead of yttrium, the material contains europium as the rare-earth element. Some of the samples were doped with ^{57}Fe corresponding to $x = 0.01$ or 0.02. Iron concentration was kept low in order to maintain the structure and superconductivity properties. Different oxygen deficiencies allow us to obtain information on the oxygen environment of the copper lattice sites. The samples were characterized by x-ray diffractometry, chemical analysis of the oxygen content, magnetic susceptibility measurements, and ^{57}Fe and ^{151}Eu Mössbauer spectroscopy.

EXPERIMENTAL

Three types of EuBa$_2$(Cu$_{1-x}$^{57}Fe$_x$)$_3$O$_{7-\delta}$ samples with $x = 0$, 0.01 and 0.02 were synthesized using conventional ceramic methods. Samples were prepared in different syntheses with slightly different annealing temperatures and durations. For $x = 0.01$, the original synthesis was as follows. Stoichiometric quantities of Eu$_2$O$_3$ (99.99 %, Kemira), BaCO$_3$ (p.a. Merck), CuO (p.a. Merck) and ^{57}Fe (95 % enriched, Techsnabexport) were carefully mixed in an agate mortar. The mixture was calcined in an alumina

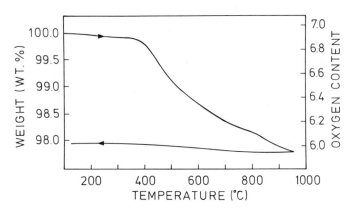

Fig. 1. Thermogravimetric data on $EuBa_2(Cu_{0.99}Fe_{0.01})_3O_{6.9}$ for heating at 2 °C/min and cooling at 3 °C/min in nitrogen. The number of oxygen atoms per unit cell is indicated on the right-hand side of the figure.

crucible for 15 h at 950 °C in air. After grinding the samples, the calcination was repeated. For the final heat treatment in oxygen, the samples were pelletized with a pressure of 7.5 kbar. The final anneal had six temperature steps at 950, 700, 600, 550, 500 and 430 °C with annealing times of 18, 6, 6, 9, 12, and 15 h, respectively. Iodometric titration gave for oxygen deficiency a value of $\delta = 0.1$.

In order to establish the preparative conditions for specimens with lower oxygen content, $EuBa_2(Cu_{0.99}Fe_{0.01})_3O_{6.9}$ samples were studied by thermal analysis. We recorded the sample weight as a function of temperature using a thermobalance (Perkin-Elmer, TGA-7). In Fig. 1, thermogravimetric data measured in nitrogen atmosphere are shown. To obtain oxygen deficiencies corresponding to $\delta = 0.6$ and 1.0, annealing temperatures of 575 and 900 °C in nitrogen were selected. The annealing time was 12 h and cooling was performed in nitrogen. The oxygen content was further confirmed with iodometric titration.

The oxygen content of the samples with $\delta \geq 0.6$ was later restored using the oxygen-annealing procedure as described above. For these reoxidized specimens, iodometric titration gave the same results as for the original samples before nitrogen annealing.

Magnetic AC susceptibility of the samples was measured using a transformer bridge, where the basic circuit consists of two identical transformers: a test frame and a compensator (Fig. 2). Changes in coil inductances due to variations of magnetic susceptibility of the specimen are measured. The magnetizing field is produced using an externally controlled current source. The pick-up signal is detected with a lock-in analyzer. This computer-based equipment was already used to characterize the Y-Ba-Cu-O samples of a neutron-diffraction study.[8]

Mössbauer experiments were carried out in standard transmission geometry using sinusoidal velocity sweep. In ^{57}Fe measurements the source was $^{57}Co:Pd$. The ^{151}Eu experiments were performed with a $^{151}Sm:Sm_2O_3$ source. Mössbauer absorbers were made by grinding the $EuBa_2(Cu_{1-x}{}^{57}Fe_x)_3O_{7-\delta}$ samples. The thicknesses of the absorbers were between 10 and 30 mg/cm^2. After Mössbauer measurements, part of the material was used for chemical analyses.

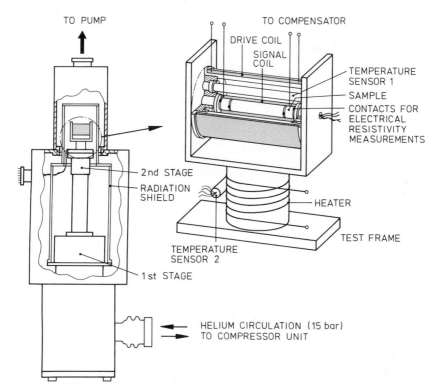

Fig. 2. Test frame and cooling system for AC susceptibility measurements. The specimen is placed on the axis of the signal coil. It is magnetized by applying a sinusoidal current at the frequency of 20 Hz through the drive coil. The effect of air flux is eliminated with a compensator coil (not seen in the figure). Electrical resistivity can be measured simultaneously with the susceptibility measurement. The test frame and the compensator are thermally connected to the cold head of a two-stage cryocooler.

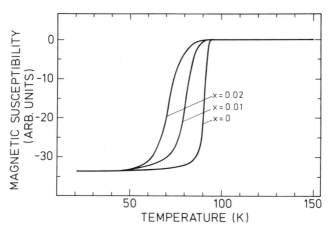

Fig. 3. Magnetic susceptibility of $EuBa_2(Cu_{1-x}{}^{57}Fe_x)_3O_{6.9}$ samples with $x = 0$, 0.01 and 0.02.

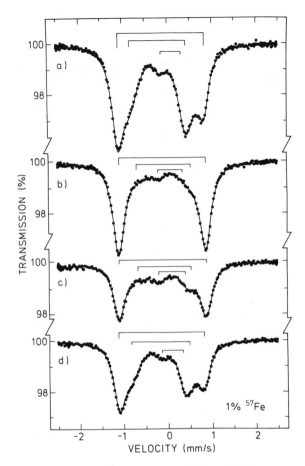

Fig. 4. Mössbauer spectra of $EuBa_2(Cu_{0.99}{}^{57}Fe_{0.01})_3O_{7-\delta}$ samples with 1 % iron concentration. The oxygen content corresponds to a) $\delta = 0.1$, b) 0.6, c) 1.0 and d) 0.1 (reoxidized from $\delta = 1.0$).

RESULTS

Measurements on crushed pellets with an x-ray powder diffractometer were carried out using $Cu(K_\alpha)$ radiation with a scanning speed of $1°/min$ (2θ). X-ray data showed pure orthorhombic phase for the samples with $\delta = 0.1$ and pure tetragonal phase for $\delta = 1.0$. For the sample with the intermediate oxygen content ($\delta = 0.6$), a small orthorhombic splitting was evident.

Temperature dependence of the real part of the AC susceptibility is shown in Fig. 3 for pelletized samples with $\delta = 0.1$. Iron doping smoothly decreases the transition temperature and increases the width of the transition range. The corresponding imaginary parts of the susceptibility are peaked at the middle of the transition ranges of Fig. 3. Magnetic susceptibility of the oxygen-deficient samples ($\delta \geq 0.6$) was close to zero in the

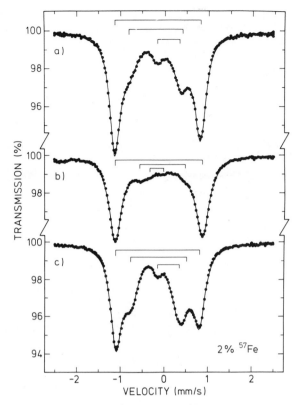

Fig. 5. Mössbauer spectra of $EuBa_2(Cu_{0.98}{}^{57}Fe_{0.02})_3O_{7-\delta}$ samples with 2 % iron concentration. The oxygen content corresponds to a) $\delta = 0.1$, b) 1.0 and c) 0.1 (reoxidized from $\delta = 1.0$).

temperature range between 20 K and 300 K. After subsequent annealing in oxygen, the reoxidized samples became superconducting, but the transition range was broad (about 40 K) and the transition was composed of two stages. In this case, the imaginary part was peaked at the lower transition temperature.

Room-temperature ^{57}Fe Mössbauer spectra of samples with $x = 0.01$ are presented in Fig. 4. The oxygen content has been varied between $0.1 \leq \delta \leq 1.0$. The sample used for the measurement of Fig. 4c was reoxidized resulting in the spectrum shown in Fig. 4d. In Fig. 5, Mössbauer spectra of samples with $x = 0.02$ are presented. Figure 5 demonstrates that the relative intensities of the spectral components can be changed by variations in the sample preparation.

The Mössbauer data have been analyzed using three quadrupole doublets. The results of least-squares fits are shown by solid lines. Values obtained for quadrupole splittings vary between 1.92 and 2.00 mm/s for the largest splitting and between 1.1 and 1.3 mm/s for the intermediate splitting. For the smallest splitting, the results are somewhat more uncertain, particularly with the oxygen-deficient samples.

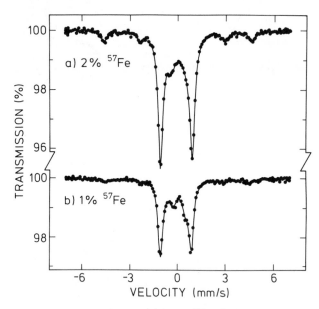

Fig. 6. Mössbauer spectra of $EuBa_2(Cu_{1-x}{}^{57}Fe_x)_3O_{6.0}$ samples with an increased maximum velocity, showing a magnetically split component.

In Figs. 4c and 5b, additional resonance lines can be seen at about -2.3 mm/s. The spectra of Fig. 6 reveal that these lines are due to a magnetically split component, which is observed only in the samples with $\delta = 1.0$. A large quadrupole splitting of this component is also evident.

Mössbauer studies with ^{151}Eu confirmed that the oxidation state of europium is three in the 1-2-3 compounds with δ between 0.1 and 1.0. No trace of Eu^{2+} was found. The effective position of the Eu^{3+} resonance line apparently depends on the oxygen content of the sample. The observed differences in the line positions are small (≤ 0.06 mm/s) but reproducible.

DISCUSSION

The $EuBa_2Cu_3O_{6.9}$ samples had orthorhombic crystal structure and they showed sharp superconductivity transitions. Also the oxygen content and ^{151}Eu Mössbauer spectra were in agreement with commonly accepted results. The change in the lattice symmetry due to variation of the oxygen content may cause the observed differences in the Eu^{3+} line position. Doping with iron allowed to use ^{57}Fe Mössbauer spectroscopy for characterization of the samples.

In some of our susceptibility measurements, a two-stage transition was observed. Mazaki et al.[9] have explained this type of behaviour to originate from growing of superconducting inclusions (first stage) and their subsequent combination (second stage). Mössbauer spectra of the samples showing one-stage (Fig. 3) and two-stage susceptibility transitions were similar. This finding supports the notion that the possible second stage of the transition describes macroscopic features of the sample.

Mössbauer spectra of the samples with $\delta = 0.1$ show three distinct components, in contrast to the oxygen-deficient samples, where only the quadrupole doublet with the largest energy splitting is clearly visible. It is seen that the shape of the spectrum is well recovered after varying the oxygen content in the compound (Fig. 4). In the samples with $\delta = 1.0$, a new magnetically split component is observed (Fig. 6). Within the measurement accuracy, the magnetic susceptibility study did not reveal an ordered magnetic structure.

On the basis of ionic radii, iron atoms are expected to substitute copper atoms at either of the two inequivalent lattice sites of the 1-2-3 compound. Variation of the oxygen content between $\delta = 0.1$ and 1.0 affects only the nearest-neighbour environment of the copper lattice site denoted by Cu(1). In our Mössbauer spectra, the quadrupole doublet with the largest energy splitting remained approximately unchanged in all samples. A straightforward interpretation then suggests that the components with smaller energy splitting originate from iron atoms at the Cu(1) sites. Also other interpretations are possible if it is assumed that iron atoms are preferentially substituted at the Cu(1) sites.[6]

In summary, [57]Fe Mössbauer spectroscopy allows to probe the local oxygen symmetry around copper lattice sites of the 1-2-3 compounds. At present, more work is needed for definite site assignments. Detailed calculations beyond the simple point-charge model should provide both qualitative and quantitative connection between the oxygen environment and the observed electric field gradients.

ACKNOWLEDGMENTS

We thank the Technology Development Centre and Academy of Finland for financial support. Two of us (K. U. and M. K.) gratefully acknowledge scholarships from the Emil Aaltonen Foundation and the Association of Finnish Chemical Societies, respectively.

REFERENCES

1. J. M. D. Coey and K. Donnelly, Z. Physik B **67**, 513 (1987).
2. H. Tang, Z. Q. Qiu, Y.-w. Du, Gang Xiao, C. L. Chien and J. C. Walker, Phys. Rev. B **36**, 4018 (1987).
3. X. Z. Zhou, M. Raudsepp, Q. A. Pankhurst, A. H. Morrish, Y. L. Luo and I. Maartense, Phys. Rev. B **36**, 7230 (1987).
4. M. Takano and Y. Takeda, Jap. J. Appl. Phys. **26**, L1862, (1987).
5. T. Tamaki, T. Komai, A. Ito, Y. Maeno and T. Fujita, Solid State Comm. **65**, 43 (1988).
6. E. R. Bauminger, M. Kowitt, I. Felner and I. Nowik, Solid State Comm. **65**, 123 (1988).
7. M. Eibschutz, D. W. Murphy, S. Sunshine, L. G. Van Uitert, S. M. Zahurak and W. H. Grodkiewicz, Phys. Rev. B **35**, 8714 (1987).
8. O. K. Antson, P. E. Hiismäki, H. O. Pöyry, A. T. Tiitta, K. M. Ullakko, V. A. Trunov and V. A. Ul'yanov, Solid State Comm. **64**, 757 (1987).
9. H. Mazaki, M. Takano, Y. Ikeda, Y. Bando, R. Kanno, Y. Takeda and O. Yamamoto, Jap. J. Appl. Physics **26**, L1749 (1987).

SEARCH FOR THE BARIUM ISOTOPE EFFECT

IN THE HIGH-T_C SUPERCONDUCTOR $YBa_2Cu_3O_{7-x}$

A.Inyushkin, N.Babushkina, V.Florentiev, A.Kopylov
V.Ozhogin, A.Kaul* and I.Graboy*

Kurchatov Institute of Atomic Energy, 123182 Moscow, USSR
*Moscow State University, 119899 Moscow, USSR

ABSTRACT

The barium isotope substitution effect on the superconducting transition temperature has been searched for in the high-T_C superconductor $YBa_2Cu_3O_{7-x}$. No shift in the transition temperature T_C was observed by ac magnetic susceptibility measurements within 0.25 K.

INTRODUCTION

Until recently, the possibility of finding superconductivity near 100 K was only hypothetical. Upper limits for T_C (30 - 40 K) are obtained in some approximations within the framework of the BCS theory with the phonon mechanism for electron pairing. Therefore, the question of possible new mechanisms for electron pairing has only arisen since the discovery of superconductivity above 90 K, in $YBa_2Cu_3O_{7-x}$[1].

The isotope substitution effect on the superconducting transition temperature provides strong support for the phonon mechanism of superconductivity. Under BCS treatment, $T_C \propto \omega_D$, where ω_D is the Debye frequency. For mono-element superconductors, $T_C \propto M^{-1/2}$, due to the relation $\omega_D \propto M^{-1/2}$, where M is the isotopic mass. This mass dependence of T_C supports the phonon mechanism for electron pairing. Generally, in the case of poly-element lattices, the mass dependences of T_C are individual for the various atoms in the compound, and depend on the contribution of each atom's vibrations to the electron-phonon coupling strength.

The effect of isotope substitution on the transition temperature of high-T_C superconductors has been studied experimentally in[2-6]. Batlogg et al.[2] and Bourne et al.[3] reported that T_C did not change significantly when ^{16}O was replaced by ^{18}O in $YBa_2Cu_3O_{7-x}$. However, more recently Leary et al.[4] have observed a non-zero oxygen isotope effect: ΔT_C = 0.3 - 0.5 K. For this system, Bourne et al.[5] reported that Cu and Ba isotope substitutions produced no measurable shift in T_C. Note that in $La_{1.85}Sr_{0.15}CuO_4$ an oxygen isotope effect was observed by Batlogg et al.[6] and Faltens et al.[7]. Thus, the available experimental data are insufficient to draw conclusions about the isotope substitution effect on T_C in high-T_C superconductors.

EXPERIMENTAL

We searched for an isotope effect in $YBa_2Cu_3O_{7-x}$ by substituting the barium isotope. Samples of $YBa_2Cu_3O_{7-x}$ were prepared by a cryochemical technique of powder synthesis[8]. Keeping in mind that the character of the superconducting transition is sensitive to sample preparation conditions, samples of different isotopic composition were processed simultaneously under identical conditions in the same furnace. The pressed ceramic pellets of 9 mm diameter and 1.5 mm thickness were heated for 12 hours in oxygen at $950^{\circ}C$ and then cooled at a rate of $200^{\circ}C/h$.

The superconducting transition was determined through the temperature dependence of the sample's magnetic susceptibility. The ac susceptibility measurements were performed on an induction magnetometer at 200 Hz with a modulation field amplitude of less than 0.1 Oe. Figure 1 shows the magnetic susceptibility data $\chi(T)$ for one of the ^{135}Ba enriched samples of $YBa_2Cu_3O_{7-x}$. The samples are paramagnetic at temperatures $T > T_c \approx 95$ K, and are almost ideally diamagnetic ($\chi \approx -1/4\pi$) below 70 - 80 K.

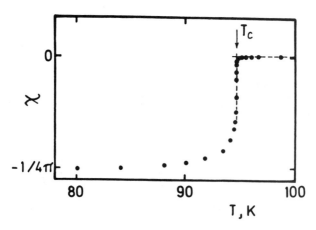

Fig. 1. Magnetic susceptibility vs temperature for $YBa_2Cu_3O_{7-x}$ prepared with ^{135}Ba.

RESULTS AND DISCUSSION

For the relevant transition temperature, T_c, we used the onset of the superconducting transition because our experimental results had shown that the deviation of the onset temperatures for various samples with a given barium isotopic composition was much smaller than when other definitions of T_c were used. In our opinion, the superconducting transition in ceramic samples starts with a transition from a normal to a superconducting phase of sample grains whose compositions are close to stoichiometric. The adopted T_c definition gives the value of T_c, irrespective of the transition width.

One can estimate the maximum expected isotope shift of T_c in the following way. According to the BCS theory, the isotope shift $\Delta T_c^{max} \approx 1.5$ K, under the assumption that barium atom vibrations are responsible for the electron coupling; and $\Delta T_c^{max} \approx 0.5$ K, when the electron-phonon coupling strength is approximately the same for all phonon modes, i.e. there are no preferred modes responsible for electron coupling.

T_c measurements of three sample groups with different Ba-isotope composition (^{134}Ba, ^{135}Ba, ^{138}Ba) have been performed. Each group consisted of seven samples with the same Ba isotope. Measured values of T_c, mean T_c values for each group, root-mean-square deviations of T_c and the mean mass of Ba-atoms (M_{Ba}) are presented in Table 1. The dependence

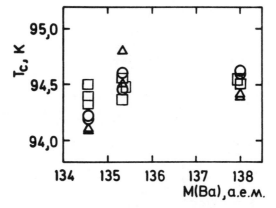

Fig. 2. T_c vs M_{Ba} for $YBa_2Cu_3O_{7-x}$.

of T_c on Ba isotope mass is plotted in Fig. 2. The differences between the mean values of the transition temperatures do not exceed 0.25 K, i.e. are equal to the sum of the rms deviations. In the case of $T_c \propto M_{Ba}^{-\alpha}$, we find $\alpha = -0.05 \pm 0.04$, with a statistical confidence of more than 0.95. We believe this result indicates the lack of the barium isotope effect in $YBa_2Cu_3O_{7-x}$.

The zero barium isotope effect itself cannot be regarded as a reason for a non-phonon mechanism of electron coupling in $YBa_2Cu_3O_{7-x}$, because the magnitude of the isotope-effect exponent, α, may be reduced (e.g. by the strong electron-electron interaction) in superconductors with phonon-mediated coupling. However, if one takes into consideration the two experimental facts - namely the high value of T_c and the zero isotope effect, then unphysically large values of electron-phonon and electron-electron interaction constants are obtained from McMillan's equation for the observed T_c[5].

Therefore, non-phonon mechanisms of superconductivity may be dominant in $YBa_2Cu_3O_{7-x}$.

Table 1. Superconducting transition temperatures, rms deviation of T_C and mean Ba-atom masses for the studied samples.

sample number	T_C, K		
	^{134}Ba	^{135}Ba	^{138}Ba
1	94.20	94.60	94.60
2	94.20	94.45	94.60
3	94.35	94.45	94.55
4	94.40	94.55	94.50
5	94.50	94.35	94.55
6	94.15	94.50	94.40
7	94.15	94.80	94.50
T_C(mean),K	94.28	94.53	94.51
rms(T_C), K	0.11	0.13	0.08
M_{Ba}, au	134.6	135.3	138.0

REFERENCES

1. M.K.Wu, J.R.Ashburn, C.J.Torng, P.H.Hor, R.L.Meng, L.Gao, Z.J.Huang, Y.Q.Wang, and C.W.Chu, Phys.Rev.Lett. 8, 908 (1987)
2. B.Batlogg, R.J.Cava, A.Jayaraman, R.B.van Dover, G.A.Kourouklis, S.Sunshine, D.W.Murphy, L.W.Rupp, H.S.Chen, A.White, K.T.Short, A.M.Mujsce, and E.A.Rietman, Phys.Rev.Lett. 58, 2333 (1987)
3. L.C.Bourne, M.F.Crommie, A.Zettl, H.-C.zur Loye, S.W.Keller, K.L.Leary, A.M.Stacy, K.J.Chang, M.L.Cohen, and D.E.Morris, Phys.Rev.Lett. 58, 2337 (1987)
4. K.J.Leary, H.-C.zur Loye, S.W.Keller, T.A.Faltens, W.K.Ham, J.N.Michaels, and A.M.Stacy, Phys.Rev.Lett. 59, 1236 (1987)
5. L.C.Bourne, A.Zettl, T.W.Barbee III, and M.L.Cohen, Phys.Rev. B36, 3990 (1987)
6. B.Batlogg, G.Kourouklis, W.Weber, R.J.Cava, A.Jayaraman, A.E.White, K.T.Short, L.W.Rupp, and E.A.Rietman, Phys.Rev.Lett. 59, 912 (1987)
7. T.A.Faltens, W.K.Ham, S.W.Keller, K.J.Leary, J.N.Michaels, A.M.Stacy, H.-C.zur Loye, D.E.Morris, T.W.Barbee III, L.C.Bourne, M.L.Cohen, S.Hoen, and A.Zettl, Phys.Rev.Lett. 59, 915 (1987)
8. A.R.Kaul, I.E.Graboy, and Y.D.Tretyakov, Synthesis of superconducting complex oxides, in "Superconductivity, 1-st issue," V.I.Ozhogin, ed., Kurchatov Institute of Atomic Energy, Moscow (1987)

PURE METALLIC BRIDGE AND POINT-CONTACT TUNNELING INTO

SINGLE- AND POLYCRYSTALLINE YBa$_2$Cu$_3$O$_{7-y}$

K.E. Gray, E. R. Moog and M. E. Hawley

Materials Science Division
Argonne National Laboratory
Argonne, Illinois, 60439, USA

INTRODUCTION

It is known theoretically and experimentally[1] that when the resistance of a point-contact is decreased the current-voltage characteristic, I(V), changes from that of an insulating barrier to that of a pure metallic bridge. Such an I(V) still shows the energy gap but, because of its much lower resistance, it is less sensitive to extraneous conduction mechanisms. Values of between 25 and 35 meV are reported here for YBa$_2$Cu$_3$O$_{7-y}$, and these are in agreement with our results for low-resistance point-contact tunneling in both single- and poly-crystal samples. Examples of I(V) and the conductance, dI/dV, will also be presented for pure metallic bridges as well as higher resistance point-contact tunneling. An electrically insulating layer is found on the surface of all samples, including single-crystals showing no evidence of impurities measured by Raman scattering[2]. Consequently, contact of the tunneling tip is necessary to mechanically scrape, and thus clean, the surface before a measurable current can be obtained. Although this represents a disadvantage, a more serious drawback of point-contact tunneling is the inability to measure the I(V) continuously through the transition temperature, T$_c$, due to thermal expansion of the mechanical apparatus. We will argue that thin film tunnel junctions are thus desirable both to unambiguously identify the energy gap with the 90 K bulk material, and to subtract the background conductance to evaluate the electron coupling mechanism in the high-T$_c$ superconductors (HTS), that is analogous to electron-phonon coupling in traditional superconductors[3].

EXPERIMENTAL

The tunneling apparatus is described in detail, together with a discussion of tests done on conventional superconductors, in a study of an organic superconductor[4]. In that case, as well as measurements[5] on La$_{1.85}$Sr$_{0.15}$CuO$_{4-y}$ and the results reported here, an insulating surface layer prevented true vacuum tunneling and the soft Au tip had to be pushed into the surface of the sample to obtain a measurable current. The Au tip is moved towards the sample, at low temperature, by the combined action of a differential micrometer and a stack of piezoelectric transducers, to which a few hundred volts is applied. The resistance of resulting junctions could be varied by adjusting the transducer voltage and hence the force between the tip and sample. All measurements were done with the apparatus cooled be exchange gas to He liquid at 4.2 K. Raising the temperature above 20 K by a heater always resulted in changes in, or complete loss of, the tunneling contact.

Measurements were made on both polycrystalline sintered pellets and single-crystals grown by solid state reaction[6]. In all cases, we found an insulating surface layer, including the flat, shiny crystalline facets in the a-b plane of YBa$_2$Cu$_3$O$_{7-y}$, which showed no evidence by Raman scattering of the impurity phases found in polycrystalline samples.

RESULTS

A variety of I(V) can be found for different tip force after a measurable current is established. Often the conductance rises monotonically with voltage and is otherwise featureless, but nonmonotonic behavior and sharp features can also be observed. In our measurements on poly- and single-crystals using both soft Au and hard W tips we have reproduced most, if not all, of the I(V) or dI/dV reported by others, a partial summary of which is contained in Refs. 7 and 8. The I(V) of very high resistance (> 1 MΩ) contacts were more likely to be very asymetrical with voltage and sometimes time dependent. Because of the high resistance, the I(V) are more susceptible to being influenced or even dominated by other, non-tunneling, conduction mechanisms[9]. For lower resistances, we have obtained the rather symmetrical dI/dV shown in Fig. 1 for poly- and single-crystal YBa$_2$Cu$_3$O$_{7-y}$ samples. These show the expected symmetrical peaks traditionally associated with the superconducting energy gap, but have an additional contribution to the conductance, which is linear in voltage. Such an additional conductance has been observed previously in tunnel junctions of traditional superconductors with a distribution of small metallic particles in the tunnel barrier[10], but has also recently been suggested as evidence for the resonating valence bond theory of superconductivity[11] being applicable to the HTS. Energy gap values of 20-25 meV are extracted from the data of Fig. 1, and these are in reasonable agreement with point-contact tunneling measurements of others, although peaks at much higher voltages have been reported[7]. Note that since the tip was pushed into the surface of the single crystal, no information on the anisotropy of the energy gap can be obtained.

A convenient way to avoid many of the problems and uncertainties of these point-contact tunneling studies is to form a pure metallic bridge by increasing the force on the tip until there is no barrier between the tip and the YBa$_2$Cu$_3$O$_{7-y}$ sample. Experimental and theoretical studies[1] have confirmed that there is a smooth transition from the I(V) of classical tunneling to that of a pure metallic bridge as the scattering strength of the barrier, and therefore junction resistance, decreases. Information about the energy gap is still contained in the pure metallic bridge since the conductance is determined by Andreev reflections at the discontinuity of the energy gap found in the region of the normal (Au) to superconductor (YBa$_2$Cu$_3$O$_{7-y}$) interface. An example of this is shown in Fig. 2 for a Au-YBa$_2$Cu$_3$O$_{7-y}$ pure metallic bridge, together with the theoretical prediction[1] for dI/dV for a gap of 25 meV. Other bridges show values of the gap up to 35 meV.

The energy gap has also been extracted from measurements of infra-red reflectivity. The most recent result[12] for a mosaic of single-crystals indicates an energy gap of about 30 meV for YBa$_2$Cu$_3$O$_{7-y}$ in reasonable agreement with the point-contact tunneling and pure metallic bridge results reported here in Figs. 1-3.

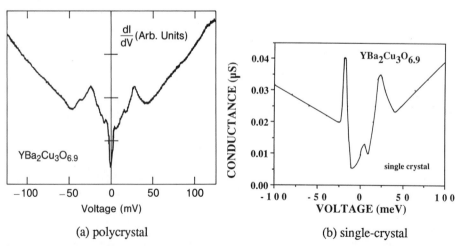

(a) polycrystal   (b) single-crystal

Fig. 1. The conductance, dI/dV, for point-contact tunneling of a Au tip into YBa$_2$Cu$_3$O$_{7-y}$.

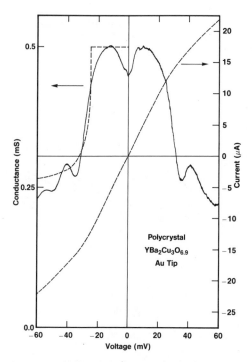

Fig. 2. I(V) and dI/dV for a pure metallic bridge between a Au tip and YBa$_2$Cu$_3$O$_{7-y}$. The theoretical conductance, dI/dV, is also shown (-------) for an energy gap of 25 meV.

DISCUSSION AND CONCLUSIONS

There consists a considerable uncertainty in the magnitude of the energy gap measured in YBa$_2$Cu$_3$O$_{7-y}$ by point-contact tunneling[7,8]. This is due to the inability to obtain BCS-type I(V) curves as in the case of traditional superconductors and the variability of results reported by various groups. Difficulties with surfaces may be the root of these problems: we know[8] that insulating layers exist on the surface of YBa$_2$Cu$_3$O$_{7-y}$ and that oxygen deficient material, possibly found at surfaces, especially in vacuum, may have a substantially reduced T$_c$ (as low as[13] 4 K) or even be semiconducting[14-16]; we also know that the short coherence length[17] (2-3 nm) requires that perfect superconducting material exists right up to a distance of a few lattice constants from the tunneling barrier. Because of this, the variability of tunneling results is not surprising. In fact, the pure metallic bridge, which has no insulator between the Au tip and YBa$_2$Cu$_3$O$_{7-y}$ sample may presently yield the most reliable measurement of the energy gap.

Unfortunately, even for the pure metallic bridge, it is not possible to determine whether the 90 K bulk superconducting phase is being probed or a lower T$_c$ oxygen deficient surface layer, because the point-contact junctions cannot be cycled through T$_c$ due to thermal expansion of the mechanical apparatus. For this reason alone, thin-film tunnel junctions would have an inherent advantage over point-contact junctions. In addition, however, such temperature excursions through T$_c$ are required to evaluate the electron coupling mechanism in HTS, that is analogous to electron-phonon coupling in traditional superconductors[3], since it is necessary to subtract the background tunneling conductance in the normal state and the critical field is very large[17].

Preparing good thin-film tunnel junctions will not be easy since the requirement for perfect superconducting material right up to a distance of a few lattice constants of the tunneling barrier is still valid. In particular, the barrier material cannot leach oxygen out of the HTS film nor can interdiffusion occur during the various processing steps of the HTS film. Such measurements represent a significant challenge to the scientific community, but their importance towards furthering our knowledge of the HTS will make it worth the effort.

ACKNOWLEDGEMENTS

The authors would like to thank J.Z. Liu and J Downey for providing the single crystals and D.G. Hinks and D.W. Capone II for providing the polycrystalline samples. This work was supported by the U.S. Department of Energy, Division of Basic Energy Sciences-Materials Sciences under contract #W-31-109-ENG-38.

REFERENCES

1.   G.E. Blonder, M. Tinkham and T.M. Klapwijk, Phys. Rev. **B25**, 4515 (1982); G.E. Blonder and M. Tinkham, Phys. Rev. **B27**, 112 (1983).
2.   R. Bhadra, T.O. Brun, M.A. Beno, B. Drabowski, D.G. Hinks, J.Z. Liu, J.D. Jorgensen, L.J. Nowicki, A.P. Paulikas, I.K. Schuller, C. Segre, L. Soderholm, B. Veal, H.H. Wang, J.M. Williams, K. Zhang and M. Grimsditch, Phys. Rev. **B37**, (1988).
3.   E.L. Wolf, Principles of Electron Tunneling Spectroscopy (Oxford Univ. Press, New York, 1985), Chaps. 4 and 6.
4.   M.E. Hawley, K.E. Gray, B.D. Terris, H.H. Wang, K.D. Carlson and J.M. Williams, Phys. Rev. Lett. **57**, 629 (1986).
5.   M.E. Hawley, K.E. Gray, D.W. Capone II and D.G. Hinks, Phys. Rev. **B35**, 7224 (1987).
6.   J.Z. Liu, G.W. Crabtree, A. Umezawa and L. Zongquan, Phys. Lett. **A121**, 305 (1987).
7.   K.E. Gray, M.E. Hawley and E.R. Moog, in Novel Mechanisms of Superconductivity, edited by S.A. Wolf and V.Z. Kresin (Plenum, New York, 1987), p. 611.
8.   E.R. Moog, M.E. Hawley, K.E. Gray, J.Z. Liu, D.G. Hinks, D.W. Capone II and J. Downey, J. Low Temp. Phys. **71**, 393 (1988).
9.   S.T. Ruggiero and J.B. Barner, Phys. Rev. **B36**, 8870 (1987); J.B. Barner and S.T. Ruggiero, Phys. Rev. Lett. **59**, 807 (1987).
10.   H.R. Zeller and I. Giaever, Phys. Rev. **181**, 789 (1969).
11.   P.W. Anderson and Z. Zou, Phys. Rev. Lett. **60**, 132 (1988).
12.   Z. Schlesinger, R.T. Collins, D.L. Kaiser and F. Holtzberg, Phys. Rev. Lett. **59**, 1958 (1987).
13.   W.K. Kwok, G.W. Crabtree, A. Umezawa, B.W. Veal, J.D. Jorgensen, S.K. Malik, L.J. Nowicki, A.P. Paulikas and L. Nunez, Phys. Rev. **B37**, 106 (1988).
14.   D.C. Johnston, A.J. Jacobson, J.M. Newsan, J.T. Lewandowski, D.P. Goshorn, D. Xie and Y.B. Yelon, Proc. Symp. on Inorganic Superconducting Materials, Amer. Chem. Soc. Nat. Mtg., New Orleans, LA, August 31-September 4, 1987.
15.   J.D. Jorgensen, M.A. Beno, D.G. Hinks, L. Soderholm, K.J. Volin, R.L. Hitterman, J.D. Grace, I.K. Schuller, C.U. Segre, K. Zhang and M.S. Kleefisch, Phys. Rev. **B36**, 3608 (1987).
16.   R.J. Cava, B. Batlogg, C.H. Chen, E.A. Reitman, S.M. Zahurak and D. Werder, Phys. Rev. **B36**, 5719 (1987).
17.   T.K. Worthington, W.J. Gallagher and T.R. Dinger, Phys. Rev. Lett. **59**, 1160 (1987).

SECTION 4: CRITICAL FIELDS – CRITICAL CURRENTS

UPPER CRITICAL FIELD OF GRAIN ORIENTED CUPRATE SUPERCONDUCTORS

D. K. Finnemore, O. B. Hyun, K. Athreya
L. A. Schwartzkopf,* and M. Damento

Ames Laboratory and Department of Physics
Iowa State University
Ames, IA 50011 USA
*Mankato State University, Mankato, MN 56001

INTRODUCTION

One of the striking structural features of the $Y_1Ba_2Cu_3O_{7-x}$ family of superconductors is the occurrence of twin planes aligned parallel to the c-axis. These materials generally form the proper phase at a temperature near 960°C with an oxygen deficiency with x approaching one. If the sample is cooled in flowing oxygen, gradually oxygen intercalates into the copper layer between the two BaO layers. At a temperature in the 600 to 700°C range, these oxygens order into chains along the b-axis with an accompanying tetragonal to orthorhombic phase transition. The strain associated with this orthorhombic distortion in turn leads to [110] twin planes with the a and b axis interchanged at each twin. Depending on the cooling rate, the grain size and possibly other factors, the twin spacing typically ranges from 20 to 200 nm. Sometimes the spacing is in the μm range or greater. Evidence of these twins is easily seen in a transmission electron microscope or by looking at a polished sample under polarized light.

These twins may affect the superconducting properties. There is considerable evidence in the literature that twin planes can enhance superconductivity in simple metallic superconductors. Enhancements of several tenths of a degree have been reported for Sn and Nb[1,2,3] and theoretical studies to explain these results have shown that this twin plane enhancement should lead to an upper critical field of the form $H_{c2}=A(1-T/T_c)^{1/2}$ where A is a constant and T_c is the transition temperature.[1]

The square root behavior for H_{c2}, in fact, is a very general feature for thin film superconductors with the field applied parallel to the film. In the regime where the film thickness, d, is greater than the coherence distance, ξ, the Ginzburg-Landau relation holds

$$H_{c2} = \frac{\Phi_0}{2\pi\xi^2} = \frac{\Phi_0}{2\pi\xi_0^2} (1-T/T_c)$$

In the regime where d<<ξ, then

$$H_{c2} = \frac{\Phi_0}{2\pi\xi d} = \frac{\Phi_0}{2\pi\xi_0 d}(1-T/T_c)^{1/2}$$

Therefore, the occurrence of the $(1-T/T_c)^{1/2}$ behavior would be an indication that the sample is behaving like a stack of thin films oriented parallel to the c-axis.

In a preliminary study[4] of grain aligned $Y_1Ba_2Cu_3O_7$ having a grain size ranging from 2 to 4 μm, it was found that $H_{c2}=30(1-T/T_c)^{1/2}$ tesla where T_c=91.8 K. For this sample, the grains behaved like a stack of thin films. In the Budzin et al.[1] model for twin plane enhancement, there are two transition temperatures, T_c for the twin planes and T_{co} for the bulk. Analyzing the data[5] in these terms gives T_0=91.8K and T_{co}=90.0K for a coherence distance in the a-b plane ξ_{oab}=1.5 nm.

The purpose of the work reported here is to see how general this result is by reporting data for $Ho_1Ba_2Cu_3O_7$ grain aligned samples and data for one single crystal.

EXPERIMENTAL

Two different kinds of grain aligned samples have been studied in this work. For the first, superconducting powder is poured into a mold containing liquid epoxy in a field ranging from 2 to 9 T and allowed to harden. For the second, the powder is poured into a mold containing toluene in a field of 9 T. The toluene is then squeezed out and a pellet is pressed. This pellet is then sintered in the usual way. In the epoxy case the alignment is essentially complete as observed both by x-rays and by optical photography of the twin planes on polished surfaces. In the pressed pellet case, approximately 20% of the grains are misaligned by large angles. Photographs taken parallel to the c-axis and perpendicular to the c-axis are shown in Fig. 1 for a sintered pressed pellet of $Ho_1Ba_2Cu_3O_7$ sample oriented in a field of 2T. As can be seen, many of the grains are aligned within a few degrees of the c-axis. A few are off by 20° or more.

RESULTS

Starting material for the $Ho_1Ba_2Cu_3O_7$ sample shows an excellent Curie-Weiss law at high temperature as shown by the inset on Fig. 2. Here the inverse of the ac susceptibility is shown to be linear in T. The superconducting to normal transition has a steep linear portion between 89 and 90 K with a small tail at the high temperature end. This sample was then ground to powder, oxygen treated and then either formed into an epoxy aligned sample or a sintered pellet. For the epoxy samples, the alignment is essentially complete. For the sintered pellet, results like those shown in Fig. 1 are typical.

In an attempt to obtain a quantitative estimate of the degree of alignment, x-ray powder patterns were taken for various sintered pellets. As shown in Fig. 3, an unaligned pellet (top spectrum) shows all the proper x-ray peaks with approximately the proper intensities. For an aligned pellet with the c-axis perpendicular to the plane of the x-rays (middle spectrum) the (00ℓ) peaks are gone. For an aligned pellet with the c-axis in the plane of the x-rays, the (00ℓ) peaks are the dominant feature. The alignment is certainly not perfect but the c-axis is predominantly along the applied field.

Figure 1. Photographs of aligned $Ho_1Ba_2Cu_3O_7$, the top picture
looking down the c-axis, the bottom picture with the
c-axis in the plane of the photo parallel to the
arrow.

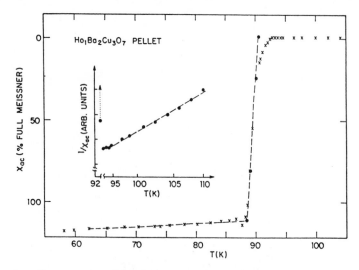

Figure 2. The ac susceptibility change at the superconducting
transition. The inset shows that the Ho moment obeys a
Curie-Weiss law.

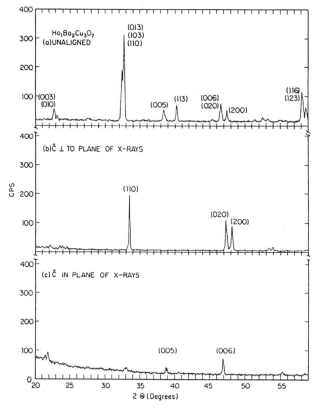

Figure 3. X-ray powder patterns to estimate the degree of alignment.
The top spectrum shows an unaligned sample. The middle
spectrum shows an aligned sample oriented to extinguish the
(001) peaks. The bottom spectrum shows an aligned sample
oriented to show only (001) peaks.

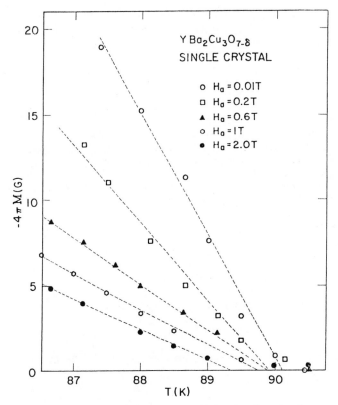

Figure 4. Magnetization data for a $Y_1Ba_2Cu_3O_7$ single crystal.

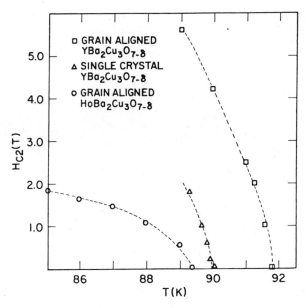

Figure 5. Upper critical field data for two grain aligned samples and
one single crystal.

Magnetization vs temperature data for both the $Y_1Ba_2Cu_3O_7$ and the $Ho_1Ba_2Cu_3O_7$ are similar to the data reported earlier.[4,5] For the Ho sample there is a rather large background to subtract but this is a very well behaved quantity. For the single crystal, the magnetization data are shown in Fig. 4. For this crystal, the magnetization is linear in temperature as shown and H_{c2} was determined by extrapolating the data to $M=0$.

The upper critical field data, H_{c2} vs T, all have a clear negative curvature in the region just below T_c and they can easily be fitted to an $H_{c2}=A(1-T/T_c)^{1/2}$ relation. The particular $Ho_1Ba_2Cu_3O_7$ sample shown here was made from freshly ground powder with no subsequent oxygen treatment. We assume that this explains the depressed T_c. Other grain aligned Ho samples which were oxygen treated after grinding had T_c above 92 K and the same negative curvature for H_{c2} vs T.

CONCLUSIONS

A wide variety of grain aligned and single crystal samples have been studied and found to have H_{c2} approximately proportional to $(1-T/T_c)^{1/2}$. This seems to be a rather general phenomenon and signals a periodic supression or enhancement of the pair potential with periodicity comparable to the twin plane spacing.

REFERENCES

1) A. I. Budzin and N. Khvorikov, Sov. Phys. JETP 62 1071 (1985).

2) W. Krah and D. Kohnlein, Z. fur Phys. B 28 19 (1977).

3) R. B. Zubeck, T. W. Barbee, Jr. and T. H. Geballe, J. Appl. Phys. 50 6423 (1979).

4) D. E. Farrell, B. S. Chandrasekhar, M. R. DeGuire, M. M. Fang, V. G. Kogan, J. R. Clem, and D. K. Finnemore, Phys. Rev. B 36 4025 (1987).

5) M. M. Fang, V. G. Kogan, D. K. Finnemore, J. R. Clem, S. Chumbley, and D. E. Farrell, Phys. Rev. B (accepted).

CRITICAL FIELDS AND CHARACTERISTIC LENGTHS

IN COPPER OXIDE SUPERCONDUCTORS

G. W. Crabtree, W. K. Kwok, and   A. Umezawa

Materials Science Division
Argonne National Laboratory
Argonne, Illinios, U. S. A. 60439

INTRODUCTION

There have been a number of studies of the critical
fields and critical currents in the two copper oxide
superconducting families $La_{2-x}Sr_xCuO_4$ and $YBa_2Cu_3O_{7-\delta}$ which
show unexpected and unusual properties. Resistive measurements
of the superconducting transition in zero and finite field
often show long tails extending to low temperatures, and the
critical field curves constructed from such data often show
unusual upward curvature, especially near T_c. Initially most
of the critical field studies were done on polycrystalline
sintered samples. With such samples there is always a
lingering question as to whether the measured properties are
intrinsic to the material or are due to poor connections
between particles or to the averaging of strongly anisotropic
behavior over many crystallite orientations. With the advent
of measurements on single crystal samples in several
laboratories it is now possible to compare measurements on
polycrystals and single crystals to determine the intrinsic
behavior and to assess the degree of sample dependence which
cannot be attributed to particle connections or to anisotropy.
In this review we characterize the typical behavior of
polycrystalline samples and compare it to results on single
crystals. We find that much of the behavior seen in
polycrystals also appears in single crystals and that among
single crystal samples there is significant sample dependence.
Nevertheless, there is sufficient agreement on the upper
critical field values to define approximately the slope at T_c,
from which the coherence length can be derived. We combine
this information with new measurements of the lower critical
field at low temperature to derive the anisotropic magnetic
penetration depth. Finally, we compare the derived
characteristic length with relevant unit cell dimensions and
with the scale of the microstructure to infer information about
the dimensionality of the superconductivity and the pinning
forces in $YBa_2Cu_3O_{7-\delta}$.

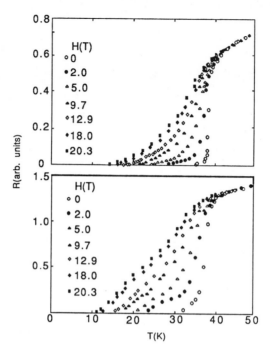

Fig. 1 Resistive superconducting transitions
in zero and finite fields for $La_{1.85}Sr_{.15}CuO_{4-y}$
(top panel) and $La_{1.85}(Sr,Ca)_{.15}CuO_{4-y}$;
Ca/Sr=3/97 (bottom panel). (Ref. 1)

POLYCRYSTALLINE MEASUREMENTS

Both the $La_{2-x}Sr_xCuO_4$ and $YBa_2Cu_3O_{7-\delta}$ families of
superconductors show unusual resistive transition curves.
Typical data[1] is shown in Fig. 1, where the resistance is
plotted versus temperature for two different samples of the
$La_{2-x}Sr_xCuO_4$ family. The top panel shows data for the sample
composition which gives the most reproducible results from
laboratory to laboratory[2]; the other panel shows data for a
sample which has not been so well studied elsewhere. In zero
field the transition is rather broad, with a gradual onset and
an obvious tail near R=0 which extends for several degrees. In
finite field, the onset temperature does not fall significantly
while the zero resistance temperature is depressed by more than
a factor of two. The same behavior can be seen in
polycrystalline samples[3] of $YBa_2Cu_3O_{7-\delta}$ as shown in Fig. 2.
The zero field curve shows its major drop at 91 K with a small
tail going to zero at 88 K, while the 8 T curve has nearly the
same onset with zero resistance occurring at 79 K. Similar
curves are seen for samples prepared at many other
laboratories.[4,5]

The field dependent tail in the resistance curves makes
the extraction of upper critical field values from the data
somewhat uncertain. A large range of critical field slopes can
be obtained depending on the value of the resistance chosen to
represent T_c. This is illustrated in Fig. 3 where the critical
field curve from the data of Fig. 2 is plotted taking the 50%

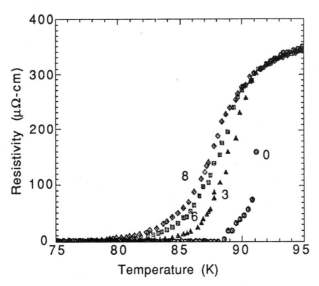

Fig. 2: Resistive transitions in polycrystalline
YBa$_2$Cu$_3$O$_{7-\delta}$ in zero and finite fields.
The fields in tesla are indicated by labels
on the curves. (Ref. 3)

and 10% values of the resistance as the definition of T_c. The
slopes dH_{c2}/dT differ by more than a factor of two, from -3.41
T/K for the 50% definition to -1.59 T/K for the 10% definition.
Thus, a characteristic critical field curve is hard to define
from polycrystalline data. However, there is one feature that
both the 50% and 10% curves have in common: both show upward
curvature near T_c. Like the long field dependent tail in the
resistive transitions, this feature appears in nearly all
measurements of the upper critical field in polycrystalline
samples. This upward curvature is directly traceable to the
relatively strong dependence of the resistivity on low magnetic
fields.

The long resistive tail and the upward curvature in the
$H_{c2}(T)$ curves might be easily explained by either of two
polycrystalline effects: the strong anisotropy in H_{c2} due to
the one- and two-dimensional character of the superconductivity
or the poor connections between particles in sintered
samples.[6,7] In both cases the sample contains material with a
range of critical field values, with the high critical field
part contributing to the transition curve near onset and the
low critical field part contributing to the curve near zero
resistivity. If anisotropy is the explanation, the high and
low critical field parts of the sample arise from the random
orientation of the crystallites, while if weak links across
particle connections are the explanation, the high and low
critical field parts correspond to the interior of the
particles and the weak links themselves. In these models, any
desired range of critical fields can be obtained by postulating
a pattern of critical field anisotropy or by assuming a range
of coupling strengths across the weak links.

If either of these explanations is correct, the long
resistive tails and the upward curvature in the critical field
curve should be absent in measurements on single crystal

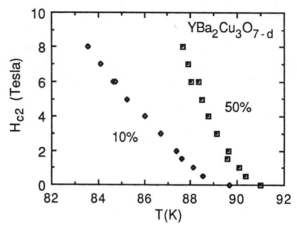

Fig. 3: Upper critical field derived from the data of Fig. 2 assuming T_C is given by 50% and 10% of the resistive transition respectively. (Ref. 3)

samples. As is shown below, these features are as obvious in single crystal results as they are in the polycrystal data, suggesting that they are intrinsic effects unrelated to the random orientation or weak links in polycrystalline samples.

SINGLE CRYSTAL UPPER CRITICAL FIELDS

There have been four measurements[8-11] of the upper critical fields of $YBa_2Cu_3O_{7-\delta}$ single crystals with the samples prepared and the measurements performed in different laboratories. The measurements were performed inductively and resistively, though to date both techniques have not been used on the same sample. Resistive measurements[8,9] are shown in Figs. 4 and 5, where the field in both cases is along the **c** direction. There is a great difference in the shape of the curves in finite field, indicating that there is a difference in sample quality. However both curves are quite sharp in zero field, with zero resistivity attained at about 91 K.

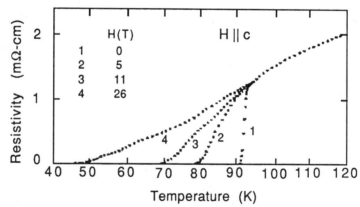

Fig. 4: Resistive transition of a single crystal of $YBa_2Cu_3O_{7-\delta}$ in zero and finite fields. (Ref. 9)

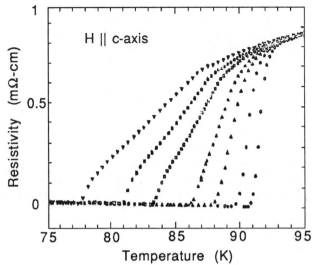

Fig. 5: Resistive transition of a single
crystal of $YBa_2Cu_3O_{7-\delta}$ in zero and finite
fields. The data are shown for fields of
0, 3, 10, 20, 40, 60, and 90 kOe. (Ref. 8)

Nevertheless, in finite field there is a long tail extending to
rather low temperature, with the onset point affected
relatively little by the field. This feature, present in both
sets of data, is qualitatively very similar to that found in
the polycrystalline data. For the field in the **ab** plane, the
critical fields are much higher and the depression of the zero
resistance point is correspondingly lower. However, the same
qualitative feature holds for this field direction as well:
there is a much stronger depression of the zero resistance
point than there is of the onset point.

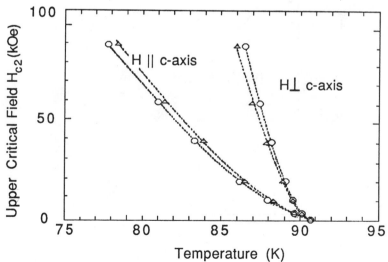

Fig. 6: Upper critical fields of two single crystal of
$YBa_2Cu_3O_{7-\delta}$. The points shown for H along **c** are
taken from Fig. 5. (Ref. 8)

The critical field curves derived from resistive measurements on single crystal samples show upward curvature as is observed in the polycrystalline samples. Fig. 6 shows the critical fields[8] taking R=0 as the definition of T_C for both field directions for two samples, one of which was used for the data in Fig. 5. Both samples show upward curvature in the $H_{C2}(T)$ curves near T_C. The data for the field in the **ab** plane closely resemble those for polycrystalline samples, with obvious curvature near T_C and more linear behavior at lower temperature. As with polycrystals it is difficult to assign a unique slope to the critical field data near T_C. Inductive measurements of H_{C2} have been performed only by one laboratory,[11] so it is difficult to assess the reproducibility of the measurements. However, there seems to be much less curvature in the data for the field in either direction except very close to T_C, where strong upward curvature appears.

A very interesting comparative study was done by Moodera et al.[10] who made resistive studies on single crystals grown by two different techniques. One set of crystals showed upper critical field curves very similar to those of Fig. 6. The other set showed curves like those of Fig. 7, with no upward curvature and, for fields along the **c** direction, a distinct downward curvature. Both crystals were measured by the same resistive technique, so the difference indicates a clear sample dependence in the critical field behavior. The origin of the sample dependence is not known, although the density of twin boundaries may play a role.[11,12]

The available data on the upper critical field in single crystal samples leads to two conclusions: the long resistive tail in the finite field superconducting transition and the upward curvature in the critical field curves are not artifacts

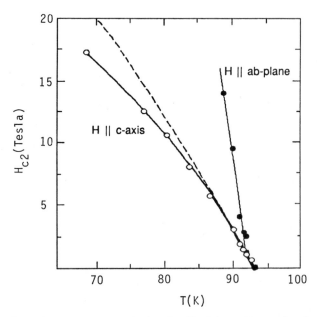

Fig. 7: Upper critical fields of a single
crystal of YBa$_2$Cu$_3$O$_{7-\delta}$ derived from
resistive measurements.(Ref. 10)

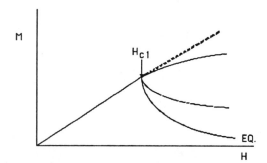

Fig. 8: Schematic representation of the
magnetization curve of a superconductor with
no flux pinning (EQ), mild flux pinning
(middle curve), and strong flux pinning
(upper curve).

of the anisotropy or the weak link connections between
particles. Rather, they are intrinsic to single crystal
samples produced in many laboratories throughout the world.
However, the upward curvature is not found in all single
crystal samples, implying that it is not necessarily a property
of the ideal defect-free material. It is likely that twin
boundaries or other defects control the degree of curvature in
the upper critical field data in a way that is not yet well
understood.

Despite the sample dependence of the critical field data,
it is possible to define a critical field slope for the two
directions below the region near T_c. For the field in the **ab**
plane, all measurements[8-11] give -3.6 T/K within about 15%.
For the field along **c**, the curvature is more pronounced and
the value of the slope depends more strongly on the field. Two
values are quoted[8-11] in the literature: -0.55 T/K and
-1.1 T/K. As shown below, this uncertainty does not severely
affect the derived values of the characteristic lengths.

SINGLE CRYSTAL LOWER CRITICAL FIELDS

The lower critical field is defined as the field where it
is energetically favorable for flux to enter the sample in the
form of a vortex lattice. In general this field must be
measured inductively as there is no distinguishing feature at
H_{c1} in resistive measurements. Normally H_{c1} is determined from
the magnetization curve as illustrated in Fig. 8. The initial
part of the magnetization curve is linear and corresponds to
perfect diamagnetism where the field is shielded from the
interior of the sample by persistent supercurrents flowing
within a penetration depth of the surface. For an ideal
sample in equilibrium there is a sharp cusp in the
magnetization curve at H_{c1} where the field first enters the
sample. However for real samples[13] containing defects which
pin flux lines, the field is prevented from entering the sample
freely and the cusp at H_{c1} is masked, as in the middle curve of
Fig. 8. If the flux pinning is very strong, very little field
enters the sample at H_{c1} and the magnetization curve hardly
deviates from the initial linear behavior characteristic of the
Meissner state, as illustrated by the top curve in Fig. 8.

Thus for many real materials there is only a subtle deviation
from linearity at the lower critical field and precise
magnetization data are required to identify H_{c1}.

A careful study of the magnetization curves for the field
along **c** and in the **ab** plane using well characterized single
crystals and a commercial SQUID magnetometer has been carried
out.[14,15] Three crystals with varying amounts of defects were
examined to explore the influence of flux pinning on the data.
The details of the measurement are reported elsewhere in this
volume.[15] After correcting for demagnetizing effects and
averaging over the three samples, the lower critical field was
determined to be 120±10 Oe for the field in the **ab** plane and
690±50 Oe for the field along the **c** direction. These values
are considerably smaller than previously estimated from earlier
magnetization data on single crystals,[16] but are in agreement
with polycrystalline results from magnetization[17] and muon spin
rotation.[18]

CHARACTERISTIC LENGTHS

One of the most useful ways of characterizing an
arbitrary superconductor is by its two length scales, the
coherence length ξ describing how fast the superconducting
order parameter varies in space, and the magnetic penetration
depth λ describing how far a magnetic field extends into the
superconductor or, equivalently, how deep the shielding
supercurrents flow. These two lengths are closely related to
the critical fields through the formulas

$$H_{c2} = \frac{\phi}{2\pi\xi^2}$$

$$H_{c1} = \frac{\phi}{4\pi\lambda^2} \ln\frac{\lambda}{\xi}$$

where ϕ is the flux quantum. The first formula follows from a
simple linearized Ginzburg-Landau argument[19] applied at H_{c2} and
can be justified intuitively from a simple geometrical
argument. The average field in the superconductor is given by
$H=n\phi \sim \phi/l^2$ where n is the density of flux lines and l the
separation between them. When $l \sim \xi$ the normal cores of the
flux lines overlap and the material is driven normal, implying
$H=H_{c2}$. The second formula follows from an energy argument[20,21]
equating the free energy of the Meissner state with the free
energy of the state with a single flux line. The energy is
composed of the core energy, the field energy, and the kinetic
energy of the supercurrents. The latter two contributions are
calculated in the London electrodynamics which is valid in the
high κ ($=\lambda/\xi$) approximation. Once formulated, the energy
calculation reduces to a geometry problem which depends only on
λ and ξ. Thus the expression for H_{c1} is independent of the
Ginzburg-Landau assumptions for the form of the free energy or
its temperature dependence.

For anisotropic superconductors the expressions for the
critical fields in terms of the characteristic lengths must be

Table 1: Anisotropic critical fields and characteristic lengths for $YBa_2Cu_3O_{7-\delta}$. The notations * and ** refer to results assuming the critical field slopes in the **c** direction are 0.55 T/K and 1.1 T/K respectively.

	H || c	H || a,b
H_{c1}	690 ± 50 G	120 ± 10 G
H_{c2}	35 T * 70 T **	230 T
$\xi_{a,b}$	31 Å * 22 Å **	
ξ_c	4.8 Å * 6.7 Å **	
κ	29 * 44 **	230
$\lambda_{b,a}$	900 Å * 950 Å **	8400 Å* 7800 Å**
λ_c		900 Å * 950 Å **

generalized to allow for anisotropy in ξ and λ. This is most easily accomplished by combining the two equations into a single equation

$$\frac{H_{c1}}{H_{c2}} = \frac{\ln \kappa}{2\kappa^2}$$

and allowing κ to take two values, one for each field direction. Following Ginzburg-Landau theory[22,23] to infer the forms for κ, we have

$$\kappa^c = \left\langle \frac{\lambda_{ab}^c}{\xi_{ab}} \right\rangle \qquad \kappa^{ab} = \left\langle \frac{\lambda_c^{ab} \lambda_{ba}^{ab}}{\xi_c \xi_{ab}} \right\rangle^{1/2}$$

where the superscript refers to the field direction and the subscript to the direction in which the coherence length or penetration depth is being measured. These equations can now be solved for the anisotropic coherence lengths and the penetration depths in terms of the anisotropic upper and lower critical fields. The solutions for the quantities ξ, κ, and λ are shown in Table 1.

It is interesting to compare the values of ξ and λ with other important lengths in the $YBa_2Cu_3O_{7-\delta}$ structure. The unit

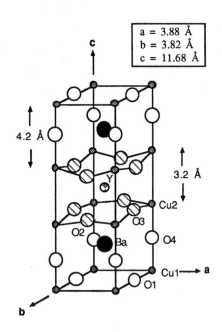

a = 3.88 Å
b = 3.82 Å
c = 11.68 Å

Fig. 9: Structure of $YBa_2Cu_3O_{7-\delta}$ showing
the distances between two dimensional
planes and one dimensional chains.
(Ref. 24)

cell[24-26] shown in Fig.9 is composed of layers of Cu-O bonds
oriented perpendicular to the **c** direction. The lower
dimensionality arises from the absence of O atoms in certain
key positions. The two dimensionality is a result of missing
O atoms along the cell edges surrounding the Y site. These
missing oxygens interrupt the Cu-O bonding chain along the **c**
direction, so that the Cu-O planes just above and below the Y
site do not overlap electronically and the metallic path along
c is destroyed. One dimensional behavior occurs in the top
and bottom faces of the cell, where O atoms along the **a**
direction are absent, leaving a system of metallic Cu-O chains
oriented along the **b** direction. The layer of chains and the
layer of planes are connected by a bridging oxygen, O4, which
provides a small electronic overlap between the one- and two-
dimensional layers. The coherence length in the **a** or **b**
direction is 20-30 Å, much larger than the unit cell
dimensions in the plane. In contrast, along the **c** direction,
the coherence length is 6-7 Å, smaller than the unit cell
dimension, leading to the possibility that the
superconductivity will be two dimensional in character.
However, the relevant comparison for determining dimensionality
is the distance between metallic or superconducting layers
rather than the cell dimension. As shown on Fig. 9, the two
planes just above and below the Y site are separated by 3.2 Å,
while the distance between the chains and the planes is 4.2 Å.
As these distances are smaller than the coherence length, it is
very unlikely that the superconductivity is less than three
dimensional. However, since the distance between planes and
between planes and chains is of the same order as the coherence
length, it is likely that the Cu-O layers are only weakly
coupled, leading to poorly developed superconducting properties
along the **c** direction. This kind of weak coupling between
two- and one-dimensional sub-systems is unusual among ordinary

superconductors and its theoretical and experimental consequences for superconducting behavior are largely unexplored.

The unusually long penetration depth in $YBa_2Cu_3O_{7-\delta}$ has potentially important implications for the Meissner effect and for flux pinning. The penetration depth of 8000 Å for the field in the **ab** plane is large compared to the thickness of thin film samples and approaches the grain size in fine-grained polycrystalline samples. Thus in these samples an external magnetic field will penetrate significantly and the Meissner effect, even in the ideal case, will be appreciably reduced. Likewise, the penetration depth is long compared to the distance between twin boundaries, which typically varies between 100 Å and 1000 Å. At the twin boundaries, the **a** and **b** axes are interchanged so that the one dimensional chains suddenly alter their direction by 90°. Because the chains play a role in determining the easy supercurrent directions and therefore the supercurrent distribution within the **ab** plane, the twin boundaries interrupt and distort the vortex supercurrents flowing within a penetration depth of the core. This distortion in the current distribution makes a significant contribution to the free energy of the vortex and therefore affects its pinning characteristics. Understanding this interaction between vortices and twin boundaries is fundamental to achieving strong pinning and high critical current densities.

CONCLUSIONS

The study of single crystals of the oxide superconductors allows two important characterizations to be made. The elimination of poor interconnections between particles in polycrystalline samples reveals the intrinsic behavior of the bulk material. Surprisingly, two of the properties of polycrystalline samples which might have been attributed to poor interconnects remain in single crystal samples: the pronounced broadening of the resistive transition in finite applied fields and the anomalous upward curvature of H_{c2} near T_c. These features are not observed in ordinary superconductors and may be characteristic of the copper oxide systems, perhaps related to the low dimensionality of the normal state crystal and electronic structure or to the weak coupling between layers in the superconducting state. Sample dependence in the shape of the upper critical field curves in single crystals of $YBa_2Cu_3O_{7-\delta}$ prepared by different techniques suggests there is an important microstructural parameter, perhaps the twin boundary density, which strongly affects the superconducting properties.

A second important result of single crystal measurements is the characterization of the anisotropy in the upper and lower critical fields. To date there is no sufficiently high quality data in single crystals to define the critical fields of the 40 K family $La_{2-x}Sr_xCuO_4$. In the 90 K family $YBa_2Cu_3O_{7-\delta}$ several measurements of the upper critical field in different laboratories are in reasonable agreement. Although curvature is observed in the upper critical field data as a function of temperature, the slope can be defined within reasonable limits for temperatures not too close to T_c. For the lower critical field, measurement is complicated by the presence of flux

pinning which obscures the characteristic cusp in the
magnetization curve. Nevertheless, with precise magnetization
measurements using a SQUID magnetometer the anisotropy in the
lower critical field can be found. The lower critical field is
much smaller than previously estimated from single crystal
measurements, with an anisotropy roughly inverse to that of the
upper critical field. The anisotropic critical fields can be
inverted to give the two anisotropic characteristic lengths
describing the superconductivity, the coherence length and the
magnetic penetration depth. A comparison of the coherence
length to the unit cell dimensions shows that the
superconductivity must be three dimensional but that there is
probably weak coupling between the chains and planes. This
weak coupling is unusual in known superconducting systems, and
its possible fundamental consequences for superconducting
properties are largely unexplored. The magnetic penetration
depth is quite large, exceeding the thickness of thin film
samples and approaching the grain size of polycrystalline
samples. This will produce a reduction of the Meissner effect
in these cases. The penetration depth is also large compared
to the scale of defects in the crystals like the separation of
twin boundary planes. This implies a strong pinning
interaction of the flux lines with the microstructure of single
crystal and polycrystalline samples.

ACKNOWLEDGEMENT

This work was supported by the U. S. Department of Energy,
Basic Energy Science-Materials Science under contract #W-31-
109-ENG-38.

REFERENCES

1. N. Kobayashi, T. Sasaoka, K. Oh-ishi, T. Sasaki,
    M. Kikuchi, A. Endo, K. Matsuzaki, A. Inoue, K. Noto,
    Y. Syono, Y. Saito, T. Masumoto, Y. Muto, Jpn. J. Appl.
    Phys. 26, L358(1987).
2. W. K. Kwok, G. W. Crabtree, D. G. Hinks, D. W. Capone, Jpn.
    J. Appl. Phys.26, Suppl.26-3, 1191(1987).
3. W. K. Kwok, Ph.d Thesis, Purdue University, 1987,
    (available from University Microfilms, Ann Arbor,
    MI. 48109).
4. T. Takabatake, M. Ishikawa, Y. Nakazawa, I. Oguro,
    T. Sakakibara, T. Goto, Jpn. J. Appl. Phys.26, L978(1987)
5. K. Takita, H. Akinaga, H. Katoh, T. Uchino, T. Ishigaki,
    H. Asano, Jpn. J. Appl. Phys. 26, L1323(1987).
6. J. R. Clem, V. G. Kogan, Jpn. J. Appl. Phys.26, Suppl.26-3
    1161(1987).
7. J. W. Ekin, A. I. Braginski, A. J. Panson, M. A. Janocko,
    D. W. Capone II, N. J. Zaluzec, B. Flandermeyer,
    O. F. de Lima, M. Hong, J. Kwo, S. H. Liou, J. Appl.
    Phys.62, 4821(1987).
8. Y. Iye, T. Tamegai, H. Takeya, H. Takei, Jpn. J. Appl.
    Phys. 26, L1057(1987).

9. Y. Hidaka, M. Oda, M. Suzuki, A. Katsui, T. Murakami, N. Kobayashi, Y. Muto, Proceedings of the Yamada Conference on Superconductivity in Highly Correlated Fermion Systems, Sendai, Japan, Aug.31-Sept.3 (1987), M. Tachiki, Y. Muto, and S. Maekawa, eds., (North Holland, Amsterdam, 1988),p329.

10. J. S. Moodera, R. Meservey, J. E. Tkaczyk, C. X. Hao, G. A. Gibson, P. M. Tedrow, Phys. Rev. B (in press)

11. W. J. Gallagher, T. K. Worthington, T. R. Dinger, R. L. Sandstrom, Proceedings of the Yamada Conference on Superconductivity in Highly Correlated Fermion Systems, Sendai, Japan, Aug.31-Sept.3 (1987), M. Tachiki, Y. Muto, and S. Maekawa, eds., (North Holland, Amsterdam, 1988), p228.

12. D. K. Finnemore, these Proceedings.

13. C. P. Bean, Phys. Rev. Lett.$\underline{8}$, 250(1962).

14. A. Umezawa, G. W. Crabtree, J. Z. Liu, T. J. Moran, S. K. Malik, L. H. Nunez, W. L. Kwok, C. H. Sowers, (submitted to Phys. Rev.).

15. A. Umezawa, G. W. Crabtree, J. Z. Liu, L. H. Nunez, these Proceedings.

16. T. K. Worthington, W. J. Gallagher, T. R. Dinger, Phys. Rev. Lett.59, 1160(1987).

17. R. J. Cava, B. Batlogg, R. B. van Dover, D. W. Murphy, S. Sunshine, T. Siegrist, J. P. Remeika, E. A. Rietman, S. Zahurak, G. P. Espinosa, Phys. Rev. Lett.$\underline{58}$, 1676(1987).

18. D. R. Harshman, G. Aeppli, E. J. Ansaldo, B. Batlogg, J. H. Brewer, J. F. Carolan, R. J. Cava, M. Celio, A. C. D. Chaklader, W. N. Hardy, S. R. Kreitzman, G. M. Luke, D. R. Noakes, M. Senba, Phys. Rev. B$\underline{36}$, 2386 (1987).

19. M. Tinkham, <u>Introduction to Superconductivity</u>, McGraw-Hill N. Y., 1975, p128.

20. V. G. Kogan, Phys. Rev.B$\underline{24}$, 1572(1981).

21. A. V. Balatskii, L. I. Burlachkov, L. P. Gor'kov, Sov. Phys. JETP63, 866(1986).

22. R. A. Klemm and J. R. Clem, Phys. Rev. B$\underline{21}$, 1868(1980).

23. D. R. Tilley, Proc. Phys. Soc.(London)$\underline{85}$, 1177(1965).

24. M. A. Beno, L. Soderholm, D. W. Capone II, D. G. Hinks, J. D. Jorgensen, I. K. Schuller, C. U. Segre, K. Zhang, J. G. Grace, Appl. Phys. Lett. $\underline{51}$, 57(1987).

25. J. E. Greedan, A. H. O'Reilley, C. V. Stager, Phys. Rev.B35, 133(1987).

26. J. J. Capponi, C. Chaillout, A. W. Hewat, P. Lejay, M. Marezio, N. Nguyen, B. Raveau, J. L. Soubeyroux, J. L. Tholence, R. Tournier, Europhys. Lett.$\underline{3}$, 1301(1987).

CRITICAL CURRENT DENSITIES OF OXIDE SUPERCONDUCTORS

O.Horigami and Y.Yamada

Toshiba R&D Center, Toshiba Corporation
4-1 Ukishima-cho, Kawasaki-ku, Kawasaki 210, Japan

INTRODUCTION

A tremendous number of papers on oxide superconducting materials have been presented in the last year.

In Japan, for instance, there were as many as 230 papers presented at the autumn meeting of the Physical Society, 130, 60 and 45 papers at meetings of the Japanese Society of Applied Physics, the Japanese Institute of Metals, and the Cryogenic Society of Japan, respectively[1].

Huge efforts have been made to characterize and fabricate high T_c materials and to increase their critical temperatures. The deciding factors for the future application possibilities of oxide superconductors are their current carrying capacity and whether or not they can be easily fabricated to a desired shape.

CRITICAL CURRENTS

Thin Films

The highest critical current density value obtained to date is 2.54×10^6 A/cm^2 [2]. It was measured using a thin film which was prepared by sputtering a Ho-Ba-Cu oxide on a MgO substrate. With Y-Ba-Cu-O, NTT (Japan) has obtained the value, 1.8×10^6 A/cm^2 at 77 K using a film prepared by sputtering on SrTiO$_3$. Both these films were grown epitaxially. The values, 2.54×10^6 A/cm^2 and 1.8×10^6 A/cm^2, have been obtained for transport currents in an a-b plane. For Y-Ba-Cu-O, the critical current density is two orders of magnitude smaller in the c-axis.

There have been many reports of Y-Ba-Cu-O materials being prepared by electron beam evaporation[4].

For a sample with the same amount of a- and c-axis orientation, the critical current density measured by transport is 5.3×10^4 A/cm^2 at 78 K without an external magnetic field. Another research group has reported on the critical current density of thin films prepared by electron beam evaporation[5]. These films have been deposited on SrTiO$_3$ and the c-axis is perpendicular to the film surface. The resulting critical current density obtained using the four-terminal method is 1.5×10^5 A/cm^2.

Wires and Bulk

Many institutes have, on the other hand, reported that the critical current density of wires is as high as around 10^3 A/cm^2 at 77 K without an external magnetic field. These wires were fabricated by a powder process using silver metal as a sheath material. Usually Y_2O_3, $BaCO_3$ and CuO powders are used as starting materials. They are mixed, calcined at around 900 °C and then mixed again. After drying, the calcined powders are put into a metal sheath and fabricated to the desired shape, e.g. wire or tape. Finally the fabricated material is heat-treated in an atmosphere of oxygen or air. As a sheath material, silver seems to be suitable because of its oxygen penetration. Fujikura Co.Ltd. has obtained a critical current density of 1.1×10^4 A/cm^2 (critical current 81 Amps) for wires which were prepared using the conventional powder process, with the exception of the heat treatment procedure[6]. Usually the wire is drawn to the desired diameter and then heat treated at a suitable temperature in oxygen or air. However, in this case, the sheathing material was etched off before putting the wire into the furnace. The aim was to avoid crack generation in the oxide material during heating, due to the different thermal expansions of the oxide material and the silver sheath.

For sintered bulk materials, 5.68×10^3 A/cm^2 was the highest critical current density value. These results were obtained using the four-terminal method.

The critical current densities are summarized in Table 1.

Table 1. Critical current densities of films and wires measured by the resistive method (A/cm^2 at 77 K)

Films

Process	Material	Critical Current Density	Institute	Ref.
Sputtering on MgO	Ho-Ba-Cu-O	2.54×10^6	Sumitomo (Japan)	2
Sputtering on SrTiO$_3$	Y-Ba-Cu-O	1.8×10^6	NTT (Japan)	3
Electron Beam on SrTiO$_3$	Y-Ba-Cu-O	5.3×10^4	U. Stanford (USA)	4
Electron Beam on SrTiO$_3$	Y-Ba-Cu-O	1.5×10^5	IBM (USA)	5

Wires and Tapes

Process	Material	Critical Current Density	Institute	Ref.
Powder process without sheath	Y-Ba-Cu-O	1.1×10^4	Fujikura (Japan)	6
Powder process with Ag sheath	Y-Ba-Cu-O	1.89×10^3 (wire) 3.3×10^3 (tape)	Hitachi (Japan)	11
Powder process without sheath	Y-Ba-Cu-O	1.24×10^3	Toshiba (Japan)	12

As seen above, the critical current density of wires is at present not as high as desired. In addition to the low current density, the critical current density decreases rapidly in an external magnetic field, as shown in Fig. 1[7].

In general, it is thought that this low critical current density is due to (1) non- or weakly superconducting materials at grain boundaries (due to segregation, and second phase materials), (2) very strong crystal anisotropy of critical current densities in these oxides.

Some interesting experimental data is being produced, which will help our understanding of current limitations.

Critical current densities can be estimated using the following formula

$$M = \mu_0 J_c d/4,$$

where M is the magnetization (Tesla), J_c is the critical current density (A/cm^2) and d is the grain diameter (cm).

According to preliminary experiments[8] on magnetization as a function of grain size, J_c is constant (4×10^4 A/cm^2 at 77 K) up to a grain diameter of about 50 μm. Above 50 μm, magnetization no longer depends on the powder size. This seems to indicate something about grain boundaries.

STABILITY OF WIRES

Metal superconductors, e.g. NbTi wires, have usually been fabricated down to a diameter of less than 50 μm to avoid flux jumping.

The diameter is determined by the following formula, derived from the relation between flux penetration and pinning force[9].

$$d < 2(3\gamma C/\mu_0 J_c(-dJ_c/dT))^{1/2} = 2(3\gamma C(T_c-T_0)/\mu_0 J_c)^{1/2}$$

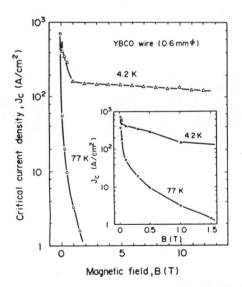

Fig. 1. Critical current density vs. magnetic field for Y-Ba-Cu-O at 4 and 77 K

If the following values are taken for Y-Ba-Cu-O, the requirement for the wire diameter will be less than 0.8 mm^{10}: $\gamma = 7 \times 10^3$ kg/m^3, $C = 158$ J/kg.K, $T_C = 94$ K, $T_0 = 77$ K and $J_C = 1.8 \times 10^{10}$ A/m^2.

This indicates that multifilamentary wires are not necessary for practical applications.

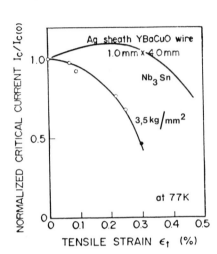

Fig. 2. Tensile strain vs. normalized critical current for YBCO

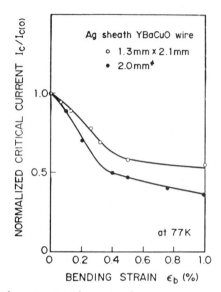

Fig. 3. Bending strain vs. normalized critical current for YBCO

MECHANICAL PROPERTIES

It is worthwhile to note the relationship between the mechanical strain and the critical current.

Figs. 2 und 3 show the normalized critical currents versus the tensile and bending strains for the wires with a silver sheath[13].

Both results show that the critical currents decrease by about 50% with strains of about 0.3%.

The rapid decrease of the critical current under a small amount of strain is one of the serious problems for magnet applications.

SUMMARY

In summary, a superconducting oxide is an interesting material, both practically and theoretically. However, from the point of view of possible applications there are stil many serious problems to be solved, including instability, anisotropy, magnetic field dependence of critical currents, mechanical properties and so on. Fortunately, with the large number of people working in this field, we can be certain that many of these problems will be solved in the future.

REFERENCES

1. K.Noto, Fifth Japan-US Workshop on High-Field Superconducting Materials for Fusion: Nov. 1987 in Japan
2. Press announcement in Japan: Jan. 1988
3. Y.Enomoto et al., Japan.J. of Applied Physics $\underline{26}$, L1248, 1987
4. B.Oh et al., Appl.Phys.Lett. $\underline{51}$, 852, 1987
5. P.Chaudhari et al., Phys.Rev.Lett. $\underline{58}$, 2684, 1987
6. Y.Ikeno et al., Fifth Japan-US Workshop on High-Field Superconducting Materials for Fusion: Nov. 1987 in Japan
7. Y.Yamada et al., Japan.J.of Applied Phys. $\underline{26}$, L865, 1987
8. D.Ito, E.Shimizu, Fall Meeting on Cryogenic Engineering in Japan: Nov. 1987
9. M.N.Wilson, Superconducting Magnets: Clarendon Press, Oxford, p.134 (1983)
10. D.Ito, Cryogenic Engineering (Japan) $\underline{22}$, 383, 1987
11. Press announcement in Japan: Oct. 1987
12. Private communication
13. M.Koizumi et al., Cryogenic Engineering Conf. in Japan (Nov. 1987)

ANISOTROPY OF THE LOWER CRITICAL FIELD IN YBa$_2$Cu$_3$O$_{7-\delta}$

A. Umezawa, G. W. Crabtree, J. Z. Liu, and L. H. Nunez

Materials Science Division
Argonne National Laboratory
Argonne, Illinios, U. S. A. 60439

INTRODUCTION

The crystal structure[1] of YBa$_2$Cu$_3$O$_{7-\delta}$ is composed of chains and planes of Cu-O bonds which are arranged in layers perpendicular to the **c** direction. The one-dimensional chains are oriented along the **b** direction and are located at the top and bottom of the unit cell shown in Fig. 1. The planes are composed of a corrugated two-dimensional network of Cu-O bonds oriented along the **a** and **b** directions and located near the center of the unit cell just above and below the Y site. The metallic character of YBa$_2$Cu$_3$O$_{7-\delta}$ is provided by strong overlap between neighboring Cu and O ions in the chains and planes. There is almost no metallic character along the **c** direction because there are no oxygen atoms along the cell edges between the Cu2 sites. Thus the Cu-O plane just above the Y site does not overlap the plane just below the Y site. In the layer of chains there is virtually no metallic character in the **a** direction because there are no oxygen atoms between the Cu1 sites along **a**. The layer of chains and the layer of planes are connected by the bridging O4 site which overlaps both the neighboring Cu1 and Cu2 sites. However, this O4 is displaced considerably toward the Cu1 site so that it is bound much more strongly to the chains than to the planes. Thus, the coupling between the chains and planes is relatively weak, and the system may be thought of as composed of two nearly independent subsystems showing one- and two-dimensional character.[2]
The low dimensionality inherent in the metallic structure of YBa$_2$Cu$_3$O$_{7-\delta}$ leads inevitably to characteristic anisotropy in the superconducting properties. To date it has been impossible to observe any one dimensional character because all single crystals of YBa$_2$Cu$_3$O$_{7-\delta}$ contain a large number of twins which are related to each other by the interchange of the **a** and **b** axes. However, the two dimensional character has been clearly observed in the anisotropy of the upper critical field H$_{c2}$. In this paper we report the first careful measurements of the two dimensional anisotropy of the lower critical field H$_{c1}$. We compare the anisotropy in our values of H$_{c1}$ with the anisotropy observed in H$_{c2}$ and derive values for the anisotropy of the effective mass.

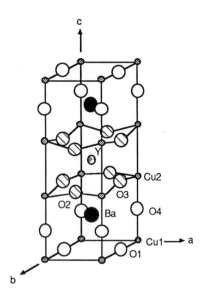

Fig.1: Structure of $YBa_2Cu_3O_{7-\delta}$

SINGLE CRYSTALS

The single crystals used in this study were grown by a solid state annealing method.[3] Y_2O_3, $BaCO_3$, and CuO powders were mixed in the stoichiometric 1:2:3 ratio, ground with a mortar and pestle, placed in a platinum crucible and heated to approximately 975 C for 3-5 days. The reaction product contained many small crystals up to typical dimensions of 0.5 mm. The crystals showed a sharp inductive onset at 91.5 K as measured in a dc shielding experiment in a field of 10 Oe. In general the crystals were not oxygen annealed before measurement. Three crystals were selected for measurements of H_{c1}. One of the crystals was used as-grown, one was annealed for 24 hours at 500 C in a pure oxygen environment, and one was irradiated with neutrons to a fluence of 8.16 x 10^{17} neutrons/cm^2. Many of the properties of these crystals have been reported in earlier studies.[4,5]

MEASUREMENTS

H_{c1} was determined from dc magnetization measurements in a SHE SQUID magnetometer at 11 K. Typical magnetization curves for as-grown single crystals are shown in Fig. 2. Because flux pinning in $YBa_2Cu_3O_{7-\delta}$ is significant, the magnetization curve does not show the characteristic sharp cusp at H_{c1} which is expected for equilibrium behavior. Instead, a subtle departure from the linear dependence of magnetization on field in the Meissner state occurs at H_{c1}. In Fig. 2 the departure appears to occur between 1 and 2 kOe for fields along the **c** direction, and near 500 Oe for fields in the **ab** plane. However, a magnification of the scale reveals that the departure occurs at

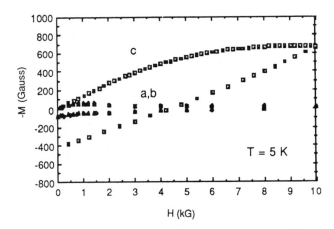

Fig.2: Magnetization curves of single crystal
YBa$_2$Cu$_3$O$_{7-\delta}$ along the **c** direction and
in the **ab** plane.

much lower field. Fig. 3 shows the difference between the
measured magnetization and a straight line fit to the low field
data for the field along the **c** direction at 11 K, expanded by
approximately a factor of 100 on the vertical axis. The first
departure from linearity can be seen to be near 500 Oe,
indicating that H$_{c1}$ is much lower than would be evident from
data on the coarser scale. A similar effect is seen for fields
in the **ab** plane, as shown in Fig. 4. There the departure from
linearity occurs near 80 Oe, a smaller value than along the **c**
direction.

While the data in Figs. 3 and 4 indicate that the values
of H$_{c1}$ are low, they do not rule out the possibility that flux
pinning plays a role in determining the point of first
departure from linearity. In this case the values of H$_{c1}$
derived from Figs. 3 and 4 would be upper limits to H$_{c1}$, and
the point of first departure would be dependent on the degree
of flux pinning. To investigate this possibility, measurements
were made on crystals where the flux pinning was significantly
altered by annealing and by neutron irradiation. If flux
pinning plays a role in determining the point of first
departure, it should be smaller and larger, respectively, in
the annealed and irradiated crystals. A comparison of the data
for the as-grown and annealed crystals is shown in Fig. 4,
where the flux enters the annealed crystal (filled squares)
more easily than the as-grown crystal (open squares). This
shows that the pinning strength was changed by the annealing
process. Despite the change in pinning, Fig. 4 shows that the
departure from linearity occurs at approximately the same field
in the two crystals. As shown below, the values of H$_{c1}$ derived
from the three crystals are approximately the same, indicating
that they are intrinsic to the material and not strongly
affected by flux pinning.

RESULTS AND DISCUSSION

Values of H$_{c1}$ were derived from the measured point of
first departure by correcting for demagnetizing effects using
the formula H$_{c1}$ = H$_{c1}$^{obs}/(1-n) where n is the demagnetizing

factor. The demagnetizing factor was determined from the
measured initial slope of the magnetization curve assuming
$M=-H/4\pi(1-n)$. The values of the demagnetizing factors derived
in this way are in reasonable agreement with those estimated
from the sample dimensions. As the sample shapes were
imperfect parallelepipeds, estimates of the demagnetizing
factors from dimensions are difficult.
    The data for the three crystals are summarized in
Table 1. There is no obvious correlation of the point of first
departure with defect concentration, as would be expected if
flux pinning played a major role in determining the low field

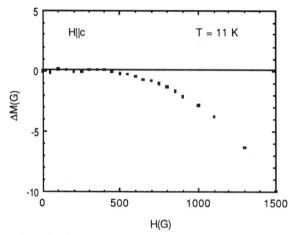

Fig.3: Difference between the measured
       magnetization and the linear fit to
       the low field points for the field
       in the **c** direction.

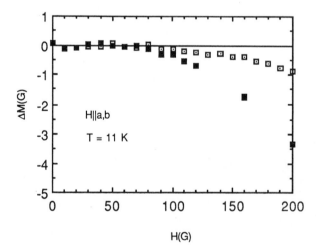

Fig.4: Difference between the measured
       magnetization and the linear fit
       to the low field data for the field
       in the **ab** plane. The open squares refer
       to the as-grown crystal and the filled
       squares refer to the annealed crystal.

Table 1. Measured values of the point of first departure of the magnetization curves from linear behavior. The two values in the first column refer to the two directions perpendicular to **c**. The third column gives the demagnetizing factors.

	H||a,b (Oe)	H||c (Oe)	n
As-Grown Sample	90 ± 10 100 ± 10	425 ± 50	.31 (H||a,b) .45 (H||c)
Annealed Sample	80 ± 10 70 ± 10	375 ± 50	.31 (H||a,b) .47 (H||c)
Irradiated Sample	90 ± 10	325 ± 50	.27 (H||a,b) .46 (H||c)
Average H_{c1}	120 ± 10	690 ± 50	

data. There is reasonable agreement for the values of H_{c1}, considering the limits on accuracy imposed by the field interval at which the data were taken. An average of the H_{c1} values for the three crystals with slightly more weight on the annealed crystal for the field in the **ab** plane gives the values shown in Table 1, 690 Oe for the field along the **c** direction and 120 Oe for the field in the **ab** plane.

The anisotropy in H_{c1} is a factor of nearly 6, approximately equal to the anisotropy of 6.5 in the measured values[6-9] of H_{c2} near T_c. Such nearly equal anisotropies are predicted by the simplest phenomenological treatments of the critical fields. One such treatment is Ginzburg-Landau theory, where all the anisotropy is attributed to one anisotropic parameter, the effective mass. Thus a convenient check on the consistency of the data for upper and lower critical fields can be made by comparing the mass anisotropies derived from them. Following Klemm and Clem,[10] the mass anisotropy can be derived from the formulas

$$\frac{H_{c2}^c}{H_{c2}^{ab}} = \left[\frac{m_{ab}}{m_c}\right]^{\frac{1}{2}}$$

$$\frac{H_{c1}^c}{H_{c1}^{ab}} = \left[\frac{m_c}{m_{ab}}\right]^{\frac{1}{2}} \frac{\ln\kappa^c + 0.497}{\ln\kappa^{ab} + 0.497}$$

$$\frac{\kappa^c}{\kappa^{ab}} = \left[\frac{m_{ab}}{m_c}\right]^{\frac{1}{2}}$$

where κ is the Ginzburg-Landau parameter and the superscripts refer to the field directions. κ may be derived from the formula

$$\frac{H_{c1}}{H_{c2}} = \frac{\ln\kappa}{2\kappa^2}$$

which is valid in the large κ approximation. The effective mass anisotropies derived from these formulas are 43, 78, and 63, respectively, assuming the value of H_{c2} extrapolated to T=0 is 35 T for H along **c** and 230 T for H in the **ab** plane as reported in the literature.[6-9] Thus the effective mass anisotropies derived from the upper and lower critical fields are of the same order, indicating that the measured anisotropy in our values of H_{c1} is consistent with that in the reported values of H_{c2}.

ACKNOWLEDGEMENT

We thank Dr. W. K. Kwok for excellent advice and assistance throughout this work. Mr. T. J. Moran assisted in the data collection and analysis.

This work was supported by the U. S. Department of Energy, Basic Energy Science-Materials Science under contract #W-31-109-ENG-38.

REFERENCES

1. M. A. Beno, L. Soderholm, D. W. Capone II, D. G. Hinks, J. D. Jorgensen, I. K. Schuller, C. U. Segre, K. Zhang, J. G. Grace, Appl. Phys. Lett. 51, 57(1987).
2. G. W. Crabtree, W. K. Kwok, and A. Umezawa, "Basic Properties of Copper Oxide Superconductors", in Quantum Field Theory as an Interdisciplinary Basis, F. C. Khanna, H. Umezawa, G. Kunstatter, and H. C. Lee, eds., World Scientific Publ. Co. Ltd., 1988.
3. J. Z. Liu, G. W. Crabtree, A. Umezawa, Li Zongquan, Phys. Lett. 121, 305 (1987).
4. G. W. Crabtree, J. Z. Liu, A. Umezawa, W. K. Kwok, C. H. Sowers, S. K. Malik, B. W. Veal, D. J. Lam, M. B. Brodsky, J. W. Downey, Phys. Rev. B 36, 4021 (1987).
5. A. Umezawa, G. W. Crabtree, J. Z. Liu, H. W. Weber, W. K. Kwok, L. H. Nunez, T. J. Moran, C. H. Sowers, H. Claus, Phys. Rev. B 36, 7151(1987).
6. W. J. Gallagher, T. K. Worthington, T. R. Dinger, R. L. Sandstrom, Proceedings of the Yamada Conference on Superconductivity in Highly Correlated Fermion Systems, Sendai, Japan, Aug.31-Sept.3 (1987), M. Tachiki, Y. Muto, and S. Maekawa, eds., (North Holland, Amsterdam, 1988),p228.

7. Y. Iye, T. Tamagai, H. Takeya, H. Takei, Proceedings of
    the Yamada Conference on Superconductivity in Highly
    Correlated Fermion Systems, Sendai, Japan, Aug.31-Sept.3
    (1987), M. Tachiki, Y. Muto, and S. Maekawa, eds.,
    (North Holland, Amsterdam, 1988),p224.
8. Y. Hidaka, M. Oda, M. Suzuki, A. Katsui, T. Murakami,
    N. Kobayashi, Y. Muto, Proceedings of the Yamada
    Conference on Superconductivity in Highly Correlated
    Fermion Systems, Sendai, Japan, Aug.31-Sept.3 (1987),
    M. Tachiki, Y. Muto, and S. Maekawa, eds., (North
    Holland, Amsterdam, 1988),p329.
9. J. S. Moodera, R. Meservey, J. E. Tkaczyk, C. X. Hao,
    G. A. Gibson, P. M. Tedrow, Phys. Rev. B, (submitted).
10. R. A. Klemm and J. R. Clem, Phys. Rev. B21, 1868(1980).

CRYSTAL GROWTH AND SOME MAGNETIC PROPERTIES OF $YBa_2Cu_3O_{7-\delta}$

H.Niculescu, A.Pajaczkowska
A.Wisniewski and M.Baran

Institute of Physics, Polish Academy of Sciences
02-668 Warsaw, Al.Lotnikow 32/46, Poland

INTRODUCTION

After the discovery of high-temperature superconductivity in the La-Ba-Cu-O system by Bednorz and Müller[1] a tremendous amount of investigations have been started on Cu-O based perovskite-type materials. Soon after superconductivity well above liquid nitrogen temperature was achieved by Wu et al.[2] in mixed phase compositions of Y-Ba-Cu-O, it was attributed to the compound with the chemical formula $YBa_2Cu_3O_{7-\delta}$[3].

It is rather easy to prepare ceramic $YBa_2Cu_3O_{7-\delta}$ (1:2:3), but extremely difficult to grow large single crystals because the 1:2:3 compound does not melt congruently.

Recently some authors have described various methods for the growth of 1:2:3 single crystals[4-17]. Takekawa et al.[12] found the lowest temperature of the oxide composition Y:Ba:Cu = 1:10:25 to be the ternary eutectic point. Balestrino et al.[15] determined the molar composition of the eutectic to be BaO 28% and CuO 72% at a temperature of $870^\circ C$. Sheel and Licci[17] obtained large size crystals for this composition. Schneemeyer et al.[11] proposed a successful method based on the existence of a region of partial melting near $1000^\circ C$ in the phase diagram of $YO_{1.5}$-BaO-CuO. By slowly cooling compositions near the ternary eutectic from $1000^\circ C$ (below the melting temperature of 1:2:3), thin platelets of 1:2:3 single crystals can be obtained. The sizes in the a-b plane of the crystals range from 1 to several millimeters. Kaiser et al.[13] developed a technique of crystal growth for compositions in the melting region of the pseudoternary subsystem 1:2:3 - $BaCuO_2$-CuO.

Some measurements on very small 1:2:3 single crystals have been performed[5-10]. They revealed anisotropies of the electrical conductivity[6,10] as well as of magnetic properties (critical fields[5-9], critical current density[7-8]).

In our experiments single crystals of 1:2:3 have been grown from nonstoichiometric compositions, rich in CuO and BaO, in the Y-Ba-Cu-O system. Cooling starts from temperatures above the temperature of incongruent melting of 1:2:3. During the cooling, reforming of the superconducting material takes place.

EXPERIMENT

The starting materials were Y_2O_3, $BaCO_3$ and CuO powders of high purity. To stoichiometric mixtures of the powders in the ratio 1:2:3, an excess of $BaCO_3$ and CuO, in a molar Ba:Cu ratio from 1:2:5 to 1:3:5 was added. The molar ratio of 1:2:3 to ($BaCO_3$+CuO) was varied from 1:9 to 1:23, respectively.

Some samples were prepared by adding to the starting materials 2-5 wt.% of Bi_2O_3 to promote crystal growth. Taking into account the fact that the presence of borate fluxes is desirable from the point of view of nucleation, some experiments were performed with 2 wt.% of B_2O_3.

Mixed powders were heated in alumina crucibles to 1030 - 1060°C, kept 5h at the chosen temperature, then slowly cooled below the eutectic temperature (880°C[11]) at rates between 1 and 4°C per hour. Next, the temperature was decreased to 500°C at a rate of 12°C per hour, held 5-7h and thereafter the power was turned off. Experiments were performed in air. Crystal growth of 1:2:3 was investigated using various materials for crucibles like MgO, ZrO_2, Pt, Au, Al_2O_3. The largest crystals were obtained in Al_2O_3 crucibles. However, the crystals were contaminated by aluminium.

RESULTS AND DISCUSSION

One typical form of crystals obtained is a thin platelet as large as several millimeters in size in the a-b plane by a few tens of micrometers in the orthogonal (c-axis) direction. To obtain a platelet in the melt plane, the lowest temperature was at the surface of the melt.

The biggest as grown crystals in crucibles have surface areas of up to 40mm^2.

Mechanical separation from solidified melts yields crystals of surface area of about 10 mm^2 and 10 - 30 μm thin.

The best results were obtained for compositions with a molar ratio Y:Ba:Cu = 1:4:10 and 2 wt.% Bi_2O_3, heated to 1050°C.

Fig. 1. Crystals of $YBa_2Cu_3O_x$ in the crucible. The size of the biggest one is 6mm x 7mm x ~30μm.

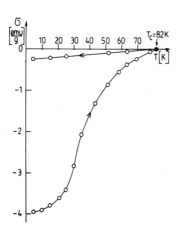

Fig. 2.
The dependence of "Meissner" signal (field cooled sample) and shielding signal (zero field cooled sample) on temperature in a field of 100 Oe.

Crystals were annealed in flowing oxygen at 380-400°C for seven days and then cooled to room temperature. The annealing at temperatures higher than 600°C during 2-3 days and then at about 400°C yields crystals with low transition temperature T_c.

No crystals were found in those samples, where the starting material was prepared by adding $BaCO_3$ and CuO to previously synthesized powdered 1:2:3 ceramics.

A small percentage of B_2O_3 had unfavourable effects, the excess of B_2O_3 supressed the growth of crystals on the surface of the melt.

The composition of the crystals was determined by electron probe microanalysis to be 1:2:3. No trace of bismuth was found down to 10 ppm. However, the contamination of crystals by aluminium is up to 0.5 wt.%. Defects, like twins and mosaic structure, were observed.

One expects that aluminium ions, entering the copper[18] and yttrium[19] sites, lead to lowered transition temperatures.

Magnetization measurements of several crystals were carried out using a VSM magnetometer (PAR-155). There were some difficulties with the measurements because of the low sample masses of about 1 mg. In all cases, it was observed that the "Meissner" to shielding signal ratio was lower than 10% (see Fig. 2).

T_c (onset) was determined from the dependence of the shielding effect signal on temperature. The magnetic field was applied along the c-axis because the signal was significantly stronger in this case, in comparison with the perpendicular orientation. Taking the influence of the magnetic field strength on the shape of the M(T) relation into account, which is probably related to the strong decrease of H_{c1} with temperature reported in[20], the measurements should be done in as low magnetic fields as possible. But, the low masses of the samples and sensitivity limitations of our VSM were the reason for applying a magnetic field of 100 Oe. We believe that this was the main cause of broadening of the superconducting to normal state transition.

For all samples, T_c (onset) was obtained in the range from 60 to 82 K. Further results are presented for the sample with the highest T_c (onset) only.

The crystal structure of $YBa_2Cu_3O_{7-x}$ suggests strong anisotropy between magnetic properties measured for the c direction and directions in the a-b plane. Also one can expect small anisotropy between the a and b directions. Since there exists strong twinning in the a-b plane, it is impossible to measure along a single $\vec{a}$ or $\vec{b}$ direction. Measurements of magnetization hysteresis loops were taken in external fields up to 16 kOe

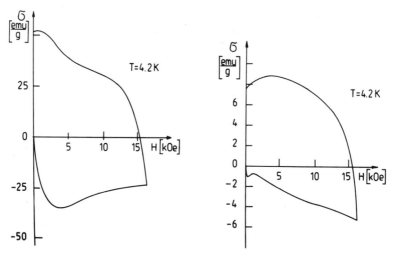

Fig. 3. Magnetization hysteresis for magnetic fields along the c-axis (a)
and perpendicular to the c-axis (b) at 4.2 K.

with $H\perp c$ and $H\|c$, treating the a-b plane as isotropic. The results
obtained at 4.2 K are presented in Fig. 3. In order to estimate the
critical current density, the formula appropriate for rectangular samples
proposed by Clem (see in[21]) was used. The critical current for a magnetic
field of 10 kOe along the c axis is roughly twice as large as in the
perpendicular orientation, both being of the order of 10^5 A/cm^2. This
very low anisotropy may be caused by the presence of Al atoms which may
be treated as additional defects. Such a suggestion may be supported by
the results of Umezawa et al.[21]. They showed that the irradiation damage
induced by fast neutrons strongly decreases the anisotropy of J_c.
    The magnetization measurements in the low magnetic field region
show, that for magnetic fields along the c-axis no hysteresis loops are
observed, a result confirmed by more sensitive SQUID measurements[22]. In
the case of the c-axis oriented perpendicular to the magnetic field, the
observed hysteresis loop[20] is connected with the very low first critical
field. The absence of extra low field hysteresis loops, observed in bulk
ceramic materials[23-25], confirms that they result only from the
intergrain connections but not from twinning boundaries.

CONCLUSIONS

    The obtained crystals of platelet form are characterized by quite a
large size, up to 40 mm^2, very seldom reported by other authors. However,
crystals of such size can be only used in investigations which can be
performed without separating them from the crucible.
    The lowering of the transition temperature, which was observed in
all samples, could be caused by the presence of some admixture of
aluminium or by an insufficient annealing process. However, all examined
samples (with different T_c) contain a very similar amount of aluminium
(about 0.5 wt.%). Therefore, it seems that an appropriate annealing
process should lead to higher T_c. This suggestion is confirmed by the
value of T_c equal to about 90 K reached for ceramic 1:2:3 samples with
the same admixture of aluminium. The presence of aluminium in crystals

can be responsible for the very small anisotropy of critical currents observed in this work and the unusual dependence of H_{c1} on temperature[20].

It seems that the addition of Bi_2O_3 and the admixture of Al_2O_3 (from the crucible) improve the crystal growth. Platinum and gold react too strongly with the melt of temperatures higher than $1000^\circ C$ and can not be used as crucible materials.

ACKNOWLEDGEMENT

We would like to thank Dr.G.Jasiolek for the electron probe microanalysis and Mr.M.Czech and Mr.J.Piechota for technical assistance.

REFERENCES

1. J.G.Bednorz and K.A.Müller, Z.Phys.B 64, 189 (1986)
2. M.K.Wu, J.R.Ashburn, C.J.Torng, P.H.Hor, R.L.Meng, L.Gao, Z.J.Huang, Y.Q.Wang and C.W.Chu, Phys.Rev.Lett. 58, 908 (1987)
3. R.J.Cava, B.Batlogg, R.B.van Dover, D.W.Murphy, S.Sunshine, T.Siegrist, J.P.Remeika, E.A.Rietman, S.Zahurak and G.P.Espinosa, Phys.Rev.Lett. 58, 1676 (1987)
4. S.Hayashi, H.Komatsu, T.Inoue, T.Ono, K.Sasaki, Y.Koike and T.Fukase, Japan.J.Appl.Phys. 26, L1197 (1987)
5. Y.Hidaka, Y.Enomoto, M.Suzuki, M.Oda, A.Katsui and T.Murakami, Japan.J.Appl.Phys. 26, L726 (1987)
6. K.Hayashi, K.Murata, K.Takanashi, M.Tokumoto, H.Ihara, M.Hirabayashi, N.Terada, N.Koshizuka and Y.Kimura, Japan.J.Appl.Phys. 26, L1240 (1987)
7. T.R.Dinger, T.K.Worthington, W.J.Gallagher and R.L.Sadstrom, Phys.Rev.Lett. 58, 2687 (1987)
8. T.K.Worthington, W.J.Gallagher and T.R.Dinger, Phys.Rev.Lett. 59, 1160 (1987)
9. R.W.McCallum, R.N.Shelton, M.A.Noack, J.D.Verhoeven, S.A.Swenson, M.A.Damento, K.a.Gachneidner, Jr.,F.D.Gobson and A.R.Moodenbaugh, preprint
10. S.W.Tozer, A.W.Kleinsasser, T.Penney, D.Kaiser and F.Holtzberg, Phys.Rev.Lett. 59, 1768 (1987)
11. L.F.Schneemeyer, J.V.Waszczak, T.Siegrist, R.B.van Dover, L.W.Rupp, B.Batlogg, R.J.Cava and D.W.Murphy, Nature 328, 601 (1987)
12. S.Takekawa and N.Iyi, Japan.J.Appl.Phys. 26, L851 (1987)
13. D.L.Kaiser, F.Holtzberg, M.P.Chisholm and T.K.Worthington, J.Cryst.Growth 85, 593 (1987)
14. S.Sato, I.Nakada, T.Kohara and Y.Oda, submitted to Acta Cryst.C
15. G.Balestrino, S.Barbanera and P.Paroli, J.Cryst.Growth 85, 585, (1987)
16. X.Z.Zhou, A.H.Morrish, J.A.Eaton, M.Raudsepp and Y.L.Luo, J.Phys.D 20, 1542 (1987)
17. H.J.Scheel and F.Licci, J.Cryst.Growth, 85, 607 (1987)
18. P.B.Kiroy, M.R.Harrison, W.G.Fracman, I.Samuel and M.J.Haines, submitted to Phys.Rev.B
19. J.P.Franck, J.Jung and M.A.-K.Mohamed, Phys.Rev.B 36, 2308 (1987)
20. J.Vanhatalo, M.Baran, R.Laiho, H.Niculescu and A.Pajaczkowska, work in progress
21. A.Umezawa, G.W.Crabtree, J.Z.Liu, H.W.Weber, W.K.Kwok, L.H.Nunez, T.J.Moran, C.H.Sowers and H.Claus, Phy.Rev.B 36, 7151 (1987)
22. J.Vanhatalo and M.Baran, private communication
23. B.Renker, I.Apfelstedt, H.Küpfer, C.Politis, H.Rietschel, W.Schauer, H.Wühl, U.Gottwick, H.Kneissel, F.Steglich, U.Rauchschwalbe and H.Spille, Z.Phys.B 67, 1 (1987)

24. S.Senoussi, M.Oussena, M.Ribault and G.Collin, Phys.Rev.B 36, 4003 (1987)
25. P.Przyslupski, M.Baran, J.Igalson, W.Dobrowolski, T.Skoskiewiecz and J.Rauluszkiewicz, Phys.Lett.A 124, 460 (1987

FLUX PINNING IN $Y_1Ba_2Cu_3O_{7-\partial}$ SINGLE CRYSTALS

L. Fruchter, C. Giovannella*, G. Collin[+], and I.A.Campbell

Physique des Solides - Universite Paris Sud - Bat. 510 -
91405 - Orsay - France
* Dip. di Fisica - II Universita' di Roma - Via O.
Raimondo - 00173 - Roma Italy
[+]UA200 - Université René Descartes - 4 av. de
l'Observatoire - 75006 Paris France

1. INTRODUCTION

We have recently shown[1] that magnetic torque is a very powerful tool to investigate flux creep in the new sintered high T_c superconducting perovskites. It has become clear[2,3,4] that these materials behave as weak coupled superconducting grains. Flux pinning and creep have been reported[1] at moderate fields (10 G < H < 10 kG) at which grains become independent.

We have studied here the behaviour of single crystal YBaCuO, in the same range of field, using a torque technique. We were able to put into evidence strong pinning of the magnetization along the C axis and, alternatively, easy creep of the flux trapped in the A-B plane.

The sample consisted of three single crystals (0.2 mm thick along C, 0.4 x 0.4 in perpendicular directions). In section 4.2, we used one larger crystal (0.25 mm x 0.8 x 0.8) which showed no other difference from the first sample than the signal magnitude. Most measurements were made at 4.2 K.

2. MAGNETIC TORQUE IN THE A-B PLANE

In this first geometry, a turning magnetic field H_t (H_t < 8.7 kG) has been rotated in the basal plane of the crystals, using two different procedures.

2.1. FIELD-CYCLED PROCEDURE (FCy)

After zero field cooling, the sample is cycled up to our maximum value (8.7 kG) and back to zero before again raising the field to a value H_t and rotating it (approx. 1°/sec.) through 360° and then back.

For H_t lower than 200 G, we observe a torque signal of the form :
$$\Gamma/H_t = K \sin(\theta)$$
reversible with reversal of the sense of rotation of the field.

We shall call it a "rigid signal", as it indicates a pinned remanent magnetization along the cycling direction whose strength is given by K.

For higher fields, the signal gradually changes to 180° periodicity, and reversal of the sense of rotation leads to inversed torque signals (Fig.1a).

2.2. ZERO-FIELD-COOLED PROCEDURE (ZFC)

After zero field cooling, the field is raised directly to H_t without any previous cycling and is then turned as in 2.1. In this case, we do not observe any rigid signal at low field, but a quasi-viscous one for fields greater than 100 G which progressively evolve as in the FCy procedure (Fig. 1b). (we shall call "viscous signal" a signal of the form $\Gamma = (1-\exp(-t/t_o))$ characteristic of a viscous moment on vortices at low angular rates).

3. MAGNETIC TORQUE THROUGH C AXIS

The sample was mounted so that the turning field lies in a plane containing the C axis. Then, during rotation, H_t could be either parallel or perpendicular to this axis.

3.1. FIELD CYCLING

We first cycled the crystals to 8.7 kG and back to zero along the C axis and then investigated the torque signal with different turning fields.

A rigid signal is observed with H_t up to 2.5 kG, progressively replaced by a 180° periodic signal (Fig. 2a). Increasing temperature to 10.6 K lowers this cross over field to about 1 kG.

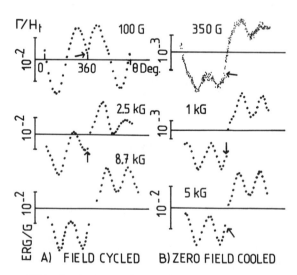

FIG. 1. Torque signal in the A-B plane

(Arrow indicates reversal of the sense of rotation of the field)

3.2. ZFC PROCEDURE

Two regimes may be roughly defined.

- For $H_t < 85$ G, no signal is observed. Such a critical field has already been observed in sintered LaSrCuO and YBaCuO and identified with the critical one for flux penetration in the grain defects.
- For 85 G $< H_t$, a quasi-reversible, 180° periodic signal is observed, which becomes anti-symmetric with increasing fields.

4. DISCUSSION

4.1. DATA INTERPRETATION

It has been shown that magnetic properties of YBaCuO single crystals are very anisotropic[5,6,7].

The torque signal can arise from i) diamagnetic geometrical form effects, ii) trapped flux lines created at $H > H_{c1}$, iii) diamagnetic anisotropy. The rigid signal observed in the FCy procedure at low fields is the simplest example of the second category.

The torque on the sample is given by $\Gamma = M \wedge H_t$, so that we essentially measure the component of the magnetization perpendicular to the turning field.

The 180° periodic signal (for which a purely geometrical interpretation cannot apply, as it is not invariant under field rotation inversion) can be understood if we neglect the magnetization along A-B.

In this hypothesis, as the field is turned, the magnetization lies along C and is determinated by the component of H_t parallel to this axis

$$M = M_c \, (H_t \cos \theta).$$

The resulting torque is :

$$\Gamma = M \wedge H_t = M_c (H_t \cos \theta) \, H_t \sin \theta.$$

We determined $M_c (H_t \cos \theta)$ by using a standard hysteresis loop[8] for a cycling field along C (Fig. 3b) and calculated the associated torque signal (Fig. 3c).

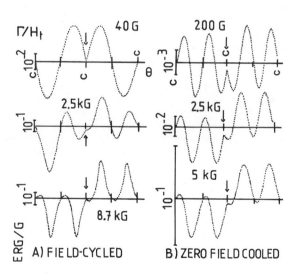

Fig. 2. Torque signal in a plane perpendicular to A-B (Turning field starts on C axis)

The excellent agreement with data confirms that we may well neglect the magnetization and any pinning of the vortices in the A-B plane, as compared to the well trapped flux lines along C.

4.2. CROSS-CYCLING

This procedure was developed in order to get an even more direct demonstration of the unpinning of vortices in the A-B plane.

The sample is first cycled at 8.7 kG along C to create trapped flux in this direction. Then, we apply a field H_{cc} (up to 8.7 kG)in a perpendicular direction (i.e. in the A-B plane). The torque signal obtained that way is almost linear with H_{cc}, shows weak hysteretic effects and no time effects at all. From this, we deduce that most flux lines remain pinned on the C axis and cannot be moved or destroyed by our perpendicular field H_{cc}.

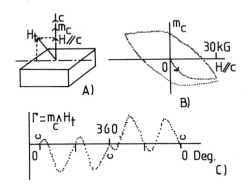

Fig. 3. Calculation procedure, using an hysteresis loop, for the torque signal through C axis

A very different behaviour is observed with the symmetric procedure, which consists of cycling the sample along A-B and then applying H_{cc} along C. The torque signal Γ/H_{cc}, in this geometry, is a measure of the magnetization along A-B. When H_{cc} is increased suddenly (typical raising time 5s) and then kept constant, Γ exhibits relaxation effects for H_{cc} greater than about 600 G (Fig. 4). We may then conclude that the magnetization along A-B is either moved to the C axis or destroyed with time. The decay of the A-B component of the magnetization is obviously not exponential as already observed in sintered samples[9].

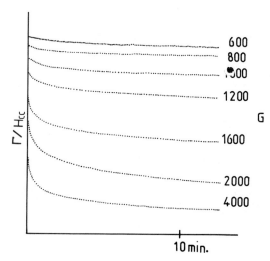

Fig. 4. Decay of the A-B component of the magnetization in the cross-cycling procedure

5. CONCLUSION

We have shown that the C axis in YBaCuO crystals is a preferential strong pinning direction for the flux lines at moderate fields (H < 10 kG), while vortices lying in the A-B plane may be quite easily moved. As a result, the behaviour of such samples in turning fields is dominated by the C axis properties.

REFERENCES

1. C. Giovannella, G. Collin, I.A. Campbell, J. Physique, 48:1835 (1987)
2. D.C. Cronemeyer, A.P. Malozemoff, in print
3. G.J. Cui, X.F. Meng, S.G. Wang et al., Phys. Rev. B, 36:16:8309 (1987)
4. V. Khachaturyan, E.R. Weyber, P. Tejedor et al., Phys. Rev. B, 36:16:8309(1987)
5. J. Hamman, M. Ocio, A. Berticotti et al., J. de Phys., 48:10: 1593 (1987)
6. J.S. Moodera R. Meservey, J.E. Tkaczyk et al., Phys. Rev. B, 37:1:619 (1988)
7. A. Umezawa, G.W. Crabtree, J.Z. Liu et al., Sub. to Phys. Rev. L.
8. S. Senoussi, M. Oussena, G. Collin, Sub. to Phys. Rev. Let.
9. A.C. Mota, A. Pollini, P. Visani et al., Phys. Rev. B, 36:7:404 (1987)

EFFECT OF NEUTRON IRRADIATION ON CRITICAL MAGNETIZATION CURRENT IN SINGLE CRYSTAL $YBa_2Cu_3O_{7-\delta}$

H.W.Weber*, G.W.Crabtree, A.Umezawa, J.Z.Liu, and L.H.Nunez

Materials Science Division
Argonne National Laboratory
Argonne, IL 60439, U.S.A.

Magnetization measurements were performed on single crystals of $YBa_2Cu_3O_{7-\delta}$ to study the effect of fast neutron irradiation on flux pinning and the critical magnetization current. Neutron irradiation up to a fluence of 8.16×10^{17} cm^{-2} (E > 0.1 MeV) always enhances the magnitude of critical magnetization currents significantly, while the transition temperatures decrease at a rate of 2.6 K per 10^{18} neutrons cm^{-2}. The temperature and field direction dependence of these enhancement factors are discussed in detail.

INTRODUCTION

Recent inductive measurements[1-3] on single crystal $YBa_2Cu_3O_{7-\delta}$ have shown large critical magnetization currents at 5 K suggesting the possibility of useful applications. At higher temperatures, however, the critical current values were found to be significantly smaller than those at 5 K. Furthermore, the anisotropy observed in the critical magnetization currents may prevent the achievement of high current densities in polycrystalline samples due to misorientations at grain boundaries. Thus an enhancement in j_c and a reduction of the anisotropy in the flux pinning would greatly improve the possibilities for using $YBa_2Cu_3O_{7-\delta}$ in applications. Fast neutron irradiation is a possible method for increasing the number of pinning centers within the system. Also, since the spatial dimensions of the pinning defects produced by the fast neutrons may be roughly isotropic, the anisotropy of the flux pinning may decrease with neutron irradiation.

EXPERIMENTAL METHODS

Single crystal $YBa_2Cu_3O_{7-\delta}$ samples were prepared using methods described earlier[4]. Their typical dimensions were about 0.2 x 0.4 mm along the c axis and the ab plane respectively and their masses ranged from 175-497 μg. Three crystals were irradiated with fast neutrons at 30 °C in the H2 position of the Intense Pulsed Neutron Source (IPNS) at Argonne National Laboratory to fluences of 1.8×10^{17}, 2.98×10^{17}, and 8.16×10^{17} cm^{-2} (E > 0.1 MeV).

Low field (20 - 50 G) shielding measurements were taken by a special SQUID designed[5] to operate in fields less than 100 G, to determine the superconducting transition temperature T_c of the irradiated crystals. T_c was defined to be the temperature where the onset of diamagnetism in the shielding curve occurred.

For determining the critical magnetization current from the Bean model[6] magnetization was measured up to 10 kG for fields along the c axis and in the basal plane using a commercial SQUID magnetometer. Magnetization measurements were taken at 5, 45, and 77 K for all three irradiated crystals for fields along the c axis as reported earlier[7] and at 5, 20, 45 and 77 K for the maximum irradiated crystal for fields in the basal plane. In this paper we report the effects of the neutron irradiation on the magnitude and anisotropy of the critical magnetization current at 45 K and 77 K.

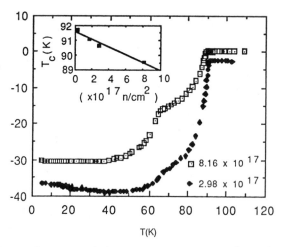

Fig. 1. dc inductive transitions for two of the irradiated samples. The inset shows the decrease in T_c with neutron fluence.

RESULTS

Fig. 1 shows the transition curve for the single crystals of $YBa_2Cu_3O_{7-\delta}$ irradiated to fluences of 8.16×10^{17} cm^{-2} and 2.98×10^{17} cm^{-2}. The transition curves taken after irradiation show varying degrees of structure implying that the crystals are not of equal quality. Increasing neutron fluence was observed to decrease the transition temperature by 2.6 K/10^{18} neutrons cm^{-2} for the single crystals. This value is comparable to the values of 2.7 K/10^{18} neutrons cm^{-2} found for irradiated polycrystalline $YBa_2Cu_3O_{7-\delta}$ (Ref.8) and 2.1 K/10^{18} neutrons cm^{-2} for polycrystalline La$_{1.85}$Sr$_{0.15}$CuO$_4$ (Ref.9). In comparison, the percentage decrease in T$_c$ observed in A15 superconductors[10] is slightly greater than this value.

Typical magnetization curves for single crystals of YBa$_2$Cu$_3$O$_{7-\delta}$ irradiated at various fluences were reported in an earlier publication[7]. Critical magnetization currents were obtained from the magnetization curves using the field independent Bean model[6] generalized for the

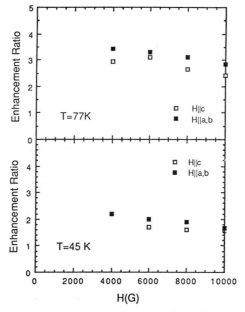

Fig. 2. Ratio of the critical magnetization current in the
irradiated crystal to that in the unirradiated crystal.

rectangular shape of the crystals[9] by Clem[11]. For all fields and temperatures neutron irradiation enhanced the critical magnetization current. The enhancement can be expressed as the ratio of the critical magnetization current for the most highly irradiated crystal to that for the unirradiated crystal. We refer to this dimensionless ratio as the enhancement ratio.

The enhancement ratio as a function of field and temperature is shown in Fig. 2. The improvement is nearly independent of field and is very similar for the field in the two directions. The improvement is slightly greater for the field in the ab plane than for the field along the c direction. Remarkably, the improvement is larger at 77 K than at 45 K, a trend not often seen in radiation damage studies. This trend applies to the data for the field along the c direction as well as for the field in the ab plane. The magnitude of the improvement, nearly 300% at 77 K, demonstrates that substantial increases in flux pinning can be achieved if the proper defects can be found.

The anisotropy of the flux pinning can be examined by the ratio of the critical magnetization current for fields along the c axis to the corresponding values for fields in the basal plane. This anisotropy ratio $J_c(c)/J_c(ab)$ is illustrated for the irradiated and unirradiated samples in Fig. 3.

Surprisingly, in both cases the crystals show anisotropies of order 6-8 at both temperatures. These values are far smaller than those observed[1-3] at 5 K, where the anisotropy at 10 kG is approximately a factor of 50. This decrease in the anisotropy of pinning with temperature is fortunate, as it means that orientation mismatch at grain boundaries in polycrystalline samples may not be so serious at the temperature where most applications are envisioned.

The data for the effect of neutron irradiation on the critical magnetization current are summarized in Tables 1 and 2 for an applied field of 10 kG. The anisotropy ratios of 5.5 - 6.5 at 77 K are not a serious limitation to high current density. The reduction of 15% in the anisotropy ratio with irradiation and the reduction of 28% with temperature suggest that even further reductions can be achieved with proper defects. The improvement in the critical magnetization current density at 77 K of 240% for the field along the c direction and of 280% for the field in the ab plane show that substantial increases of the current carrying capability are possible.

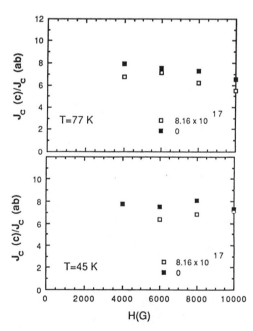

Fig. 3. Ratio of the critical magnetization currents for the field along c to that for the field along ab.

Table 1. Critical current densities in $YBa_2Cu_3O_{7-\delta}$ single crystals prior to and following neutron irradiation (applied field: 10 kG).

	$j_c(H\|\|c)$, A cm^{-2}		$j_c(H\|\|ab)$, A cm^{-2}	
	Irradiated sample	Unirradiated sample	Irradiated sample	Unirradiated sample
45 K	1.12×10^5	6.90×10^4	1.54×10^4	9.27×10^3
77 K	1.01×10^4	4.30×10^3	1.84×10^3	6.48×10^2

Table 2. Summary of neutron irradiation effects on the critical magnetization current density (applied field: 10 kG)

	$j_c(c)/j_c(ab)$		Enhancement Ratio	
	Irradiated sample	Unirradiated sample	$H\|\|ab$	$H\|\|c$
45 K	7.27	7.25	1.67	1.67
77 K	5.49	6.48	2.83	2.40

CONCLUSIONS

In summary, we have found that critical magnetization currents can be increased through irradiation with fast neutrons. Radiation fluences of up to 8.16×10^{17} cm^{-2} give critical magnetization current enhancements of up to 280% at 77 K. The critical current enhancement at 77 K is greater than that at 45 K at all fields. Above 45 K both the irradiated sample and the unirradiated sample show fairly low anisotropy in their flux pinning, less than a factor of eight. The dependence of the superconducting transition temperature with neutron fluence was found to be -2.6 K/10^{18} cm^{-2} implying that higher fluences can be used to study the effect of defects on the critical magnetization currents in single crystals.

ACKNOWLEDGEMENT

We thank Professor H.Claus for the inductive T_c measurements. One of us (H.W.W.) wishes to thank his colleagues at ANL for their hospitality during an extended visit in summer 1987. This work was supported by the U.S.Department of Energy, BES-Material Sciences under contract number W-31-109-ENG-38.

REFERENCES

1. G.W.Crabtree, J.Z.Liu, A.Umezawa, W.K.Kwok, C.H.Sowers, S.K.Malik, B.W.Veal, D.J.Lam, M.B.Brodsky, J.W.Downey, Phys.Rev.B36, 4021 (1987)

2. T.R.Dinger, T.K.Worthington, W.J.Gallagher, R.L.Sandstrom, Phys.Rev. Lett. 58, 2687 (1987)

3. J.L.Tholence, H.Noel, J.C.Levitt, M.Potel, P.Gougeon, Solid State Commun. (in press)

4. J.Z.Liu, G.W.Crabtree, A.Umezawa, Li Zongquan, Phys.Lett.A121, 305 (1987)

5. H.Claus, G.W.Crabtree, J.Z.Liu, W.K.Kwok, A.Umezawa, Proceedings of MMM Conference, Nov. 9-12, 1987, Chicago, IL., J.Appl.Phys.(in press)

6. C.P.Bean, Rev.Mod.Phys.36, 31 (1964)

7. A.Umezawa, G.W.Crabtree, J.Z.Liu, H.W.Weber, W.K.Kwok, L.H.Nunez, T.J.Moran, C.H.Sowers, H.Claus, Phys.Rev.B36, 7151 (1987)

8. J.R.Cost, J.O.Willis, J.D.Thompson, D.E.Peterson, Phys.Rev.B, (in press)

9. S.T.Sekula, D.K.Christen, H.R.Kerchner, J.R.Thompson, L.A.Boatner, B.C.Sales, Jpn.J.Appl.Phys.26, (Suppl. 26-3), 1185(1987)

10. A.R.Sweedler, D.L.Snead, Jr., D.E.Cox, in Treatise on Materials Science and Technology, edited by T.Luhman and D.Dew-Hughes (Academic Press, New York, 1979), Vol.14, p.349

11. J.R.Clem (private communication).

* Permanent address: Atominstitut der Österreichischen Universitäten, A-1020 Wien, Austria

MAGNETIZATION AND X-RAY STUDIES OF $YBa_2Cu_3O_{7-x}$ IRRADIATED BY FAST NEUTRONS

A.Wisniewski, M.Baran, P.Przysłupski, H.Szymczak
A.Pajączkowska, L.Dobrzyński*, J.Waliszewski*, M.Biernacka*
K.Pytel**, and B.Pytel**

Institute of Physics, Polish Academy of Sciences
Al.Lotników 32/46, PL 02-668 Warszawa, Poland
* Faculty of Physics, Warsaw Univ. Branch at Białystok, Poland
** Institute of Atomic Energy, Swierk, Poland

Abstract

The influence of fast neutron irradiation on magnetic properties and structure of $YBa_2Cu_3O_{7-x}$ was studied. Samples irradiated with five neutron fluences between $4*10^{16}n/cm^2$ and $8.7*10^{17}n/cm^2$ were investigated. The irradiation caused significant changes in the magnetic hysteresis of the samples. The creation of defects within the superconducting grains caused an increase of pinning forces, which enhance critical magnetization currents However, irradiation strongly damages the weak links between grains, which was observed through the decrease of low field hysteresis loops. The transport measurements showed a systematic decrease of transport critical current with increasing irradiation dose. It was noticed that only the highest fluence of neutrons ($8.7*10^{17}n/cm^2$) caused a small decrease of the transition temperature.

Introduction

The $RE_1Ba_2Cu_3O_7$ superconducting compounds exhibit transition temperatures, which offer the possibility of applications at liquid nitrogen temperature [1,2]. Today, the main problem is to find the best processing conditions for achieving materials with high critical current densities at 77K. Recently, current densities of the order $10^5A/cm^2$ at 77K in zero magnetic field have been achieved in epitaxially grown thin films of $YBa_2Cu_3O_7$ [3]. Sintered samples exhibit currents, which are three orders of magnitude lower. However, this type of material may also be suitable for some applications. Therefore, it is important to improve its critical current. The critical current is determined by the average interaction of flux lines with pinning centers. Pinning can strongly be influenced by radiation damage. Fast neutrons create point defect clusters, which may collapse to form dislocation loops interacting with the flux lines through their stress fields. Following this conclusion we performed a series of neutron irradiations of $YBa_2Cu_3O_{7-x}$ samples to investigate their magnetic and structural properties.

Experiment

The samples were prepared according to the processing technology
applying three stages of thermal treatment in an atmosphere of flowing
oxygen. In each stage, the samples were annealed at $940^{\circ}C$ for 15 hours,
slowly cooled to about $700^{\circ}C$, kept at this temperature for a few hours,
slowly cooled to about $500^{\circ}C$, kept at this temperature for 24 to 48
hours and finally slowly cooled to room temperature. The samples were
pressed into a suitable form before the last thermal treatment.

The T_c onset of the diamagnetic signal was determined from
magnetization measurements and found to be $92.5\pm0.5K$. From resistivity
measurements, T_c midpoint was determined to be $90.5\pm0.2K$, the transition
width (10%-90%) being equal to 1.3K.

The samples were irradiated in the reactor EWA at the Institute of
Atomic Energy at Swierk, Poland. The thermal neutrons, which are the main
cause of activation of the material, were shielded by a cadmium screen.
The presence of epithermal neutrons caused, however, some activation of
Ba and Cu atoms. Therefore, the samples could only be safely measured
after some days or weeks following irradiation. Since some of the RE
elements give long-life isotopes, the number of $RE_1Ba_2Cu_3O_7$ compounds
which can be studied after neutron irradiation is rather limited.

The samples were irradiated with fluences of 4×10^{16}, 1.2×10^{17},
2.3×10^{17}, 4.5×10^{17} and 8.7×10^{17} n/cm^2. The values of fluence were
estimated with the accuracy of 10% and correspond to neutron energies
above 1 MeV. The neutron flux for this range of energy was about
$2.1\times10^{13}n/s\times cm^2$.

The magnetization measurements were performed with a vibrating
sample magnetometer (PAR, model 155) in fields up to 16kOe and in a
range of temperatures from 4.2K up to 100K.

The Carl Zeiss x-ray diffractometer HZG-4 was used for obtaining
diffraction patterns. A bent quartz primary beam monochromator ((1011)
reflection - Johansson type) was applied. The room temperature
measurements of samples irradiated with fluences of 4×10^{16} and
$1.2\times10^{17}n/cm^2$ and of a reference (nonirradiated) sample were performed using a step scanning method (0.05° steps, 200 seconds counting
time per point).

Results and discussion

From the magnetization measurements it was found, that the transition
temperature T_c being equal to $92.5\pm0.5K$, measured as an onset of
diamagnetic signal, was not affected by irradiation up to the fluence of
$4.5\times10^{17}n/cm^2$. Irradiation with the highest applied fluence (8.7×10^{17}
n/cm^2) lowered the T_c to $91.0\pm0.5K$. Preliminary x-ray studies of this
sample seem to indicate some differences in comparison to the remaining
samples.

The magnetization hysteresis measurements were carried out at low
fields (below H_{c1}; see Fig.2) and at fields up to 16 kOe (Fig.1).
Hysteresis loops in fields increasing up to 16 kOe and then decreasing to
zero were recorded at 4.2, 42 and 77K. The observed hysteresis is
consistent with the critical state of a hard superconductor. The
penetrating flux lines are pinned by structural defects. The increase in
magnetic hysteresis (Fig.1) is a consequence of stronger flux pinning
following irradiation.

The formula proposed by Fietz [4]

$$j_c(H) = [M_+(H) - M_-(H)]\times15/R$$

where M_+ and M_- are values of magnetization measured for increasing and
decreasing magnetic field, respectively, and R is the sample radius
perpendicular to the magnetic field direction, can be applied to calculate the

critical currents. The values obtained for a field of 10 kOe and at 4.2K, 42K and 77K are presented in the Table 1. It should be noted, that there is a significant increase in critical currents with fluence up to 2.3×10^{17} n/cm^2, where saturation is reached. It is also important to note that the ratio of critical currents at 4.2 and 77K, being equal to 290 for the non-irradiated sample, is only equal to 46 for samples irradiated with doses of about 5×10^{17}n/cm^2. On the basis of the above results it may be concluded that irradiation improves the strength of pinning and raises the critical magnetization currents.

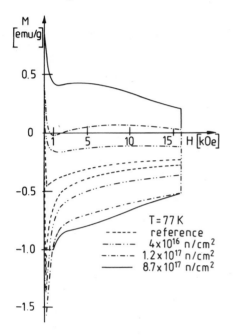

Fig. 1
The growth of hysteresis loops with incresing neutron fluences (0, 4×10^{16}, 1.2×10^{17}, 8.7×10^{17} [n/cm^2]). Results for two more fluences are omitted for better readibility of the figure. Before the measurements each sample was cooled down in nearly zero magnetic field.

Table 1

Critical current densities in [A/cm^2] at H = 10 kOe

Fluence [n/cm^2]	4.2K	42K	77K
0	1.1×10^4	5.0×10^2	38
4×10^{16}	1.2×10^4	1.4×10^3	1.5×10^2
1.2×10^{17}	1.6×10^4	2.5×10^3	3.2×10^2
2.3×10^{17}	2.2×10^4	4.1×10^3	4.7×10^2
4.5×10^{17}	2.1×10^4	4.0×10^3	4.8×10^2
8.7×10^{17}	2.2×10^4	4.3×10^3	4.8×10^2

It is necessary to realize that the bulk polycrystalline sample is an array of superconducting grains connected via weak links. In a magnetic field of 100 Oe or more, the critical currents of weak links are exceeded and critical magnetization currents for the whole superconducting sample are connected with their intragrain properties. Because the transport current is limited by weak intergrain links, the influence of irradiation on these links must be checked.

The measurements in the low field region showed, that the "Meissner" signal relative to the shielding signal (both measured in the field of 100 Oe) decreased after the irradiation (see Table 2). Since the field of 100 Oe is sufficiently above the low field loop (see below and Fig.2), this tendency reflects the behaviour of the superconducting grains. A decrease of the "Meissner" to shielding ratio shows that, after irradiation, the flux lines can penetrate the grains more easily during cooling in the field, which is consistent with the behaviour in higher fields. It is worthwhile to notice that the first critical field also decreases with irradiation (see Table 2).

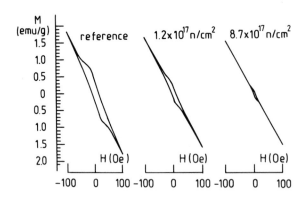

Fig. 2
The low field hysteresis loops at 4.2K for three of the measured samples The samples were chosen to present the general tendency of decreasing the loops with increasing fluences.

Table 2.

The dependence of "Meissner" effect to shielding effect ratio and of the first critical field on applied fluences.

Fluence $[n/cm^2]$	"Meissner"/shielding	H_{c1} [Oe]
0	21%	380
$4*10^{16}$	16%	300
$1.2*10^{17}$	16%	320
$2.3*10^{17}$	12%	315
$4.5*10^{17}$	12%	270
$8.7*10^{17}$	9%	210

The hysteresis loops for fields   far below H_{c1} were measured for all the samples at 4.2K (Fig.2). It should be noted, that the loops decrease with increasing irradiation dose. The existence of these loops reflects the response of the weak links to the magnetic field. The first virgin curve (after zero field cooling) has a linear initial part reflecting shielding effects of the weak links. After exceeding the critical currents of some links by shielding currents, the flux starts to penetrate into regions between grains and the curve passes to a nonlinear part. When the shielding currents exceed the critical currents for all the links, the second linear part begins, with a smaller slope reflecting the shielding effect of "isolated" grains. Reversing of the field in this part leads to the hysteresis loop reported and discussed earlier in [5,6,7]. The decrease of loops after the irradiation results from the damage of the intergrain connections. This fact indicates that the transport critical current should decrease also after irradiation and this prediction was confirmed recently [8].

In all analysed x-ray diffraction patterns (see for example Fig.3) the observed lines were indexed assuming an orthorhombic unit cell, although the ratios of peak intensities suggest the presence of small amounts of the tetragonal phase (about 10%). The unit cell parameters obtained by using the least-square Cohen method are given in Table 3.

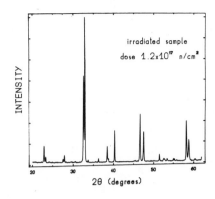

Fig. 3
An example of obtained x-ray
diffraction pattern s

Table 3. Lattice parameters

Fluence [n/cm^2]	a [A]	b [A]	c [A]
0	3.8221(1)	3.8926(1)	11.6790(1)
4×10^{16}	3.8133(8)	3.8858(8)	11.6697(8
1.2×10^{17}	3.8182(4)	3.8865(4)	11.6623(4)

It is seen that the length of the c axis systematically decreases with irradiation dose, whereas no such systematic dependence is found for the unit cell dimensions along the a and b axes. Some of the observed peak intensities are appreciably different from those given in [9]. Brief inspection of our data indicates that the atom and vacancy arrangement is different from the one quoted in [10,11]. The x-ray patterns of all 3 samples showed only minor differences and are being subjected to a careful analysis at present.

Conclusions

The behaviour of low field hysteresis loops in irradiated samples reflects the destruction of weak intergrain links by neutrons. This destruction has to lead to a lowering of critical transport currents and this fact was confirmed recently [8]. It seems that this tendency for the polycrystalline samples will be preserved even for samples with significantly improved transport parameters.

From another point of view, it must be pointed out that the increase of critical magnetization currents with irradiation up to a certain dose, reported also in [12,13], reflects an improvement of the intragrain pinning. From this fact one should conclude that the proper irradiation of monocrystalline samples (bulk or thin films) might provide prospects for improving their transport parameters. Recently, such an investigation was reported by Umezawa et al [15] and their results are in agreement with this conclusion.

There is also the important problem concerning the influence of fast neutron irradiation on transition temperature. For fluences up to 4.5×10^{17} n/cm^2 we did not observe (in the limit of error) changes in transition temperature, which is in agreement with results of Kupfer et al [14]. The highest fluence applied in our experiment (8.7×10^{17} n/cm^2) caused a decrease of T_C of about 1.5K; higher neutron doses (see [14]) affect T_C much stronger. The influence of irradiation on T_C was also reported for polycrystalline samples in [13] and for single crystal samples in [15]. The average decrease obtained from the initial slope is 3.0K/10^{18}n/cm^2 for polycrystals and $2.6/10^{18} n/cm^2$ for single crystals. These higher decreases of T_c might be connected with slightly different intrinsic properties of nonirradiated samples.

References

[1] E.M.Engler , IBM Res.Report (preprint)

[2] K.E.Gray, Third Workshop on RF Superconductivity, Argonne Nat.Lab. Sept. 1987

[3] P.Chaudhari, R.H.Koch, R.B.Laibowitz, T.R.McGuire and R.J Gambino, Phys.Rev.Lett. **58** (1987) 2684

[4] W.A.Fietz and W.W.Webb, Phys.Rev. **178** (1969) 657

[5] B.Renker, I.Apfelstedt, H.Kupfer, C.Politis, H.Rietschel, W.Schauer, H.Wuhl, U.Gottwick, H.Kneissel, U.Rauchschwalbe, H.Spille and F.Steglich, Z.Phys.B, Condensed Matter, **67** (1987) 1

[6] S.Senoussi, M.Oussena, M.Ribault and G.Collin, Phys.Rev.B **36** (1987) 4003

[7] P.Przysłupski, M.Baran, J.Igalson, W.Dobrowolski, T.Skoskiewicz and J.Raułuszkiewicz, Phys.Letters A **124** (1987) 460

[8] P.Przysłupski, A.Wiśniewski, S.Koleśnik, W.Dobrowolski, A.Pajączkowska, K.Pytel, B.Pytel, presented at HTSC-M^{2}S, Interlaken , Feb. 1988

[9] A.Reller, J.G.Bednorz, K.A.Muller, Z.Phys.B, Condensed Matter, **67** (1987) 285

[10] F.Beech, S.Miraglia, A.Santoro, R.S.Roth, Phys.Rev.B **35** (1987)

[11] M.Francois, E.Walker, J.L.Jorda, K.Yvon, P.Fischer, Solid St. Comm. **63** (1987)

[12] A.Wiśniewski, M.Baran, P.Przysłupski, H.Szymczak, A.Pajączkowska, B.Pytel, K.Pytel, Solid St. Comm. **65** (1988) 577

[13] J.R.Cost, J.O.Willis, J.D.Thompson and D.E.Peterson, Phys.Rev.B **37** (1988)

[14] H.Kupfer, I.Apfelstedt, W.Schauer, R.Flukiger, R.Meier-Hirmer, H.Wuhl and H.Scheurer, Z.Phys.B, Ccondensed Matter, **69** (1987) 167

[15] A.Umezawa, G.W.Crabtree, J.Z.Liu, H.W.Weber, W.K.Kwok, L.H.Nunez, T.J.Moran, C.H.Sowers and H.Claus, Phys. Rev. B **36** (1987) 7151

SUPERCONDUCTING PROPERTIES OF OXYGEN DEFICIENT $YBa_2Cu_3O_{7-\delta}$

W. K. Kwok, G. W. Crabtree, and B. W. Veal

Materials Science Division
Argonne National Laboratory
Argonne, Illinois, U.S.A. 60439

H. Claus

Department of Physics
University of Illinois at Chicago
Chicago, Illinois, U.S.A. 60680

M. Guillot

Laboratoire Louis Neel
BP 166X
38042, Grenoble Cedex, France

INTRODUCTION

An important and interesting feature of the high temperature superconductor $YBa_2Cu_3O_{7-\delta}$ is the dramatic effect of oxygen vacancies on its superconducting properties.[1-10] As oxygen is systematically removed from the structure, the orthorhombic distortion decreases and the system undergoes a structural transition to a tetragonal phase. Simultaneously, T_C falls monotonically, with superconductivity disappearing at approximately the stoichiometry where the orthorhombic to tetragonal structural transition occurs. The curve of T_C versus δ shows two regions of stable superconductivity where T_C is not a strong function of δ. For δ near zero, there is a nearly flat region where T_C is close to 90 K and for $.35 < \delta < .45$ there is a plateau where T_C remains constant at approximately 60 K. The width of the inductive transition in these two regions is narrow suggesting that the superconductivity is due to a homogeneous well defined phase. In order to investigate the mechanisms and properties of superconductivity in $YBa_2Cu_3O_{7-\delta}$ it is necessary to characterize and compare the superconducting behavior in these two phases. In this paper we present resistivity and magnetization data from which we determine the upper critical field H_{c2} and compare the relative pinning strengths of polycrystalline samples with $\delta = 0.04$ and 0.4.

SAMPLE PREPARATION

The samples were made by mixing powders of Y_2O_3, $BaCO_3$, and CuO that had been individually prescreened to 200 mesh. The powders were then pressed into pellets of approximately 5 grams and placed in a platinum crucible which was heated to 960 °C for calcination and sintering. The temperature was maintained at 960 °C for 24 hours after which the samples were air cooled, reground and screened to 200 mesh and remixed. The mixed powder was repressed at 2.79×10^8 Pa into pellets of 2 to 5 grams, sintered in 960 °C for 24 hours, held at 670 °C for an additional 24 hours after which the sample was cooled to ambient temperature at a rate of approximately 50 °C per hour. For one of the samples, the oxygen stoichiometry was reduced by heat treatment at 500 °C in an O_2/Ar mixture containing 1.67% O_2 flowing over the sample for 22 hours. The sample was then quenched in liquid nitrogen to prevent any change in the oxygen stoichiometry upon cooling. The difference in the oxygen content of the two samples was determined by weighing.

MEASUREMENTS

Superconducting transition temperatures were determined by magnetic shielding measurements in a low field SQUID magnetometer[11] in a field of 5 Gauss. The measurements were carried out by cooling the sample to 4.2 K in zero field, applying the measuring field and taking data on warming. Resistivity measurements were made by the four probe AC technique at frequency of 100 Hz and with a measuring current of 1.5mA in fields of 0 to 8 Tesla. Magnetization measurements were taken in an integrating magnetometer by the extraction method in fields up to 18 Tesla at the Grenoble High Field Magnet Facility (SNCI/MPI).

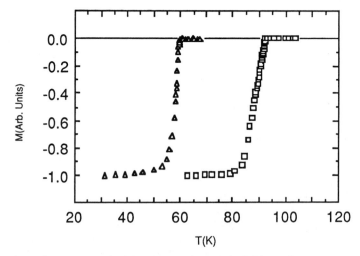

Fig. 1: Normalized magnetic shielding in a field of 5 Gauss versus temperature for two $YBa_2Cu_3O_{7-\delta}$ samples: □, $\delta = 0.04$; Δ, $\delta = 0.4$.

RESULTS AND DISCUSSION

The inductive transitions for the two samples are shown in Fig.1. The shielding curves are smooth and show no structure indicative of any impurity superconducting phases. The superconducting transition temperature is defined as the intersection of two straight line fits to the data above and below onset of diamagnetism. The δ =.04 and 0.4 samples show transition temperatures of 92.0 K and 59.5 K respectively. Resistivity measured in fields up to 8 Tesla is shown in Fig.2 for the sample with δ = 0.04. The superconducting resistive onset of approximately 93 K is defined as the temperature at which the resistivity in the presence of an applied magnetic field first deviates from the zero field curve (Fig.2). In H=0, the midpoint of the transition is at 91.3 K followed by a small tail starting at 91 K and zero resistivity at 88 K. In the presence of a magnetic field there is a nonuniform shift of the curve to lower temperature. The onset temperature moves down slightly with field while the zero resistivity point decreases considerably. This feature is unlike that observed in ordinary superconductors where the entire resistivity curve shifts uniformly to lower temperatures with field. This could be due to a strong anisotropy in the critical fields or to a strong field dependence of the coupling between the grains which appears primarily near R=0. An unusual field dependence is observed near the onset where the shift is independent of the field for fields above 2 Tesla.

The magnetoresistance of the sample with δ = 0.4 is shown in Fig.3. The onset of superconductivity as defined above is at 90 K with a sharp downturn signifying the main transition occurring at approximately 68 K and zero resistance at 56 K. The field dependence of the resistance is qualitatively similar to that of the sample with δ = 0.04 except that near the sharp downturn the shift in the resistance with field is more uniform. In addition, a shoulder appears in the low resistance values as the field is increased suggesting that either the coupling between grains or the critical field anisotropy is different in the two materials.

The upper critical field values obtained from the zero resistance points for the two materials are shown as insets in Figs.2 and 3. The zero resistance points were chosen because they correlated with the midpoint of the sharp inductive transition (Fig.1). Except for the point at H_{c2}=0, there is a well defined linear dependence of the critical field on the temperature. The apparent change in slope for temperatures near T_c is a common feature of both single crystal and polycrystalline samples of this material.[12-17] There is a change of 56% in the critical field slopes compared to a change of 35% in the transition temperatures between the two samples. These two relative changes are similar in magnitude, suggesting that the character of superconductivity near 90 K and 60 K in oxygen deficient $YBa_2Cu_3O_{7-\delta}$ is similar.

Magnetization curves at 4.2K for the two samples are shown in Fig.4. The sample with δ = 0.04 shows significant hysteresis while the sample with δ =0.4 shows no hysteresis on the scale of Fig.4. Even if the vertical scale of Fig.4 is expanded by a factor of 20, the virgin and return curves overlap within the scatter of the data for fields larger than 1 kG. Therefore in this range, within experimental accuracy, the sample with δ =0.4 shows no measurable hysteresis or flux pinning.

There are two possible explanations for the large difference in pinning in these samples. It has been suggested

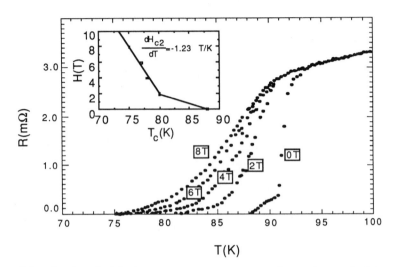

Fig 2: Resistivity vs.temperature for the δ=0.04 sample in various magnetic fields as indicated. Inset: H_{c2} vs temperature determined from the points where the resistivity becomes zero.

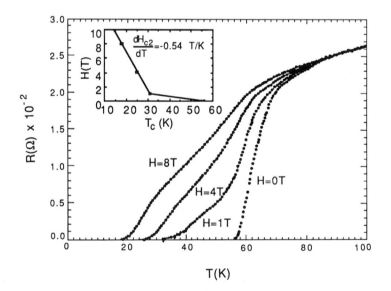

Fig 3: Resistivity vs.temperature for the δ=0.4 sample in various magnetic fields as indicated. Inset: H_{c2} vs temperature determined from the points where the resistivity becomes zero.

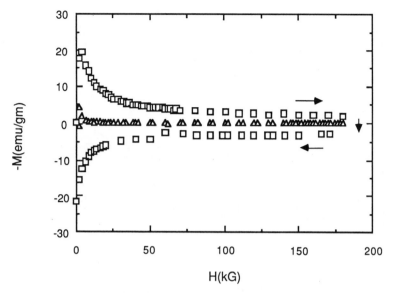

Fig.4: Magnetization with increasing magnetic
field (virgin curve) and decreasing field for
the two samples □, δ=0.04; Δ, δ=0.4. For the
δ=0.04 sample,the increasing and decreasing
branch of the hysteresis curve is connected by
a nearly vertical drop (not shown) at the
highest field.

that the pinning is due to the presence of twin boundaries
in the orthorhombic structure.[18] With decreasing oxygen
concentration the samples approach the tetragonal phaseboundary
and the reason for twin formation is removed. However, at δ =
0.4 the orthorhombic distortion is hardly reduced from its
value at δ = 0 (Ref. 10), and the degree of twin formation
should be the same. The main difference between the two
samples is the oxygen occupation in the Cu1-O1 chains.[1,2,10]
In the sample with δ = 0.4 the occupancy of the
O1 site is 36% less than in the sample with δ = 0.04. Thus the
Cu1-O1 chains are strongly disrupted so that the
superconducting properties no longer depend on their
orientation and the flux lines can move freely across the twin
boundaries.

CONCLUSION

We have examined the differences in the superconducting
properties between the 90 K and 60 K superconductors in
$YBa_2Cu_3O_{7-\delta}$ using magnetization and resistive measurements of
the superconducting transition, the upper critical field, and
the flux pinning. We find the resistive and inductive
superconducting transitions to be qualitatively similar for the
two samples. The resistive curves in both samples show a
strong field dependence near R=0 but a much smaller dependence
near the onset. The upper critical field slopes scale
approximately with T_C suggesting that there is no essential
difference in the nature of superconductivity of the 90 K and
60 K materials. Magnetization curves at 4.2 K reveal no

hysteresis indicative of pinning for the 60 K material. We attribute this absence of pinning to the disruption of the Cu1-O1 chains making superconducting properties independent of the chain direction. Therefore the twin boundaries are ineffective as pinning sites for the oxygen deficient 60 K material.

ACKNOWLEDGEMENTS

We thank J. L. Piccoche for his generous technical assistance in the magnetization measurements at the Grenoble High Field Magnet Laboratory (SNCI/MPI). One of us (W.K.K.) thanks HML/MPI for their hospitality and support during his visit. This work was supported by the U. S. Department of Energy, Basic Energy Sciences-Material Sciences under contract No.W-31-109-ENG-38.

REFERENCES

1.   W. K. Kwok, G. W. Crabtree, A. Umezawa, B. W. Veal, J. D. Jorgensen, S. K. Malik, L. J. Nowicki, A. P. Paulikas, L. Nunez, Phys. Rev. B 37 (Jan. 1988).

2.   J. D. Jorgensen, B. W. Veal, W. K. Kwok, G. W. Crabtree, A. Umezawa, L. J. Nowicki, A. P. Paulikas, Phys. Rev. B 36, 5731 (1987).

3.   J. D. Jorgensen, M. A. Beno, D. G. Hinks, L. Soderholm, K. J. Volin, R. L. Hitterman, J. D. Grace, I. K. Schuller, C. U. Segre, K. Zhang, M. S. Kleefich, Phys. Rev. B 36, 3608 (1987).

4.   J. van den Berg, C. J. van der Beek, P. H. Kes, G. J. Nieuwenhuys, J. A. Mydosh, H. W. Zandbergen, F. P. F. van Berkel, R. Steens, and D. J. W. Ijdo, Europhys. Lett., 4, 737 (1987).

5.   D. C. Johnston, A. J. Jacobson, J. M. Newsam, J. T. Lewandowski, D. P. Goshorn, D. Xie, and W. B. Yelon, Chemistry of High-Temperature Superconductors, edited by D. L. Nelson, M. S. Whittingham, and T. F. George, (American Chemical Society, Washington, DC, 1987), p.136.

6.   P. Monod, M. Ribault, F. D'Yvoire, J. Jegoudez, G. Collin, and A. Revcolevschi, J. Physique 48, 1369 (1987).

7.   E. Takayama-Muromachi, Y. Uchida, M. Ishii, T. Tanaka, K. Kato, Jpn. J. Appl. Phys. 26, L1156 (1987).

8.   R. J. Cava, B. Batlogg, C. H. Chen, E. A. Rietman, S. M. Zahurak, and D. Werder, Phys. Rev. B 36, 5719 (1987).

9.   M. Tokumoto, H. Ihara, T. Matsubara, M. Hirabayashi, N. Terada, H. Oyanagi, K. Murata, Y. Kimura, Jpn. J. Appl. Phys. 26, L1565 (1987).

10.   G. W. Crabtree, W. K. Kwok, H. Claus, B. W. Veal, J. D. Jorgensen, L. H. Nunez, A. Umezawa, A. P. Paulikas (to be published)

11.   H. Claus, G. W. Crabtree, J. Z. Liu, W. K. Kwok, A. Umezawa, Proceedings of the MMM Conference, Nov.9 - 12, 1987, Chicago, IL., Journ. of Appl. Phys., (in press).

12.   T. Tamegai, A. Watanabe, I. Oguro, Y. Iye, Jpn. J. Appl. Phys. 26, L1304 (1987)

13. K. Hayashi, K. Murata, K. Takahashi, M. Tokumoto,
    H. Ihara, M. Hirabayashi, H. Terada, H. Koshizuka,
    Y. Kimura, Jpn. J. Appl.Phys.$\underline{26}$, L1240(1987)

14. Y. Iye, T. Tamegai, H. Takeya, H. Takei, Jpn. J. Appl.
    Phys.$\underline{26}$, L1057(1987)

15. T. Takabatake, M. Ishikawa, Y. Nakazawa, I. Oguro,
    T. Sakakibara, T. Goto, Jpn. J. Appl. Phys.$\underline{26}$,
    L978(1987)

16. J. S. Moodera, R. Meservey, J. E. Tkaczyk, C. X. Hao,
    G. A. Gibson, P. M. Tedrow (submitted to Phys. Rev. B)

17. T. K. Worthington, W. J. Gallagher, T. R. Dinger, Phys.
    Rev. Lett.$\underline{59}$,1160(1098)

18. G. W. Crabtree, J. Z. Liu, A. Umezawa, W. K. Kwok,
    C. H. Sowers, S. K. Malik, B. W. Veal, D. J. Lam,
    M. B. Brodsky, and J. W. Downey, Phys. Rev. B
    $\underline{36}$, 4021 (1987).

AC SUSCEPTIBILITY AND INDUCTIVE CRITICAL CURRENT MEASUREMENTS IN

POLYCRYSTALLINE YBa$_2$Cu$_3$O$_7$

H. Küpfer, I. Apfelstedt, R. Flükiger, R. Meier-Hirmer, W. Schauer
T. Wolf, and H. Wühl
Kernforschungszentrum Karlsruhe, Institut für Technische Physik, and
Universität Karlsruhe P.O. Box 3640, D-7500 Karlsruhe, FRG

Polycrystalline sintered specimens of YBa$_2$Cu$_3$O$_7$ with densities up to 95% and resistivities as low as 150 µΩcm at 100 K were investigated by means of ac susceptibility, resistivity and inductive critical current measurements. The tremendous increase of the transition width in an applied field observed by ac susceptibility measurements is attributed to the anisotropy of the upper critical field of the intragrain material and to the intergrain weak links which become superconducting at different fields and temperatures. The screening and loss behavior of the inter- and intragrain current system determines the ac susceptibility from which weak link properties are obtained by comparison of various specimens prepared following different procedures.

The critical current density of polycrystalline material is severely affected by the weak intergrain coupling and the anisotropy of the upper critical field. We have investigated the critical current density by an inductive measurement which allows to distinguish between intergrain- and intragrain contributions in the same measurement. In addition, the intergrain volume of the individual sample was also derived from the same measurement.

INTRODUCTION

Since the 90 K superconducting transition in the system Y-Ba-Cu-0 has been observed[1] many papers were published concerning the dependence of the superconducting properties on preparation conditions. In polycrystalline bulk material of YBa$_2$Cu$_3$O$_7$, the weak intergrain coupling is recognized as being the reason for the disappointingly low transport critical current densities.[2,3] Recently it was shown by Jin et al.[4] that this weak link problem may be solved in the near future. By means of a new fabrication technique they obtained a crystal with well aligned grains and probably low impurity content at the interfaces. The weak link structure makes the interpretation of the measurements difficult because the assignment of the measured properties to the superconducting grain material or to the weak link material connecting the grains becomes difficult. In this paper we discuss ac susceptibility measurements and inductive measurements of the critical current densities performed on three specimens prepared by two different sintering techniques.

The first method is the common or indirect one. The appropriate powder is calcined at 950° in air, ground and mixed several times before being compressed and deformed into a pellet or wire geometry as reported by Flükiger et al.[5] In the direct method described in detail by Wolf et al.[6] the carefully mixed powder is pressed into a pellet and calcined only once. The subsequent sintering at 950° C and oxidation at < 500° C is the same for both methods. Table 1 shows some properties of an indirectly prepared (1) and of directly prepared samples (2,3). The density of indirectly prepared specimens is between 70% and

Table 1. Some properties of an indirectly (1) and of two directly prepared specimens (2,3). The critical current density determined both resistively j_c^{res}, and inductively, j_c^{ind}, is given for 77 K in zero field (see text).

sample	$\rho(100\text{ K})$ $[\mu\Omega\text{cm}]$	density [%]	mean grain size $[\mu\text{m}]$	$j_c^{res}(77\text{ K, 0T})$ $[\text{A/cm}^2]$	$j_c^{ind}(77\text{ K, 0T})$ $[\text{A/cm}^2]$
1	195	75	20	700	650
2	270	90	40	230	220
3	150	95	26	130	100

80% of 6.37 g/cm^3 whereas directly prepared specimens show densities above 90%. The grain structure of the directly prepared specimens is polycrystalline with essentially plane contact surfaces between the grains in contrast to point like contacts between the granules in indirectly prepared specimens. Sharper X-ray diffraction lines indicate a better homogeneity for the directly prepared specimens. Susceptibility and critical current measurements on various $Y_1Ba_2Cu_2O_7$ samples are presented and correlated to the different preparation methods.

AC SUSCEPTIBILITY MEASUREMENTS

The interpretation of the ac susceptibility measurements is complicated due to the different screening behavior of the grains (intragrain) and of the weak links between the

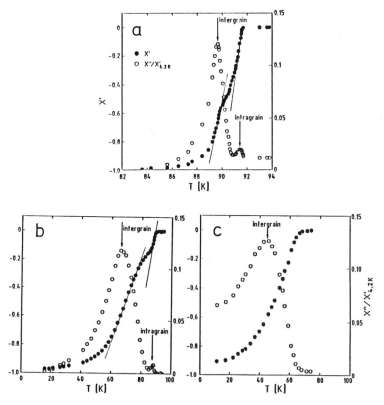

Fig. 1: Real and imaginary part of the ac susceptibility, X' and X", versus temperature for sample 2 at different magnetic fields: (a) B = 0T, (b) B = 0.5 T and (c) B = 8 T. The ac loss component X" is normalized by dividing through the value of X' at 4.2 K..

grains (intergrain). The screening behavior with respect to magnetic field, ac field amplitude, frequency and temperature differs between intra- and intergrain material.[7] Fig. 1a shows X' and X" of specimen 2 measured with a frequency of 11Hz and an ac amplitude of 10^{-4} T in zero field. There are two drops in X' versus T which correspond to the two peaks in X". After grinding the specimen only the smaller peak at higher temperature is measured as already shown in Ref. 8 and 9. Obviously, this peak corresponds to intragrain hysteresis losses within the grains whereas the larger peak is related to losses within the weak link structure which is destroyed after grinding. The two drops in the transition are therefore not caused by two superconducting phases but simply correspond to the inter- and intragrain current systems.

In a non-zero dc field the intergrain peak broadens and is shifted to a considerably lower temperature. In addition, a remarkable low temperature tail in X' appears, as shown in Fig. 1b for an applied field of 0.5 T. The peak due to intergrain losses is shifted by about 23 K whereas the intragrain peak is lowered by 3 K only indicating the transition of the grains into the superconducting state. With increasing dc field the intragrain peak and the corresponding shoulder in X' become completely buried by the intergrain contribution as shown in Fig. 1c for 8 T. Therefore information about the transition of the grain material is only reliable if the intragrain peak can still be detected. This condition depends strongly on the properties of the weak link structure and is correlated to the measured normal state resistivity ρ. For specimens with ρ larger than about 0.5 mΩcm at 100 K the intergrain contribution to X' and X" dominates with increasing resistivity.

The presence of the intragrain peak demonstrates that the intergrain losses at this temperature are negligibly small which means that the major part of the weak links is still in the about loss-free normal conducting state. With decreasing temperature the number of weak links becoming superconducting increases due to a broad distribution of the weak link quality, whereas the ac losses of an individual weak link are assumed to decrease with decreasing temperature. The combination of both effects causes the maximum of the intergrain ac losses. This maximum corresponds to the second drop of X' and masks the intragrain transition governed by the B_{c2} anisotropy and by possible inhomogeneities. The

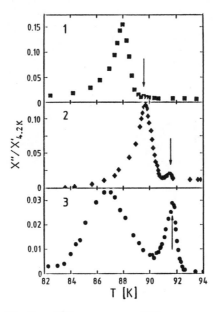

Fig. 2. AC loss component X" normalized to X' at 4.2 K versus temperature. The arrows mark the intragrain loss peak.

weak link losses cause a shift of the transition in X' to lower temperatures which is roughly parallel to the intragrain X' transition as indicated by the thin lines in Fig. 1 a,b. For this reason, at best, only the slope of the upper critical field $B_{c2}(T)$ can be obtained from X' (T) for low ρ specimens.

Figure 2 shows X"(T,B = 0) vs. temperature for the three specimens. The intragrain loss peaks are marked by the arrows. The ratio between the inter- and intragrain peak heights decreases from 20 in specimen 1 to about 1 in specimen 3. This is in full agreement with the X-ray patterns showing the sharpest diffraction lines for the directly prepared specimens 2 and 3 and with the intergrain volume discussed later.

In zero field, the intergrain volume - consisting of intergrain porosity, normal conducting or poorly superconducting material - is shielded up to temperatures at which the transition into the normal state starts. In an applied field, the tail of X' in the low temperature region of Fig. 1 b,c is caused by weak links which have become normal conducting and by enhanced ac losses of the superconducting weak links. Now the intergrain volume becomes partly magnetically visible by the susceptibility measurement. The tail in X' is sample dependent. It increases with decreasing quality of the weak links, i.e. with increasing resistivity ρ and with increasing porosity.

In conclusion, the susceptibility X'(B,T) is determined by the intragrain transition partly masked by the ac losses of the weak links and by the unshielded part of the intergrain volume. The intergrain behavior determines dominantly the lower part of the X'(B,T) transition. This is the main reason for the pronounced curvature in B(T) taken from X' = - 0.9 as shown in Fig. 3. The strength of the curvature is strongly dependent on intergrain properties, i.e. on preparation conditions whereas at high fields the sample independent slope dB/dT is representative for the grain material. This slope can be attributed to the lowest upper critical field of the anisotropic material, $dB_{c2}/dT \approx 0.9$ T/K.

INDUCTIVE MEASUREMENT OF THE INTER- AND INTRAGRAIN CRITICAL CURRENT DENSITY IN BULK SPECIMENS

The inductive measurement of the critical current density j_c in a bulk sample is described by Rollins et al.[10]. An ac field b(t) superimposed to a dc field aligned parallel to the axis of the cylindrically shaped sample causes a periodic change in the critical flux line lattice profile B(X). Here B is the mean local induction and X is the distance from the

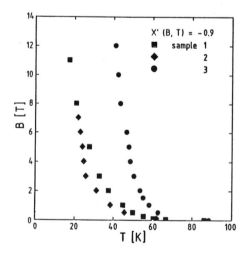

Fig. 3: Magnetic field B at which X' = - 0.9 is obtained versus temperature.

surface. From the induced voltage in a pick up coil around the specimen the penetration depth X of the ac field as a function of time is obtained. A plot of X versus the magnetic field at the corresponding time reproduces the flux profile up to the maximum penetration depth. A radial distribution of the local magnetic induction and its slope which is proportional to j_c can be detected by this measurement. In addition, one obtains from the induced voltage at small ac fields the reversible penetration depth λ' of the pinned flux line lattice.[11] In order to neglect skin effects, i.e. ac losses in the normal conducting state, and to obtain quasistationary flux profiles, i.e. no losses in the flux flow state, the frequency has to be chosen sufficiently small.

Using this method in a polycrystalline material with weak link character at the grain boundaries one gets B versus X/R flux profiles as shown in Fig. 4a where R is the sample radius. The induced signal from which this flux profile is obtained results from both the intergrain volume and from the grains themselves. For small b(t), however, this last contribution can be neglected since the intragrain j_c, i.e. the shielding capability is larger by 2 to 3 orders of magnitude. After the field has fully penetrated the intergrain volume of the specimen the corresponding contribution to the induced signal becomes reversible. With further increasing field the irreversible part of the signal results solely from the penetration into the grains. After subtracting the reversible sinusoidal voltage attributed to the intergrain material already penetrated, one obtains in addition the B(X) profile within the grains as shown in Fig. 4b. The value which has to be subtracted from the measured signal is determined from the position $(X/R)_0$ in the flux profile. This kink indicates full field penetration into the intergrain volume of the specimen. From Fig. 4b one obtains a value $(X/R)_0 \simeq 0.11$ and a corresponding part of the intergrain volume of $1 - (1-(X/R)_0)^2 = 0.21$, i.e. 21% of the normal state signal has to be subtracted for the evaluation of the intragrain B(X) beyond $(X/R)_0$.

The critical current density obtained from the intergrain part of the B vs. X/R profile and taking the sample radius is in good agreement with the resistively measured critical transport current density given for B = 0 and T = 77 K in Table 1 and as a function of field in Fig 5a. From the slope of the intragrain profile and the mean grain radius R_{Grain} the intragrain critical current density is obtained. These values are in rough accordance with those obtained from magnetization measurements and measurements made after powdering the sample as shown in Fig. 5b.

The intergrain volume V_{inter} obtained from $(X/R)_0$ is given in Table 2 for 4.2 K and 77 K at zero field. These values represent the entire sample volume except the high current

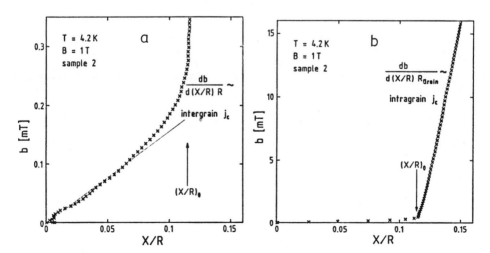

Fig. 4a,b. Magnetic flux profile b (X/R) at T = 4.2 K and B = 1 T. R is the sample radius and X/R is the normalized distance from the sample surface. $(X/R)_0$ corresponds to complete penetration of the intergrain volume by the field b.

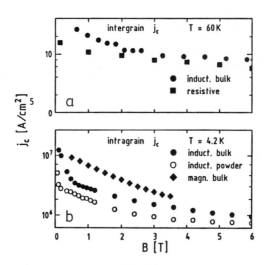

Fig. 5a,b.   Comparison of the critical inter- and intragrain current density j_c obtained
from different methods.

Table 2. Intergrain volume V_{inter} of the samples in zero field for 4.2 and 77 K.

sample	V_{inter} (4.2 K, 0T)	V_{inter} (77 K, 0T)
1	33%	39%
2	19%	28%
3	4%	4.5%

carrying grains contrary to the results of ac susceptibility measurement from which only
the magnetically visible part of the intergrain volume is obtained. The intergrain volumes
of Table 1 are in satisfying agreement with those expected from the actual current carrying
weak link material ($< 1\%$) the intergrain porosity and the non superconducting phase
content ($< 5\%$). The difference of the intergrain volume between 4.2 K and 77 K at $B = 0$
may be caused by the temperature dependence of the magnetic penetration depth λ resulting
in a suppression of the magnetic visibility for decoupled grains as shown by Clem and
Kogan[12]. Using a penetration depth of $\lambda(0K) = 0.12$ µm[13] and a mean grain radius of 10 µm
we obtain from Fig. 1 of Ref. 12 a supression of about 6% for the grain volume from 4.2 K to
77 K.

The explanation for the temperature dependence of the intergrain volume does not hold
for the observation of its striking field dependence shown in Fig. 6 for sample 2. Similar
results were observed in Ref. 8 and 14. At 4 K V_{inter} is almost field independent as expected
from the constant contribution from porosity and second phase. At 77 K, however, V_{inter}
increases with increasing field especially in the low field region, for instance from 28% at 0T
to 74% at 0.5 T. This large difference cannot be explained   either by a growing intergrain
layer thickness   or by the B_{c2} anisotropy causing the presence of normal conducting grains
once the applied field exceeds the lowest B_{c2} of the material. The reversible penetration
depth λ' of the intragrain material may be proposed as another possibility for an apparent
increase of V_{inter}. The field dependence of $\lambda' \sim (B/j_c)^{1/2}$ [11] is not in contradiction with the
measurement of V_{inter} versus B. Another explanation for V_{inter} (B) dependence would be to
assume that not all grains are decoupled, i.e. some grain boundaries are able to carry the
intragrain j_c in zero field but become weak links for small applied fields. This could result in
a screening of intergrain material in zero field as observed at 77 K.

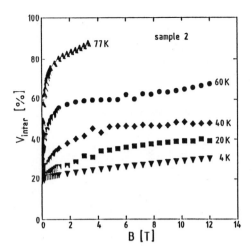

Fig. 6.   Intergrain volume V_{inter} versus
magnetic field B for different
temperatures.

CONCLUSION

Measurements of the ac susceptibility on polycrystalline specimens obtained by different sintering processes with weakly coupled grains show a X'(B, T) transition related to the intragrain properties as long as the weak links do not carry a supercurrent. After the weak links have become superconducting their   properties and the intergrain volume govern X' (B, T). The dependence of the intergrain ac loss peak in X" on B and T and its height in comparison to the intragrain hysteresis peak are a very sensitive measurement for the correlation between weak link structure and preparation conditions.

By means of the inductive measurement, it was possible to determine not only the intergrain and intragrain j_c but also the entire intergrain volume. For specimens with a low resistivity ($<$ 300 $\mu\Omega$cm at 100 K) no correlation between intergrain j_c, resistivity or density is found.[15] But the intergrain j_c increases with decreasing ratio of the total intergrain volume to the unshielded part of it which shows the dominating influence of specific weak link properties on the transport current.

REFERENCES

1. M.K. Wu, J.R. Ashburn, C.J. Torng, P.H. Hor, R.L. Meng, L. Gao, Z.J. Huang, Y.Q. Wang, C.W. Chu: Phys. Rev. Lett. 58, 908 (1987)

2. D.C. Larbalestier, M. Daeumling, X. Cai, J. Seuntjens, J. McKinnell, D. Hampshire, P. Lee, C. Meingast, T. Willis, H. Muller, R.D. Ray, R. G. Dillenburg, E.E. Hellstrom, and R. Joynt, J. Appl. Phys. 62, 3308 (1987)

3. S.B.Newcomb, B.A. Glowacki, A.M. Campbell, J.E. Evetts, W.M. Stobbs, to be published in British Ceramic Proceedings 40 (1988)

4. S. Jin, T. Tiefel, R. Sherwood, B. van Dover, presented at the Materials Research Society Fall Meeting at Boston, 30 November to 5 December 1987

5. R. Flükiger, T. Müller, T. Wolf, I. Apfelstedt, E. Seibt, H. Küpfer, W. Schauer, to be presented at the Int. Conf. on Materials and Mechanisms of Superconductivity at Interlaken, 29 February to 3 March 1988

6. T. Wolf, W. Goldacker, I. Apfelstedt, H. Küpfer, R. Flükiger, ibid. [5]

7. R.B. Goldfarb, A.F. Clark, A.I. Braginski, A.J. Panson, Cryogenics 27, 475 (1987)

8. H. Küpfer, I. Apfelstedt, W. Schauer, R. Flükiger, R. Meier-Hirmer, H. Wühl, Z. Phys. B 69, 159 (1987)

9. H. Mazaki, M. Takano, Y. Ikeda, Y. Bando, R. Kanno, Y. Takeda, O. Yamamoto, Jap. J. of Appl. Phys. 26, L1749 (1987)

10. R.W. Rollins, H. Küpfer, W. Gey, J. Appl. Phys. 45, 5392 (1974)

11. A.M. Campbell, J. Phys.C2, 1492 (1969)

12. J.R. Clem, V.G. Kogan, Proceedings LT 18, 1161 Kyoto 1987

13. T.K. Worthington, W.J. Gallagher, T.R. Dinger, Phys. Rev. Lett. 59, 1160 (1987)

14. T. Matsushita, B. Ni, Y. Sudo, M. Iwakuma, K. Funaki, M. Takeo, K. Yamafuji, presented at the Japan - US Workshop on High-Field SC. Mater. Nov. 1987

15. H. Küpfer, I. Apfelstedt, R. Flükiger, R. Meier-Hirmer, W. Schauer, T. Wolf, H. Wühl, ibid. [5]

THE COMPLEX AC SUSCEPTIBILITY - CRITICAL CURRENT RELATIONSHIP IN OXIDE SUPERCONDUCTORS

L.E. Wenger*, W. Win*, C.J. McEwan*, J.T. Chen*

E.M. Logothetis# and R.E. Soltis#

*Department of Physics, Wayne State University, Detroit, MI 48202, USA
#Ford Motor Company, Dearborn, MI 48121, USA

The critical currents of the high-temperature superconducting oxides (LaBaCuO & YBaCuO) in either bulk or thin film form are limited by the Josephson coupling between the superconducting grains. This coupling is also reflected in the complex ac susceptibility as a function of the ac magnetic field, h. The inductive component χ' actually consists of an ac field dependent contribution that reflects the coupling strength between the grains and an essentially field independent contribution at larger fields which represents the Meissner state of the grains. The resistive component χ'' shows a maximum that shifts to lower temperatures with increasing h and correlates with the temperature dependence of the critical current. Furthermore, the ac susceptibility data can be analyzed by assuming that these granular materials are composed of a multiconnected Josephson network which behaves coherently as a single, weakly-connected superconducting loop.

INTRODUCTION

Since the discovery of the high-temperature superconducting oxide materials of LaBaCuO and YBaCuO, a flurry of research activity has ensued. This activity has encompassed a spectrum of work ranging from fundamental research into the nature of the superconductivity to the construction of superconducting devices from these materials. Throughout this period of research, the granularity in these materials whether in bulk ceramic or thin film form has significantly affected the superconducting properties, and consequently has created some concern as to the future applicability of these superconducting oxides. One problem has been for researchers to increase the critical current density as determined by electrical measuring techniques to levels comparable to the densities deduced from magnetization-loop measurements. Thus the critical current densities may be limited by the nature of the coupling between the superconducting oxide grains and not the grains themselves.

Although current-voltage characteristics and correspondingly critical current measurements provide a measure of the strength of this coupling, measurements of the ac magnetic susceptibility as a function of temperature and magnetic field are a useful macroscopic tool for investigating the coupling in more detail, in addition to determining the critical temperature T_c. In general for a bulk superconductor (e.g. Pb or Nb), the inductive (or real) part of the susceptibility χ' goes from its diamagnetic limit of -1 (in SI units) just below T_c to zero or a small positive value above T_c indicating a weak normal paramagnetic state. The resistive (or imaginary) part of the susceptibility χ'' goes from zero just below T_c

---

* Supported by the National Science Foundation, the Office of Naval Research, and the WSU Institute for Manufacturing Research.

through a peak near T_c, to zero in the normal state. This χ'' signal is usually associated with resistive losses occuring in the vicinity of T_c as the superconductor is composed of multiconnected superconducting and normal regions. In addition, for highly conducting materials, one may actually observe negative χ' and nonzero χ'' signals even in the normal state due to eddy current effects.[1]

Since these oxide materials are granular in nature, an analysis of the magnetic properties based on the similarities to inhomogeneous superconductors would seem to be more appropriate. Previously, Ishida and Mazaki[2] studied the magnetic response in the ac susceptibility of multiconnected superconducting Tc in a porous-alumina substrate and found that both the inductive χ' and resistive χ'' components of the susceptibility were very sensitive to the ac magnetic field h. The transition width of χ' broadened as h increased although the onset temperature did not change, while χ'' had an asymmetric peak that shifted to lower temperatures as h increased. These features were well-reproduced by a phenomenological model that assumed the multiconnected network behaves like a single superconducting loop due to the coherent nature of the specimen. Since the ac susceptibility data[3-7] on the high-temperature superconducting oxide materials show a similar field sensitivity, the magnetic properties should be discussed in a similar fashion. In this paper, we report the results from our ac susceptibility measurements on a variety of oxide materials, demonstrate the dependence between the coupling and the sintering conditions, show a direct correlation between the critical current and the complex susceptibility, and extend the single loop concept to account for some additional features in the susceptibility signals.

EXPERIMENTAL

ac susceptibility

The differential (ac) susceptibility was measured by a mutual inductance technique utilizing a two-coil secondary arrangement at a driving frequency of 250 Hz and ac magnetic fields ranging from 4.2 mOe to 5.8 Oe. By using a two-phase lock-in amplifier, both the inductive component χ' and resistive component χ'' of the susceptibility could be measured simultaneously. A double-dewar apparatus permitted the temperature of the sample to be varied from 4 to 150 K while maintaining the primary-secondary coils at liquid helium, thus ensuring a constant inductive phase relationship with respect to the ac field current. A pneumatic system for raising and lowering the sample between the two secondary coils permits a determination of the absolute value of the susceptiblity. From the calibration of the susceptometer with different paramagnetic salts, the accuracy of the absolute susceptibility is approximately 1% and the relative accuracy an order of magnitude better. The susceptibility, in some cases, has been normalized into SI units, where the volume is considered to be that of the sample measured, and not of the ratio of the sample mass to the oxide's theoretical density. Since the present samples vary from bars to irregular shapes, the diamagnetic limit is attained when χ' reaches a constant value as the temperature is lowered. Also note that in the diamagnetic limit, χ'' must be zero.

Electrical characteristics

The electrical resistances of all samples were measured by a standard four-probe technique using dc or ac currents of 10 µA with zero resistance occurring at the resolution of 20 nV. This technique was readily adaptable for performing I-V characteristic measurements and the determination of the critical currents.

RESULTS

Figure 1 shows the ac magnetic field variation (fields expressed in mOe) in the in-phase χ' and the out-of-phase χ'' components of the ac susceptibility $\chi'-i\chi''$ on a $La_{1.8}Ba_{0.2}CuO_4$ sample made by the standard solid-state reaction technique. The x-ray diffraction pattern confirmed that this sample was single-phase with the K_2NiF_4 type tetragonal structure.

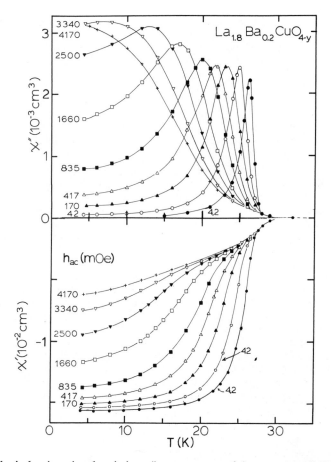

Figure 1. The inductive χ' and resistive χ'' components of the ac susceptibility for single phase $La_{1.8}Ba_{0.2}CuO_4$ as a function of the temperature. The ac magnetic fields are expressed in mOe (peak).

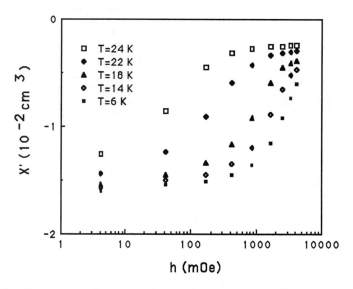

Figure 2. The magnetic field dependence of χ' for $La_{1.8}Ba_{0.2}CuO_4$ at various temperatures.

SEM photographs revealed a very porous, granular structure, with grains being typically oblong platelets and varying from 1 to 10 microns in size. The lower portion of Fig. 1 (see Fig. 2 as well) show that the inductive component χ' consists of two features: one which is extremely sensitive to the amplitude of the ac magnetic field, and the other which does not depend strongly upon the amplitude. This latter field-independence is most discernible in the temperature range of 20 to 32 K where the data converges into a single curve at the largest fields. Also note that the onset temperature of the diamagnetic χ' signal is unaffected by the field. At the lowest temperatures, the χ' signal (h=4.2 mOe) approaches the sample's diamagnetic limit; while at the largest field, the χ' value is only 1/4 of the diamagnetic limit. This value in the large-field limit is reasonable since the sample density is approximately 60% of the theoretical density for $La_{1.8}Ba_{0.2}CuO_4$, the demagnetization factors for the bulk specimen and the grains may be different, and the local field may vary due to the proximity of the grains to one another. Thus one interprets the χ' signal in the large amplitude limit as the Meissner effect of the individual superconducting material in the grains and the amplitude-sensitive signals as resulting from the shielding effect being limited by the super-current (i.e. Josephson current) between the grains. Thus the transition width reflects the strength of the coupling and the temperature dependence of the super-current.

This interpretation is consistant with the peaks observed in the resistive χ'' component. The χ'' peaks broaden and shift to lower temperatures as the field h increases although the magnitude only increases slightly. Since the χ'' signal is usually associated with energy dissipation, the maxima should occur when the induced current from the ac field exceeds the super-current and the resistive losses across the Josephson junctions are maximized. As will be shown later, the magnetic field-temperature dependence of the χ'' maxima is similar to the temperature dependence of the critical current for the bulk sample.

To further demonstrate the microscopic probe capability of the ac susceptibility measurements in relating the microstructure to the coupling in these ceramic materials, Figure 3 shows the inductive χ' component measured for a pressed ceramic composite with a 50:50 mass ratio of superconducting $Y_1Ba_2Cu_3O_7$ to insulating $Y_2Ba_1Cu_1O_5$ sintered at various temperatures. A 900°C sintering resulted in a decoupled system of the superconducting 1-2-3 grains as evidenced by a field-independent χ' for all measuring fields. (Only the 4200 mOe field result is shown.) However, a 2 hr sintering at 950°C in air produced sufficient coupling that a field-sensitive χ' is observed below 60 K. Clearly the large-amplitude χ' limit approaches that of the decoupled grains.

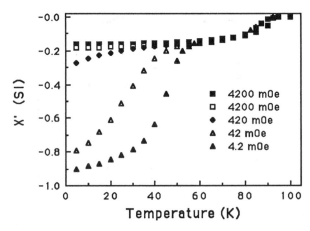

Figure 3. The inductive χ' susceptibility for a 50:50 composite of $Y_1Ba_2Cu_3O_7$ and $Y_2Ba_1Cu_1O_5$. The solid squares (■) represent χ' for the sample sintered at 900°C for 2 hrs and the remaining symbols are χ' for the 950°C/2 hr sintered sample.

MULTICONNECTED JOSEPHSON NETWORK MODEL

To understand the features described in the preceding section, we will utilize the model of a multiconnected Josephson coupled network. In a multiconnected network, a number of superconducting loops may appear for inducing current to shield against the ac magnetic field. Thus a distribution of Josephson currents should be taken into account. However, in practice, a number of such loops can be replaced by a single loop to study the magnetic field variation.[2] This is equivalent to considering that the sample behaves coherently as previously shown for 3-D granular superconductors.[8] Thus the magnetic flux ϕ generated inside a loop is given by

$$\phi = \phi_{ext} + LI$$

where ϕ_{ext} is the externally applied flux due to the ac magnetic field, L is the inductance of the ring, and I is the current through the loop. If one considers the multiconnected network to behave coherently as a single superconducting loop with a weak-link or microbridge, then the current is given by

$$I = - I_J \sin 2\pi\phi/\phi_o$$

where I_J is the maximum Josephson current and ϕ_o the magnetic flux quantum.

The solution of this nonlinear equation for ϕ is found in many textbooks usually in connection with the theory of the rf SQUID. In the limit of $2\pi LI_J >> \phi_o$, the result[2] is that the fundamental frequency susceptibility has two components and that the maximum in χ'' occurs at the midpoint of the χ' transition. Furthermore, the maximum value of χ'' occurs at a temperature where the induced current from the ac magnetic field is identical to the maximum Josephson current, I_J. Thus the magnetic field is proportional to the critical current at the temperature of the χ'' peak with the proportionality constant be determined by the inductance L and the area of the loop. In Figure 4, we show the critical currents for the $La_{1.8}Ba_{0.2}CuO_4$ sample and a mixed phase $Y_{1.8}Ba_{0.2}CuO_4$ sample. The magnetic field-temperature loci for the χ'' maxima are also displayed. By scaling the h and I_o data at one temperature, the agreement between the two sets of data is quite good.

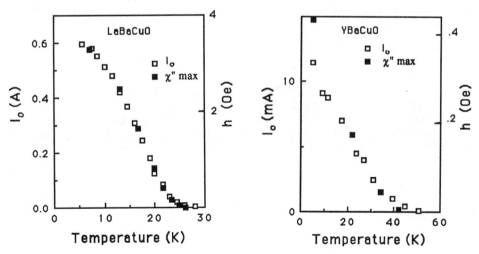

Figure 4. The critical current (□) as a function of the temperature for single phase $La_{1.8}Ba_{0.2}CuO_4$ and mixed phase $Y_{1.8}Ba_{0.2}CuO_4$. The solid squares (■) represent the temperature dependence of the χ'' maxima for various ac magnetic fields, h.

Overall, this model qualitatively describes the magnetic behavior observed in these high-temperature superconductors. Recently, we have studied the waveforms produced from the susceptilibity coils in the vicinity of the χ'' maxima. The waveform consisted of a distorted sine wave similar to that described by Ishida and Mazaki[3] but the waveform also showed a significant phase shift as compared to a purely inductive signal. Thus there appears to be two contributions to the χ'' signal: one due to Fourier components of the nonsinusoidal waveform and the other due to the phase shift arising from hysteretic behavior. This latter feature can be incorporated into this model by treating the Josephson junction as a current source shunted by a resistance (1/G) and capacitance C. The current through the junction now becomes

$$I = -I_J \sin 2\pi\phi/\phi - G\dot{\phi} - C\ddot{\phi}.$$

It is quite apparent that the $\dot{\phi}$ term results in a 90° phase-shifted signal with respect to a purely inductive signal. The effect of these additional terms will be elaborated more fully in a subsequent paper.

CONCLUSIONS

The coupling between the superconducting grains in the high-temperature oxides materials is reflected in the field dependence of the complex ac susceptibility. These features can be interpreted within the framework of granular superconductors and are in qualitative agreement with a model assuming the multiconnected network behaves as a coherent, Josephson coupled superconducting loop.

REFERENCES

1. R.A. Hein, *AC magnetic susceptibility, Meissner effect, and bulk superconductivity*, Phys. Rev. B 33:7539 (1986).
2. T. Ishida and H. Mazaki, *Superconducting transition of multiconnected Josephson network*, J. Appl. Phys. 52:6798 (1981).
3. Y. Oda, I. Nakada, T. Kohara, H. Fujita, T. Kaneko, H. Toyoda, E. Sakagami, and K. Asayama, *AC susceptibility of superconducting La-Sr-Cu-O system*, Jpn. J. Appl. Phys. 26:L481 (1987).
4. R.B. Goldfarb, A.F. Clark, A.J. Panson, and A.I. Braginski, *AC susceptibility measurements near the critical temperature of a Y-Ba-Cu-O superconductor*, in: "High Temperature Superconductors," D.U. Gubser and M. Schluter, eds., MRS, Pittsburgh (1987) EA-11, 261; *Evidence for two superconducting components in oxygen-annealed single-phase Y-Ba-Cu-O*, Cryogenics 27:475 (1987).
5. K.V. Rao, D.-X. Chen, J. Nogues, C. Politis, C. Gallo, and J.A. Gerber, *Frequency and field dependences of the ac susceptibility: a probe to the microstructure of high T_c superconducting materials*, in: "High Temperature Superconductors," D.U. Gubser and M. Schluter, eds., MRS, Pittsburgh (1987) EA-11, 133.
6. L.E. Wenger, J.T. Chen, E.M. Logothetis, W. Win, C.J. McEwan, R. Soltis, and D. Ager, *Josephson-coupled high temperature granular oxide superconductors*, in: "High-Temperature Superconductivity," Drexel Conference, in press.
7. H. Mazaki, M. Takano, Y. Ikeda, Y. Bando, R. Kanno, Y. Takeda, and O. Yamamoto, *Complex susceptiblity in $YBa_2Cu_3O_{6.86}$*, Jpn. J. Appl. Phys. 26:L1749 (1987).
8. J. Rosenblatt, P. Peyral, and A. Raboutou, *Phase ordering and structural disorder in bulk superconductors*, in "Inhomogeneous Superconductors-1979," D.U. Gubser, T.L. Francavilla, S.A. Wolf, and J.R. Leibowitz, eds., AIP, New York (1980) 33.

JOSEPHSON PHASE COHERENCE AND COUPLING IN HIGH T_c GRANULAR

SUPERCONDUCTORS

J.F.Beaudoin, F.Weiss, and J.P.Senateur

INPG-LMGP-ENSPG-CNRS UA 1109
B.P.46, 38402 Saint Martin d'Heres, France

INTRODUCTION

In order to understand the superconductive behaviour of the new high T_c superconductors, many experiments have been carried out. Among these, the measurements of a.c. complex susceptibility $\chi = \chi' + i\chi''$ as a function of temperature are of great interest[1]. The in-phase component χ', corresponding to inductive variations, indicates perfect diamagnetism and flux shielding. The out-of-phase component χ'', corresponding to resistive losses, can behave differently depending on which loss mechanism is involved. In the (RE) $Ba_2Cu_3O_{7-\delta}$ oxides, the observed magnetic behaviour has been analyzed by several authors in terms of percolation, junctions, disorder or superconducting inclusions[4,7] with the physical data apparently varying from one sample to another, depending on the synthesis methods and on the different heat treatments used. We have, therefore, performed inductive a.c. complex susceptibility as well as d.c. resistive measurements on sintered bulk and powdered samples of $HoBa_2Cu_3O_{7-\delta}$ and on two types of $YBa_2Cu_3O_{7-\delta}$ samples in order to correlate the resistivity, the morphology and the diamagnetic behaviour in these granular high T_c superconductors. The observed phenomena are directly dependent on the contact resistance between the grains (i.e. the ability of carrying supercurrent) and on the surface to volume ratio of the grains in a given material.

EXPERIMENTAL

The (RE) $Ba_2Cu_3O_{7-\delta}$ samples were prepared by sintering a mixture of $BaCO_3$, CuO and (RE) oxides in air at 850 °C. After this treatment the samples were ground into a fine powder, pressed into pellets and sintered again at 920 °C for 8 hours. For $HoBa_2Cu_3O_{7-\delta}$ and for the first $YBa_2Cu_3O_{7-\delta}$ sample (sample a) a final annealing was conducted in oxygen at 920 °C. The second $YBa_2Cu_3O_{7-\delta}$ sample (sample b) was annealed at 980 °C. Finally the three specimens were slowly cooled down in the furnace at a rate of 100 °C/hour. The three samples were found to be single phase orthorhombic by x-ray powder diffraction. The average grain size was determined by scanning electron microscopy and was found to be 5 μm for the $HoBa_2Cu_3O_{7-\delta}$ and $(YBa_2Cu_3O_7)a$ samples and 20 μm for the $(YBa_2Cu_3O_7)b$ sample, respectively. The resistive measurements were made on thin discs using a van der Pauw geometry. Current and voltage contacts were made with silver paste. The measuring apparatus for a.c. susceptibility consists of a phase sensitive detector connected to the usual arrangement of two pick-up coils wound in opposition, one of them containing the sample. The a.c. field amplitude can be changed by almost 3 orders of magnitude, from 2.86 Am^{-1} (36 m Oe) to 286 Am^{-1} (3.6 Oe). The real and imaginary parts of the complex susceptibility are alternately recorded by measuring the in-phase and the out-of-phase signal, respectively. The frequency was set to 800 Hz. The results do not depend on frequency within the range 1 to 20 kHz. The χ' signal of the sample was calibrated against a Nb rod and Nb powder samples in the same form. Taking into account that the different samples have a compactness, ranging from 50% to about 90% of the theoretical density, the measured signal coincides well with the reference one. The temperature of the sample was measured with a calibrated platinum thermometer connected to an a.c. resistance bridge. The samples were measured in two different shapes, as powder filled in little plastic tubes and as sintered cylinders.

RESULTS

In Fig. 1 we present the resistivity vs. temperature and the transition curve of the $HoBa_2Cu_3O_7$ sample for several values of the

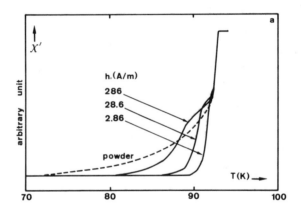

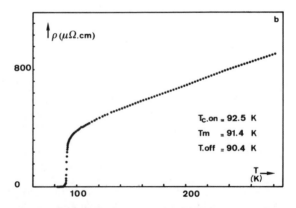

Fig. 1. a) In-phase susceptibility of $HoBa_2Cu_3O_{7-\delta}$ for several applied magnetic fields.
b) Resistivity versus temperature.

applied magnetic field. Variations in the results can be seen, according to whether the sample is in a compact or powder form.

The powder sample presents a single transition with a steep slope at high temperature, which extends smoothly to lower temperature. On the contrary, the compacted specimen presents two transitions, one at high temperature, which coincides with the high temperature part of the powder sample, and a second at lower temperature which is very sensitive to the applied magnetic field, ho: the transition temperature is lower and its variation broader for the highest applied field.

Fig. 2 shows the resistivity vs. temperature and the χ' transition curve of the $(YBa_2Cu_3O_7)a$ sample, for three values of the applied magnetic field. There are again two superconductive transitions, but the upper one is not as steep as that of the former sample.

In Fig. 3 we have presented resistivity vs. temperature and the χ' variations of the second $(YBa_2Cu_3O_7)b$ sample, for several values of ho. A comparison between the two Y-based high T_c samples leads us to the following conclusions: the upper transition is common for the two samples, the lower transition can be divided into two groups obtainable by a translatory movement.

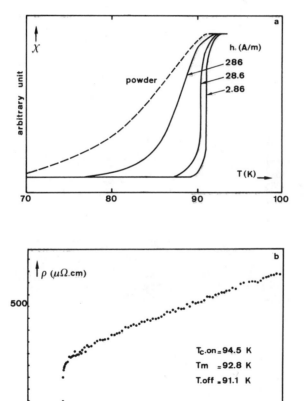

Fig. 2. a) In phase a.c. susceptibility of $YBa_2Cu_3O_{7-\delta}$ annealed at 920°C, for several applied magnetic fields.
b) Resistivity versus temperature.

For the analysed samples, the behaviour of the complex susceptibility as a function of applied magnetic field is typical of granular superconductors. Measurements on powdered samples show a broadening of χ' at low temperature, appearing after the network of links between the different grains have been to a large extent destroyed. The superconducting properties of this collection of grains do not necessarily coincide with those of bulk samples. Each grain must be

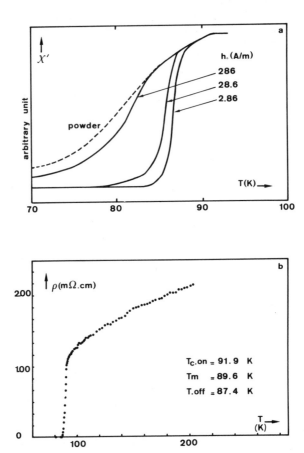

Fig. 3. a) In-phase a.c. susceptibility of $YBa_2Cu_3O_{7-\delta}$ annealed at $980°C$, for several applied magnetic fields.
b) Resistivity versus temperature.

considered as a piece of bulk, since the coherence length in the $(RE)BaCu_3O_{7-\delta}$ oxides is small, $\xi < 50$ Å, in comparison with the grain size. As long as the applied magnetic field, ho, is lower than H_{c1}, the flux explusion in a grain due to the Meissner effect enhances the field seen by its neighbours. These diamagnetic interactions[2] lead to the penetration of vortices, as soon as the local field at one point of the superconducting particle exceeds H_{c1}. It is clear that the apparent value of H_{c1} is smaller than in isolated grains and that vortices are present in the granular high T_c oxides below H_{c1}, giving rise to a broadening of χ'. This phenomenon is related to the grain size and to the particular shape of the particles and can explain the difference between the two $YBa_2Cu_3O_{7-\delta}$ samples, which have different particle sizes (5 μm (sample a) and 20 μm (sample b)) leading to higher diamagnetic interactions and to a broader transition in the former sample than in the latter. In comparison with $YBa_2Cu_3O_{7-\delta}$, the broadening of χ' in $HoBa_2Cu_3O_{7-\delta}$ (for a sample with the same grain size) is considerably reduced: this may essentially be due to a higher value of H_{c1} in the second case.

In the bulk sintered sample, the intrinsic properties of the grains are shown to be partially masked by the induction of screening currents at the sample surface, leading to larger apparent superconducting volumes including the holes, due to the porosity of the samples. The superconducting transition can commonly show two steps (Fig. 1), with a lower transition temperature depending on the applied magnetic a.c. field. This dependence can be explained in terms of multi-connected grains by Josephson interactions. Rosenblatt[3] has shown that an ordering of superconducting phases in different grains coupled by Josephson contacts could occur, which leads to a pseudo-superconductive transition of the susceptibility. The transition temperature increases when the contact between grains becomes better. This is illustrated in Fig. 3, where the $(YBa_2Cu_3O_7)b$ sample, sintered at higher temperature, has a lower resistivity than the $(YBa_2Cu_3O_7)a$ sample (Fig. 2); this leads to a better granular contact.

Ishida and Mazaki[4] have studied the influence of an a.c. magnetic field on granular superconductors and their results, analysed in terms of the weakly connected loop model, are similar to the results in Fig. 2.

The above discussions emphasize the difficulty to interpret the a.c. susceptibility measurements. Some authors have shown the existence of a second superconducting phase in the Y-Ba-Cu-O system[5,6]. It seems to be more appropriate to speak of a "pseudophase" or of a Josephson phase.

It is also very important to specify the sample form (powder or compact) and the measuring parameters, especially the value of the applied magnetic a.c. field, ho.

Nevertheless, granular behaviour is very important for the current carrying ability, which is improved when the critical temperature, T_0, of the Josephson coupled phase is higher and the intergranular contact better, as seen previously in Nb_3 (AlGe) compounds[8].

REFERENCES

1. E.Maxwell and M.Strongin, Phys.Rev.Lett 10, 212 (1963)
2. G.Waysand, Europhys.Lett. 5, 73 (1988)
3. J.Rosenblatt, Rev.Phys.Appl. 9, 217 (1974
4. T.Ishida and H.Mazaki, J.Appl.Phys. 52, 6798 (1981)
   Jap.J.Appl.Phys. 26, L1296 (1987)
5. R.B.Goldfarb, A.F.Clark, A.J.Panson and A.I.Braginski,
   Cryogenics 27, 475 (1987)
6. J.Garcia, C.Rillo, F.Lera, J.Bartolome, R.Navarro, D.H.A.Blank and
   J.Flokstra, J.Mag.Mat. 69, L225 (1987)
7. D.C.Larbalestier, M.Daeumling, X.Cai, J.Seuntjens, J.McKinnell,
   D.Hampshire, P.Lee, C.Meingast, T.Willis, H.Müller, R.D.Ray,
   R.G.Dillenburg, E.E.Hellstrom and R.Joynt, J.Appl.Phys. 62, 3308
   (1987)
8. J.F.Beaudoin, F.Weiss, J.P.Senateur, T.Verhaege, P.Dubots,
   to be published at the Interlaken HTSC-M2S-Conference, March 1988

ELECTRONIC AND MAGNETIC PROPERTIES OF $YBa_2Cu_3O_{7-x}$ SUPERCONDUCTOR

E.Babić [a,b], M.Prester [a] and G.Leising [c]

a-Institute of Physics of the University, POB 304, Zagreb
   Yugoslavia
b-Department of Physics,Faculty of Science,Zagreb,Yugoslavia
c-Institut f. Festkörperphysik, T.U.Graz, A-8010 Graz, Austria

ABSTRACT

The a.c. susceptibility and electrical resistivity of a very dense $YBa_2Cu_3O_{7-x}$ sample have been investigated. The initial susceptibility reveals a rather weak intergrain coupling and yields an estimate for $dH_{c1}/dT = 6$ Oe/K. The resistive transition is rather narrow ($\Delta T_c \approx 1.5$ K) and the measurements in the magnetic field yield $dH_{c2}/dT \approx 5$ kOe/K. The resistivity variation just above T_c agrees well with 3-D Aslamazov-Larkin theory.

INTRODUCTION

After the discovery of high temperature superconductivity in LaBaCuO[1] and YBaCuO[2] ceramics an extensive investigation of their properties has been initiated. This extraordinary activity is largely motivated by the bright outlook for the technological applications of the new superconductors. However, it was soon realized that although the upper critical fields (H_{c2}) of ceramic samples are very high, the critical transport currents (J_c) are low and strongly dependent on the magnetic field.

The measurements on the corresponding monocrystals have shown that low J_c are not intrinsic to these compounds but are rather due to a weak intergrain coupling and inhomogeneity of the ceramic samples.

Recently it was suggested that the density may have a significant influence on the intergrain coupling in a single phase sintered $YBa_2Cu_3O_{7-x}$ sample [3]. This was later confirmed by the investigation of the a.c. susceptibility of the same compound[4]. Here we report the results of the systematic investigation of a.c. susceptibility and electrical resistivity of a very dense YBaCuO sample.

EXPERIMENTAL

YBaCuO oxides were synthesized starting with 0.5:2:3 mixture of Y_2O_3, $BaCO_3$ and CuO as described earlier [5]. Sintering at 950°C under pure oxygen gas flow resulted in a single phase $YBa_2Cu_3O_7$ compound as verified by the electron microprobe analysis and X-ray diffraction. The scanning electron microscopy showed that the sample consists of dense packed fine grains (the average grain size of about 3 μm). The direct measurements yielded the density of a pellet to be 0.94 of an ideal one (6300 kg/m^3). The rec-

tangular sample with approximate dimensions 6x0.9x0.55 mm³ and a mass of
18 mg was cut from the pellet and used for the subsequent electrical and
magnetic measurements.

The initial magnetic susceptibility was measured with a standard a.c.
method [6] using a field along the long axis of the sample.   For practical
reasons we used a rather low frequency (28.4 Hz) but no dependence on frequency up to 3kHz was observed. The electrical resistivity was measured by
a standard a.c. four probe technique. The a.c. current amplitude was 0.1 mA
but even the ten times larger current did not change the resistivity variation.

RESULTS AND DISCUSSION

The superconducting transitions of our sample obtained from the a.c.
susceptibility and electrical resistivity measurements are shown in figs.1
and 3 respectively. Whereas the resistive transition does not depend on the
current through the sample (as long as the selfheating of the sample due to
the contact resistances is avoided) the "inductive" transition depends
strongly on the magnitude of the exciting a.c. field (fig.1). This latter
feature is usual for the ceramic superconductors and apparently arises from
the weak coupling between some individual grains and between some more macroscopic parts of the sample. Hence in order to compare the inductive and
resistive transition one has to find out the lower limit for the a.c. field

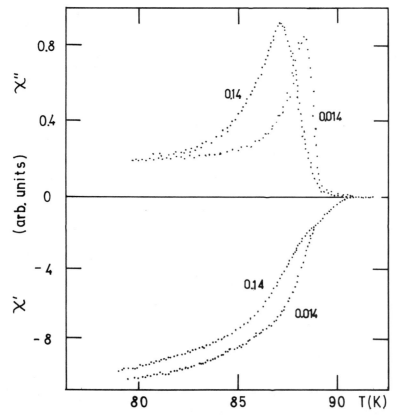

fig.1 The imaginary (χ") and real (χ') susceptibility
of $YBa_2Cu_3O_{7-x}$ for two a.c. fields. Numbers give
the field value in Oe

below which the width and the shape of a transition do not depend on the magnitude of the a.c. field ("intrinsic" transition). For our, as well as for many other sintered YBaCuO samples, this limiting field has an amplitude of about 20 mOe (14 mOe r.m.s.).

Comparing figs. 1 and 2 we see that even the intrinsic width of the inductive transition ($\Delta T_c'$) is several times larger than the resistive one (ΔT_c). Indeed, considering the microstructure of the ceramic sample it can be expected that the diamagnetism sets in gradually and that under unfavourable conditions the transition may not be completed down to well below T_c. For that reason $\Delta T_c'$ of a real part (χ') of the initial susceptibility is certainly not the best measure of the width of a "bulk" transition in a ceramic superconductor. A more useful quantity for that purpose is the halfwidth (Γ) of the peak in the imaginary part (χ'') of the magnetic susceptibility which is associated with the a.c. loss in the sample. For our sample it turns out that the intrinsic Γ (fig.1) is about the same as ΔT_c (fig.3).

However, the position of the maximum in χ'' (for 14 mOe r.m.s.) is close to the end of the resistive transition (about 1 K below its midpoint). This shows rather clearly the insensitivity of the resistive measurements of the superconducting transition on the volume effects. In other words, zero-resistance state can be achieved while a sizeable fraction of a sample is still non-superconducting (χ'' should vanish in a superconductor).

Since the low field a.c. loss is absent in a superconductor, the onset of χ'' at low temperature (left side in figs. 1 and 2) on increasing

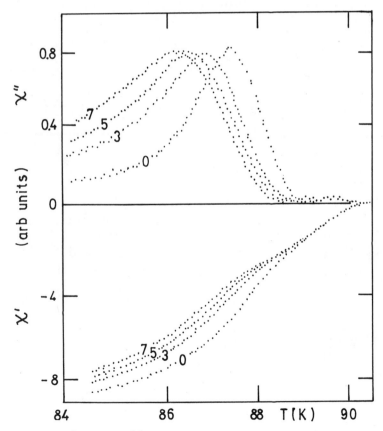

fig.2 χ'' and χ' of $YBa_2Cu_3O_{7-x}$ for different d.c. fields (numbers give the value in Oe)

magnetic field is due to the flux penetration, thus yielding the dependence of the lower critical field (H_{c1}) on temperature. We note however that $H_{c1}(T)$ of a ceramic sample is probably not the same as that of a corresponding homogeneous compound (monocrystal) since the flux penetration will depend on its microstructure. Furthermore, for a ceramic sample, different effects of the d.c. and a.c. magnetic fields (of the same amplitude) on $\chi"$ can be expected. Indeed from figs. 1 and 2 one can see that an a.c. field is over ten times more effective in broadening $\Delta T_c'$ and shifting the maximum of $\chi"$, than the d.c. field of the same amplitude. Hence the effect of an a.c. field on $\chi"$ is probably not due to the flux penetration only but also reflects the interaction of the a.c. field with a given array of (weak) junctions between the grains of the material. Namely, the increasing shielding currents drive some junctions normal.

Because of this we believe that the effect of d.c. field on $\chi"$ represents rather better the flux penetration in a ceramic superconductor than that of an a.c. field [4,7]. (We note however that $H_{c1}(T)$ variation for sintered YBaBuO sample is probably not the same as that for the monocrystal). From the measurements of $\chi"$ in different static magnetic fields (some of which are shown in fig.2) we obtain the initial slope of H_{c1}, $dH_{c1}/dT \approx 6$ Oe/K. This value is close to the values deduced by other technique [8] and places the sintered $YBa_2Cu_3O_{7-x}$ compounds among the extreme type II superconductors. ($H_{c1} \ll H_{c2}$).

Before discussing the electrical properties we wish to emphasize the importance of the a.c. susceptibility measurements for the characterization of the ceramic superconductors. We believe that this technique is very suitable for the study of the intergrain coupling and the homogeneity of these materials. As an example we analyze the origin of the two "steps" in χ' (fig.1) and the appearance of the additional peaks in $\chi"$ (fig.2) at higher magnetic fields [4,7]. To our opinion the first step in χ' (and corresponding very small peak(s) in $\chi"$) is related to the superconducting transitions in the individual grains and the clusters of well connected grains,

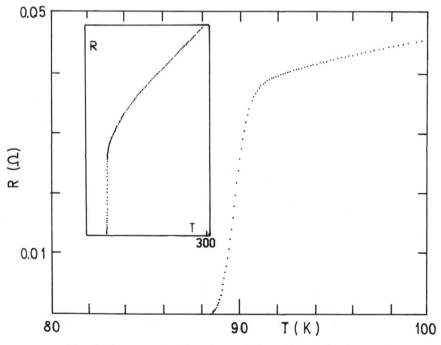

fig.3 Superconducting transition of $YBa_2Cu_3O_{7-x}$
Inset: Resistance variation up to 300 K.

whereas the larger step in χ' (and a large peak in χ'') reflects the development of the magnetic shielding throughout the sample. Therefore, these
steps (χ') and corresponding peaks (χ'') reflect the intra- and intergrain shielding currents respectively. A very weak magnetic field dependence
of the first step in χ' (fig.2) of the ceramic samples and the single step
pattern in monocrystals support our hypothesis.

In fig.3 we show the electrical resistivity of our sample. As expected [3] from its rather large density the sample has the relatively low resis-
sivity which varies rapidly with temperature (fig.3, inset). The resistivity
at 94 K is about 500 $\mu\Omega$ cm and the ratio $R_{293}/R_{94} \simeq 2.5$. A convex curvature
of resistivity below about 170 K and strong linear variation above are not
well understood at present. Since a linear resistivity variaiton is also
observed in a-b plane of monocrystals it is probably intrinsic to $YBa_2Cu_3O_{7-x}$
compound. The site and range of the convex curvature depends however on
the sample preparation.

The superconducting transition (fig.3), although narrow, has a characteristic s-shape. This probably arises from the fluctuation in the oxygen
content as indicated by the somewhat lower T_c (89.7 K, midpoint). Due to
their high normal state resistivity and a short coherence length, the high
temperature superconductors are suitable for investigations on the effects
of superconducting fluctuations. It is therefore of interest to compare the
resistivity variation just above T_c with that expected to arise due to
superconducting fluctuations [9]. We observed rather good $T^{-0.5}$ variation of
the conductivity for $0.01 \leq t \leq 0.05$ ($t=(T-T_c)/T_c$). The deviation from $T^{-0.5}$
dependence below t=0.01 is also observed in amorphous superconductors and
probably arises from inhomogeneities. Since both the Aslamazov-Larkin (AL)
and Maki-Thomson (MT) contributions to fluctuation conductivity have the
same initial temperature dependence (but with different coefficients), we
had to determine the coherence length (ξ) in order to find out whether AL
or MT term dominates the excess conductivity. By the measurement of the
shift of T_c in applied magnetic field ($H \leq 5$ kOe) we obtained the initial
slope of the upper critical field (H_{c2}), $dH_{c2}/dT \simeq 5.8$ kOe/K which corresponds to $\xi = 25$ Å. In fig.4 we show the ratio of our excess conductivity
to the calculated from AL expression for the three dimensional (3D) case
by using our value of ξ. Very good agreement between the calculated and
measured excess conductivity indicates that MT contribution is small or
absent in our sample. This in turn implies that the pair-breaking parameter
is rather large. Since even the mechanism of the high temperature super-

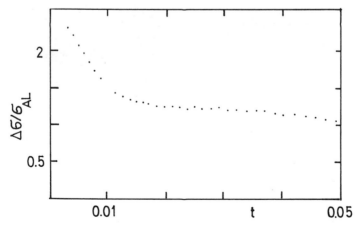

fig.4 Ratio of excess and AL conductivity vs t
       (t = $(T-T_c)/T_c$) for $YBa_2Cu_3O_{7-x}$ sample

conductivity in YBaCuO compounds is not clearly established it is perhaps premature to discuss the origin of the pair-breaking parameter. We note however that the unusually strong linear resistivity variation at higher temperatures may arise due to the strong inelastic electron-phonon scattering which thus becomes a good candidate for a pair-breaking effect. The measurements of the fluctuation conductivity in the high magnetic fields may elucidate this question.

CONCLUSION

A detailed investigation of the superconducting and normal state properties of a dense $YBa_2Cu_3O_{7-x}$ ceramic confirms that the density is an important controlling parameter. The electrical resistivity and initial susceptibility measurements yield the slopes of the upper and lower critical fields consistent with the extreme type II superconductivity. In addition, the a.c. susceptibility appears to be uniquely sensitive contactless probe of the microstructure and homogeneity of the ceramic superconductors. The superconducting transition reveals a slight oxygen deficiency and some fluctuation in the oxygen content. The variation of the conductivity just above T_c agrees both qualitatively and quantitatively with the 3D AL expression. Rather strong pair-breaking effects are tentatively ascribed to the inelastic electron-phonon scattering. Detailed study of the density dependence of I_c as well as of the intra- and intergrain currents has been reported by Küpfer et al [10].

Acknowledgement

We thank to T.Rudez and Z.Marohnić for the help in the data analysis and A.Hamzić for the assistance in resistivity measurements. The support from NBS is also acknowledged.

REFERENCES:

1. J.G.Bednorz and K.A.Müller,Z.Phys.B,64,189(1986)
2. M.K.Wu,J.R.Ashburn,C.J.Torng,P.H.Horr,R.L.Meng,L.Gao,Z.J.Huang,YQ.Wang C.W.Chu,Phys.Rev.Lett.,58,908(1987)
3. E.Babić,Z.Marohnić,M.Prester and N.Brnicević,Phil.Mag.Lett.,56,91(1987)
4. E.Babić,Z.Marohnić,D.Drobac,M.Prester and N.Brnicević, Int.J.Mod.Phys.B, 1,973(1987)
5. G.Leising et al. in "High T_c superconductors and potential applications", Proceedings (J.Vilain and S.Gregoli eds.), C.E.C.,Genova,271(1987)
6. D.Drobac and Z.Marohnić, Rapidly quenched metals (S.Steeb and H.Warlimont eds.), Elsevier Sc, Publ.B.V.,1133(1985)
7. R.B.Goldfarb,A.F.Clark,A.I.Braginski and A.J.Panson,Cryogenics,27,475(1987)
8. A.Bezinge,J.L.Jorda,A.Junod and J.Muller,Sol.State Comun.,64,79(1987)
9. W.J.Skocpol and M.Tinkham,Rep.Prog.Phys.,38,1049(1975)
10. H.Küpfer,I.Apfelstedt,R.Flükiger,R.Meier-Hirmer,W.Schauer,T.Wolf and H.Wühl, this conference (1988)

COEXISTENCE OF MAGNETISM AND SUPERCONDUCTIVITY IN A $YBa_2Cu_3O_{7-x}$ COMPOUND

D.Rassi* and F.Chovanec[+]

*Department of Physics, University College of Swansea
 Singleton Park, Swansea SA2 8PP, U.K.
+ Elektrotechnicky Ustav, CEFV SAV, Dubravska
 Cesta 8, 842 39 Bratislava, Czechoslovakia

INTRODUCTION

Since the discovery of the superconductor $YBa_2Cu_3O_{7-x}$ with transition temperatures above 90 K, similar compounds have been prepared with a number of rare-earth elements substituting the Y[1]. Surprisingly, no depression of critical temperature was observed for magnetic ions with high magnetic moment such as Gd and others. Antiferromagnetic ordering has been reported for $GdBa_2Cu_3O_{7-x}$ at a temperature of 2.3 K[2]. Furthermore, it has been found that changes in oxygen content of these materials strongly affect their electrical and magnetic properties. The question of possible interplay between superconductivity and magnetism therefore requires close scrutiny.

In this work we report on measurements of the remanent magnetic moment in multiphase YBaCuO material at 77 K and 300 K. Measurements were performed on sintered, as well as powder, specimens using a SQUID magnetometer. It was found that the remanent magnetic moment at 77 K is a combination of the diamagnetism of the superconducting part of the sample and a small remanent moment of the normal part. The normal state remanent moment observed at 300 K is superimposed on the diamagnetic moment created at 77 K by superconducting screening currents. A reversal of the normal state moment has been observed during the initial cool-down of the magnetised specimen to 77 K in the Earth's magnetic field. This effect implies the presence of magnetically oriented domains within the specimen.

EXPERIMENTAL DETAILS

Measurements of the magnetic moment of the samples were carried out with a BTi model 601 SQUID magnetometer. A second-order gradiometer pick-up coil arrangement was used; this and the relative sample position and orientation are schematically shown in Fig. 1. The mass of the powder specimens was 0.1 g and the dimensions of the bulk specimens was 2 mm x 2 mm x 4 mm. Magnetisation measurements were performed by (1) cooling the specimen to 77 K in zero magnetic field (2) applying the magnetising field (of 0 to 0.8 T) and (3) removing the specimen from the field and measuring its remanent magnetic moment with the SQUID magnetometer.

Steps 2 and 3 were also carried out at room temperature. Further measurements were made by reversing the order of steps 1 and 2.

The measured remanent magnetic moment allows us to estimate the critical current density of the screening supercurrents. In type II superconductors the critical current density is proportional to the width of the magnetisation loop as indicated in Fig. 2. The remanent magnetic moment in specimens magnetised in magnetic fields higher than the field of full penetration is proportional to the critical current density at zero magnetic field. The remanent magnetic moment for magnetisation in fields higher than the field of full penetration (in our case about 0.1 T) then should have a constant value.

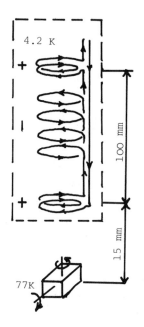

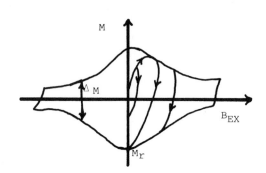

Fig. 1: Sample position and orientation with respect to the pick-up coil of a SQUID magnetometer.

Fig. 2: Typical magnetisation loop of type II superconductors.

CHARACTERISATION OF SPECIMENS

Fine powders of pure Y_2O_3, $BaCO_3$ and CuO (all 99.9% purity) in the appropriate proportions were thoroughly mixed and pulverised. The mixtures were then heated in air at 900°C for 8 hours, homogenised and (either in powder form or as pressed pellets) annealed at 920°C for a further period in oxygen as detailed in Table 1.

Preparation details of samples A,B,C and D

Sample	Form	Duration of first firing (in air)	Duration of second firing (in oxygen)
A	Powder	8 h	0
B	Powder	8 h	4 h
C	Powder	8 h	10 h
D	Bulk	8 h	10 h

The resistivity versus temperature curve, as measured with the standard four-probe technique, is given in Fig. 3. The superconducting transition starts at 93 K and is completed at 88 K. The step in the transition curve may be attributed to inhomogeneity in material with regions of different oxygen deficiency and consequently different critical temperatures. Spot spectroscopy showed inhomogeneous distribution of Y,Ba and Cu: more Y and less Cu have been found inside the specimen compared to the surface.

RESULTS

The remanent magnetic moments of samples A,B,C and D at 77 K are plotted in Fig. 4 as a function of the magnetising field. At fields greater than 0.1 T, where a constant value of remanent moment is expected, a small linear decrease can be observed. Also in this Figure, the strong dependence of superconductivity on sample preparation conditions is evident illustrating, in particular, the crucial role of oxygen in these materials.

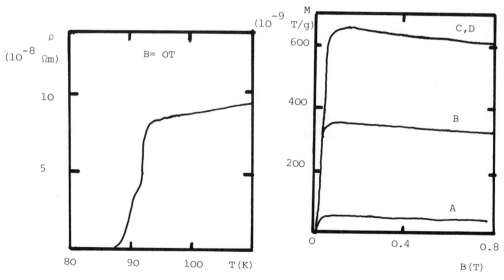

Fig. 3: Resistivity vs. temperature of the YBaCuO samples.

Fig. 4: Magnetic moment due to screening supercurrents at 77 K in samples A,B,C and D as a function of applied magnetic field.

There is no difference between the observed moment of the powder specimen C and that of the sintered bulk specimen D. This may be attributed to the existence of two types of magnetisation current: macroscopic intragrain currents and microscopic intergrain currents[3]. Intergrain currents in polycrystalline materials are limited by grain boundary interfaces (thought to behave like Josephson junctions) and are much smaller than intragrain currents which are the same in powder and bulk specimens. Direct electrical measurement of the intergrain current density in the bulk specimen D in zero magnetic field gives the value of 7 A/cm^2. Magnetisation measurements, on the other hand, lead to intragrain current densities of more than 10^3 A/cm^2, assuming a mean grain diameter of about 3μm[4].

Following thermal transition of the specimens from superconducting to normal state, a small remanent magnetic moment was detected, the magnitude of which depended on its magnetisation history. This normal state remanent moment is much smaller than that created at 77 K by screening supercurrents. The room temperature remanent magnetic moment of samples C and D as a function of magnetising field, obtained by the procedure described above, is presented in Fig. 5. It can be seen that the bulk specimen D has a remanent moment which is much higher, and a hysteresis loop which is much narrower than the powder specimen C. No easy direction of magnetisation for the normal state magnetic moment was found, implying the absence of anisotropy.

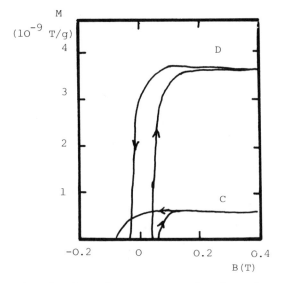

Fig. 5: Normal state magnetisation of samples C and D as a function of applied field at room temperature.

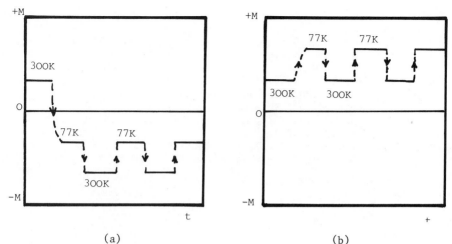

(a)   (b)

Fig. 6: Magnetic moment of sample D during thermal cycling between 300K
and 77 K in the Earth's magnetic field. The direction of applied field in
(a) is perpendicular to that in (b).

The cooling down, from 300 K to 77 K, of the specimens with a
remanent magnetic moment, produced some unexpected results. During the
initial cool-down, in the Earth's magnetic field, a reversal of the
remanent magnetic moment was observed. Subsequent thermal cycling,
however, did not produce any further reversal of the moment (see Fig.6(a)).
A change of 90° in the direction of the magnetising field produces the
results shown in Fig. 6(b). In this case there is no difference in the
initial and subsequent thermal cycles except for the longer time taken
for the establishment of screening supercurrents during the initial cool-
down. Change of magnetic moment during cool-down in a magnetic field,
known as exchange anisotropy, has been observed in inhomogeneous ferromagnetic and ferrimagnetic materials[5].

CONCLUSIONS

The coexistence of normal state permanent magnetic moment and
superconductivity has been observed in inhomogeneous YBaCuO material.
Various interpretations such as the existence of magnetically oriented
normal state domains, existence of "holes" in the antiferromagnetic
ordering of Cu-O chains as well as others can be put forward as
explanations for this effect.

A possible alternative explanation may be in terms of ferromagnetic
impurity particles in the sample. Although the starting materials were
pure, the possibility of contamination during pelletisation cannot be
ruled out. Attempts to measure the Curie temperature of the observed
ferromagnetic behaviour have produced inconclusive results and need to be
repeated. The magnetic measurements will also be carried out on new, pure-
phase materials. We aim to present a detailed account of our findings in
the near future.

REFERENCES

1. P.H.Hor, R.L.Meng, Y.Q.Wang, L.Gao, Z.J.Huang, J.Bechtold,
    K.Foester, C.W.Chu, Phys.Rev.Lett., 58: 1891 (1987).
2. J.M.Tarascon, W.R.McKinnon, L.H.Green, C.M.Hull, E.M.Vogel,
    Preprint, Bell Communication Research, New Jersey (1987).
3. H.Küpfer, I.Apfelsted, W.Schauer, R.Flükiger, R.Meier-Hirmer,
    H.Wühl, Submitted to Z.Phys.B. (Aug. 1987).
4. M.Polak, F.Hanic, I.Hlasnik, M.Majoros, F.Chovanec, Submitted
    to Czech.J.Phys.B, (1988).
5. W.M.Meiklejohn, J.Appl.Phys., 33: 1328 (1962).

STUDIES OF HTS CERAMICS-COPPER INTERACTION

A.D.Nikulin, V.Y.Fil'kin, I.I.Davydov, A.K.Shikov
N.V.Shishkov, E.V.Antipova, V.S.Sergeev, V.A.Zheleznyakov
N.I.Kozlenkova, V.I.Panzyrny, and V.V.Medkov
All Union Research Institute for Nonorganic Materials
Moscow, USSR

Consideration is given to the interaction of Y-Ba-Cu-O compounds and copper in a single filament $YBa_2Cu_3O_{7-x}$ powder conductor. The structure and composition of a HTS compound is shown to change, depending on the heat treatment conditions within 400 - 900 °C. It is shown that there is no diffusion-induced redistribution of metal atoms in the ceramics-copper interaction zone after annealing at a temperature of up to 900 °C.

The discovery of high temperature superconducting (HTS) Y-Ba-Cu-O ceramics, with a transition temperature above that of liquid nitrogen[1], issued the challenge to produce Y-Ba-Cu-O based conductors for practical usage. One of the first methods developed to produce the brittle intermetallic compound-based superconductors, was to deform the Y-Ba-Cu-O powder in a ductile can[2]. This method was employed to produce wire specimens based on A15 compounds (Nb_3Sn)[2], NaCl-(NbN)[3,4], and Chevrel-phases ($PbMo_6S_8$)[5]. The deformation stability problem of such conductors is not related to the crystallographic structure or the superconducting properties of an enclosed powder but is rather determined by factors such as powder dispersivity, mechanical properties of the canning material and the ratio between the canning and the powder-core areas[4,6].

Using the method of cold drawing we have produced a single filament conductor, 1.0 mm - 1.5 mm dia., with an $Y_1Ba_2Cu_3O_{7-x}$ powder core, the cross-section of which is shown in Fig. 1.

The initial powder for the HTS compound was produced by a solid-phase method using initial yttrium and copper oxides and barium carbonate. The structure of a HTS ceramic powder with a grain size, prior to deformation, of < 50 μm, is shown in Fig. 2a. During deformation the ceramic grains decrease in size. The sizes of individual crystals become equal to 5 - 10 μm. The ceramic structure is uniform throughout the whole filament section.

X-ray diffraction studies showed that the structure of the deformed $Y_1Ba_2Cu_3O_{7-x}$ phase remains orthorhombic; however, the broadening of the line profile indicates a high level of residual stress. In this case, as in Nb_3Sn[8] and $PbMo_6S_8$[9] powder deformation, a degradation of the superconducting properties is observed. To regain the initial transition temperature level and to provide a structure with a high current-carrying capacity, a heat treatment is required which, along with a residual stress relief, may result in an interaction with the canning material.

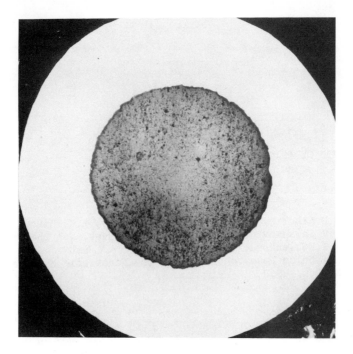

Fig. 1. Cross section of a single filament conductor containing HTS
ceramic powder in a copper can

e.g., during the heat treatment of Nb_3Sn in the high temperature range,
in contact with copper or low tin Cu-Sn alloys, the atoms of tin leave
Nb_3Sn, which results not only in the restoration of T_c, but in a sharp
decrease of T_c and Nb_3Sn lattice parameter[10]. A similar process, with S
leaving the compound, takes place when producing $PbMo_6S_8$-based
superconductors[8].

For $Y_1Ba_2Cu_3O_{7-x}$, oxygen is the most reactive element. Under the
action of temperature, the oxygen content of the compound changes
reversibly from 6.9 (400 °C) to 5.8 (900 °C). In this case, the critical
temperature of Y-Ba-Cu-O is lowered until the superconducting properties
are completely lost.

During heat treatments between 400 and 900 °C the ceramic powder,
which is in contact with a copper can, interacts with the copper. A
400 °C anneal in air for tens of hours will not restore the
superconducting properties of the ceramics and according to the data of
metallographic and micro x-ray spectral analyses, the width of the
interaction zone at the copper-core interface does not exceed 10 - 15 μm.
An increase in the heat treatment temperature to 700 - 900 °C resulted in
a drastically changed structure and composition of the ceramics. The
$YBa_2Cu_3O_{7-x}$ compound decomposed to form large crystals of copper oxides
and complex oxides of yttrium and barium (Fig. 3a,b), primarily due to a
considerable loss of oxygen in the superconducting phase.

Therefore, in the production of composite HTS ceramic-based
superconductors, special attention should be paid to conditions which
provide for the retention of oxygen in the superconducting phase. For
this purpose one should select a matrix material adequately compatible
with $YBa_2Cu_3O_{7-x}$ and not exceed the temperature range within which the
compound decomposes[11], regulate the composite cooling and heating rates
and take other measures conducive to the retention of the stoichiometric
composition of HTS ceramics in the composite.

Fig. 2a. Structure of the initial $Y_1Ba_2Cu_3O_{7-x}$ powder, prior to the onset of deformation

Fig. 2b. ... and after deformation to the final diameter.

1 μm

Fig. 3. Structure of a ceramic filament when heat treated in air for one
hour at a) 700 °C and b) 800°C

REFERENCES

1. M.K.Wu, J.R.Ashburn, C.J.Torng, P.H.Hor, P.L.Meng, L.Gao, Z.J.Huang,
    Y.Q.Wang, C.W.Chu: Phys.Rev.Lett. $\underline{58}$, 908 (1987)
2. E.M.Savitzky, J.F.Efimov, N.D.Kozlov, B.P.Michailov, L.F.Myzenkova,
    E.D.Doron'kich: "Superconducting Materials", M.Metallurgia (1976)
3. Patent USA N 4411953
4. E.M.Savitsky, V.V.Baron: "Powder metallurgy and superconducting
    materials" in "I.P.Bardeen and national metallurgy"
5. T.S.Luhman, O.Horigami, D.Dew-Hughes: Bull.Amer.Phys.Soc. $\underline{20}$, 343
    (1975)
6. V.F.Suchkov, V.I.Svetlova, E.E.Finkel: "Thermoresistable cables with
    magnesia insulation", M.Energia 1969, p.96
7. A.S.Nikiforov, A.D.Nikulin, V.J.Fil'kin, N.V.Shishkov, I.I.Davydov,
    A.K.Shikov, E.V.Antipova, N.A.Chernoplekov, E.J.Klimenko:
    "La-Sr-Cu-O and Y-Ba-Cu-O composite superconductors", M.Atomnaya
    Energia $\underline{63}$, 421 (1987)

8. N.M.Michailov, M.N.Smirnova, I.N.Groznov, V.A.Zhorin: "High pressure
    and deformation modification of structure and superconducting
    temperature of A-15 compounds", Proceedings of the 2nd scientific
    meeting on "Metallophysics of superconductors", 1983, Kiev, 134-135
9. T.Luhman, D.Dew-Hughes: J.Appl.Phys. _49_, 936 (1978)
10. A.D.Nikulin, V.A.Kovaleva, E.G.Kasakov, A.I.Skvortsov, V.I.Panzirny,
    V.A.Vargin: "Investigation of interaction between Cu-Sn alloy melts
    and Nb_3Sn layer", Proceedings of the 2nd scientific meeting on
    "Metallophysics of superconductors" 1983, Kiev, 268-269
11. Y.Takagi, R.Liang, Y.Inaguma, T.Nakamura: Jpn.J.App.Phys. _26_, 1266
    (1987)

SECTION 5: THIN FILMS

HIGH T_c SUPERCONDUCTING OXIDE THIN FILMS: FABRICATION

APPLICATION, AND CHARACTERIZATION

R.B.Laibowitz

IBM Research Center, P.O.Box 218
Yorktown Heights, NY 10598, U.S.A.

Vapor-deposited thin films of the high T_c superconducting compound, $YBa_2Cu_3O_y$, have been fabricated using several thin film systems including e-beam deposition and sputtering. In the e-beam system three independent sources are used for the metals and oxygen is added to the system during the deposition at a pressure of about $8x10^{-4}$ Torr. Available film thicknesses range from less than a tenth μm to several μm and transition temperatures as high as 91 K have been observed. Important parameters for the fabrication of these films include choice of substrate and temperature during the deposition, oxygen incorporation during the deposition and the annealing conditions. While critical currents well in excess of 10^6 A/cm^2 at 4 K have been measured, typically on epitaxial films, the polycrystalline films generally show much lower values. These results will also be discussed with particular reference to anisotropic effects in the superconducting films.

In order to use these materials in a variety of thin film applications, techniques for lithographic patterning and processing had to be developed. Several techniques including photolithography, laser etching and ion implantation have been used and linewidths to about 2 μm have been achieved. Utilizing these techniques, applications such as SQUIDs operating above 77 K, tunnel junctions and transmission lines will be discussed.

STOICHIOMETRY AND DEPOSITION RATE OF DC MAGNETRON SPUTTERED Y-Ba-Cu-O THIN FILMS AS A FUNCTION OF TARGET PRESPUTTERING TIME

T.I.Selinder, G.Larsson, U.Helmersson, P.Olsson and
J.-E.Sundgren

The Thin Film Group, Department of Physics, Linköping
University, S-581 83 Linköping, Sweden

S.Rudner and L.D.Wernlund

FOA 3, Box 1165, S-581 11 Linköping, Sweden

ABSTRACT

Y-Ba-Cu-O thin films have been grown from stoichiometric compound
targets using dc magnetron sputtering. Initially the film composition was
strongly off-stoichiometric and the deposition rate very low. However,
after extensive presputtering stoichiometric films, with respect to the
metal content, were deposited at a rate of 2-2.5 Å/s. The long presputtering time (20-30 h) needed is due to a slowly changing oxygen
content on the target surface. This is caused not only by preferential
sputtering effects but also by bulk diffusion in the target.

INTRODUCTION

The recent discovery of high critical temperature superconductivity
in $YBa_2Cu_3O_{6+\delta}$ and other metallic oxides with modified perovskite
structure presents new challenges in the synthesis of superconducting thin
films. There has been considerable progress in the fabrication of thin
films of these materials and a number of groups have successfully applied
a whole range of deposition techniques, e.g., electron beam co-evaporation
(1), laser beam ablation (2), molecular beam epitaxy (3) and various
sputtering techniques (4,5). Also several types of substrates have been
tried and the best results, so far, have been attained by depositing films
on single crystalline substrates of $SrTiO_3$, MgO and ZrO_2 stabilized with
9% Y_2O_3.

Of the deposition methods mentioned above, sputtering from a single
target normally has the advantage of a well defined deposition rate and
reproducible composition over a large substrate area. Several papers on
sputter-deposited films have been published recently, e.g., Enomoto et al
(5) have produced thin sputtered films with a critical temperature
T_c = 84 K and a critical current density of 1.8×10^6 A cm^{-2} at 77.3 K.

In this paper, target presputtering effects on the deposition of
$YBa_2Cu_3O_{6+\delta}$ thin films are discussed. The films have been grown on single
crystalline (100)-oriented MgO substrates by dc magnetron sputtering
utilizing a compound oxide target. A strong dependence of the metallic

composition and the deposition rate of the films on target presputtering time is observed. This dependence is caused not only by preferential sputtering effects but also by oxygen bulk diffusion in the target.

EXPERIMENTAL

The sputtering system was pumped by a diffusion pump giving an ultimate pressure of 2.6×10^{-4} Pa. The sputtering was performed at an Ar pressure of 3.0 Pa. The target current was constantly 750 mA.

Targets used in the experiment were ceramic disks of 50 mm diameter and with a thickness of 4 mm. They were prepared by the usual sintering and annealing procedure from powders of yttrium and copper oxides and barium carbonate. The targets had the $YBa_2Cu_3O_{6+\delta}$ composition, a density of 4.69 g cm^{-3} and T_c = 92 K.

Energy Dispersive X-ray Spectroscopy (EDS) was carried out to determine the metallic composition of the as-deposited films using the LINK analytical systems AN 10 000. Yttrium, copper, barium fluoride and a stoichiometric $YBa_2Cu_3O_{6+\delta}$ bulk sample were used as elemental standards when calibrating the EDS analyses. Compositions were calculated using a ZAF-4/FLS (LINK) program. All films described in this paper were analyzed in as-deposited condition without any post annealing. However, after post annealing in a continuous flow of oxygen at a temperature of 950 $^\circ$C, films with T_c values ranging from 70 to 75 K were obtained. Typical sample thickness was 1.4 μm.

RESULTS AND DISCUSSION

Films were grown using a high argon pressure to avoid effects from energetic oxygen neutrals (6) and the temperature of the substrates was held low to obtain a sticking coefficient close to unity for all atoms. This should give nearly identical composition of the films as compared with the flux of sputtered atoms from the target. Figure 1 shows the film composition as a function of accumulated sputtering time. The film composition is initially strongly off-stoichiometric with Ba being deficient and Cu and Y in excess. However, after 20 - 30 hours of sputtering stable conditions are achieved and the film composition is practically identical

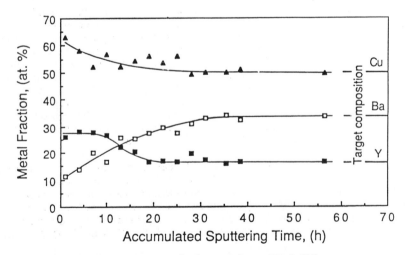

FIG. 1. Composition of the as-deposited films as a function of total target sputtering time. The target material is stoichiometric $YBa_2Cu_3O_{6+\delta}$

to the target composition. The films are still stoichiometric after 60 hours of sputtering. An initial off-stoichiometric composition is normal when sputtering from an alloy or compound target. A layer of altered composition is formed on the target surface if the sputtering yields of the constituent atoms are different. However, the time needed for the composition of this layer to stabilize is normally much less than the above mentioned times and dependent on factors such as differences in sputtering yield, ion energy and ion flux (7).

The composition of the target surface will also influence the total deposition rate. This is particularly noticeable for variations in the oxygen content on the target surface since oxides normally have a very low sputtering yield.(For example, the sputtering yield of Ba is reduced by more than a factor of six after exposure to oxygen (8).) Thus, if the surface composition of the oxide target is changed, a change in the total deposition rate can be expected. An example of this can be seen in Figures 2a and 2b, which show the deposition rate and the target potential as a function of total target sputtering time. The fact that the target voltage also varies with the status of the target surface is due to the higher electron and negative ion yields from an oxidized surface compared to those from a more metallic like surface. A larger amount of negative

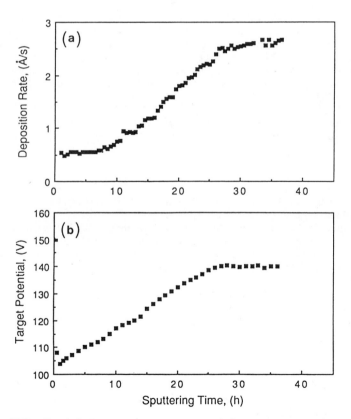

FIG. 2. (a) Deposition rate and (b) target potential as a function of sputtering time from a new target.

charge carriers in the plasma will consequently reduce the current carried by the positive Ar ions (9) and also reduce the impedance of the plasma. Since the target current is held constant the target potential will be reduced in proportion to the oxide content on the target surface. Furthermore, it can be seen in Figure 2, that steady state conditions for a new target are reached first after ~ 25 hours of sputtering in agreement with the results shown in Figure 1.

However, in spite of the low sputtering yield that oxides normally exhibit the time required for a steady state target composition to develop due to preferential sputtering effects alone, is only in the range of a few minutes (10). The very long times observed for obtaining steady state conditions in the sputtering cases described in this paper are, however, caused by a diffusion of oxygen from the bulk of the target. Due to the bad thermal conductivity of these materials the impinging ions might, if the flux is high enough, give rise to an increase in the temperature of the target. At higher temperatures oxygen bulk diffusion is rapid (11). Thus when sputtering from a composite oxide target in pure Ar discharges, the oxygen depletion of the target surface (both through sputtering and thermal desorption) will not reach a steady state value until a large region of the target has been depleted of oxygen.

The described effects giving non-stoichiometric film compositions can be one of the explanations for problems some groups have had in obtaining stoichiometric films. Shah and Carcia (12) for example, observed Ba deficiency in films grown on room temperature substrates using rf magnetron sputtering. A Ba deficiency is also observed by us during the first 10-20 hours of sputtering from a new target.

The time to establish steady state conditions on the target can be reduced by keeping the target at low temperature where oxygen diffusion is negligible. However, due to the bad thermal conductivity of these materials this might be difficult, especially when high deposition rates are desired. A high but stable oxygen content on the target surface will automatically imply a low deposition rate due to the low sputtering yield of highly oxidized targets.

The authors gratefully acknowledge the financial support from The Swedish Board for Technical Development (STU). One of the authors (G.L.) would like to extend his acknowledgements to Ericsson Radar Electronics.

REFERENCES

1 See for example, P.Chaudhari, R.H.Koch, R.B.Laibowitz, T.R.McGuire and R.H.Gambino, Phys.Rev.Lett. 58, 2684 (1987).
2 D.Dijkkamp, T.Venkatesan, X.D.Wu, S.A.Shaheen, N.Jisrawi, Y.H.Min-Lee, W.L.McLean, and M.Croft, Appl.Phys.Lett. 51, 619 (1987).
3 C.Webb, S.-L.Weng, J.N.Eckstein, N.Missert, K.Char, D.G.Schlom, E.S.Hellman, M.R.Beasley, A.Kapitulnik, and J.S.Harris,Jr, Appl.Phys.Lett. 51, 1191 (1987).
4 See for example, T.Aida, T.Fukazawa, K.Takagi, and K.Miyauchi, Jpn.J.Appl.Phys. 26, L1489 (1987).
5 Y.Enomoto, T.Murakami, M.Suzuki, and K.Moriwaki, Jpn.J.Appl.Phys. 26, L1248 (1987).
6 R.E.Somekh, J.Vac.Sci.Technol.A 2, 1285 (1984).
7 P.S.Ho, J.E.Lewis, H.S.Wildman and J.K.Howard, Surface Sci. 57, 393 (1976).
8 D.Grischkowsky, M.L.Yu, and A.C.Balant, Surface Science 127, 315 (1983)
9 S.M.Rossnagel, and J.J.Cuomo, presented at the AVS fall meeting, Anaheim, CA, USA, Nov. 6, 1987.

10 Preliminary Auger Electron Spectroscopy measurements on amorphous $YBa_2Cu_3O_x$ films show that the time required to reach steady state during sputtering with Ar ions is in the order of seconds or minutes depending on the ion current and ion energy.

11 K.N.Tu, S.I.Park and C.C.Tsuei, Appl.Phys.Lett. **51**, 2158 (1987).
12 S.I.Shah and P.F.Carcia, Appl.Phys.Lett. **51**, 2146 (1987).

SUPERCONDUCTING FILMS OF YBaCuO

P. R. Broussard, J. H. Claassen, M. S. Osofsky
S. A. Wolf, C. R. Gossett, and R. J. Soulen, Jr.

Naval Research Laboratory
Washington, DC 20375

INTRODUCTION

The purpose of this article is to report on progress at our institution on the fabrication of thin films made from high transition temperature materials. Of particular interest was the effect of substrate material on the quality of the superconducting transition.

SAMPLE PREPARATION

Films of YBaCuO were prepared by evaporation of the components from three different sources while in the presence of oxygen. Our evaporator system (shown in Fig. 1) has the capability of depositing materials simultaneously from two electron-beam guns and one Knudsen thermal source. Many different configurations of the elements in these sources were tried, and currently we place Y and Cu in the electron-beam guns and either Ba metal or BaF_2 in the Knudsen cell. Rate control for the electron-beam guns is provided by crystal rate monitors mounted at an angle to the line connecting the sources to the substrates. The Knudsen cells have no rate control, per se, but are provided with a temperature stabilizing circuit, which keeps the temperature constant to within $\pm$ 0.1 °C. Rates for the individual sources are in the range of 1-10 Å/s, and the oxygen partial pressure is typically 10^{-6} mbar (measured at the bottom of the chamber). The oxygen is introduced via a 2 mm diameter tube aimed at the substrates and located ≈ 5 cm away. Before the deposition, the substrates are baked to 500 °C to remove water vapor and organics from the surface, and then cooled to 50-100 °C. The sources are then opened with O_2 present, and deposition proceeds. The substrates rotate during deposition in order to improve homogeneity. After deposition, the films can be analyzed in situ by RHEED or XPS. Upon removal from the system, the samples are given an oxygen anneal (typical, or type A procedure: samples at 850 °C in flowing O_2 for 30 minutes followed by a ramp to room temperature at 1 °C/min).

SAMPLE CHARACTERIZATION

We have used substrates made from Al_2O_3, $LiNbO_3$, MgO, and $SrTiO_3$. As found by previous researchers, our best results occur for films grown on $SrTiO_3$. To characterize the films, we have examined

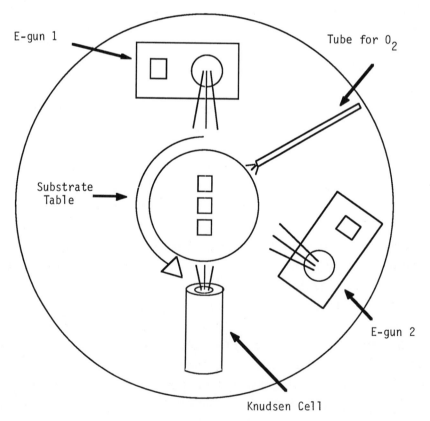

Figure 1: Schematic of deposition system, viewed from above.

them by resistivity, T_c, x-ray diffraction, and Rutherford Backscattering measurements.

Independent of the source configuration, the samples are amorphous immediately following deposition (as determined by RHEED) and, are also insulators. After annealing, films deposited using Ba metal in a Knudsen cell were usually found to be of poor quality: Only a few films had a sharp T_c, metallic behavior, etc as shown in Fig. 2.

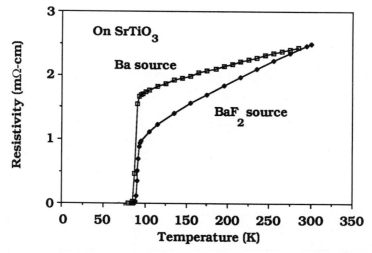

Figure 2: Comparison of resistivity traces for YBaCuO samples deposited on SrTiO3 using either Ba metal or BaF2 in the Knudsen cell.

Our best films (on SrTiO3) with this method had a T_c of 85-90K (complete to onset), a resistivity ratio (R(300 K)/R(100 K)) of 1.4, and a resistivity of 1.5 mΩ-cm at 100 K. The x-ray diffraction profiles showed the films to be polycrystalline and consist of small grains (< 1000 Å). Typically though, films on SrTiO3 and MgO had broad T_c's and resistivity ratios of ≈ 1. We attribute this to poor Ba rate control in the Knudsen cell as determined by rate checks of the Ba rate before and after deposition. This has been seen by other researchers.[1] Films on Al2O3 had strongly activated resistivity traces, while films on LiNbO3 typically came out insulating.

We substituted BaF_2 for Ba in the Knudsen cell as per the results of Mankiewich et al.[2] and found that our rate control of BaF_2 was substantially improved over that of Ba. Using our in situ XPS analysis, we have seen that films deposited using BaF_2 instead of Ba uptake approximately 1/3 less oxygen during film growth. This is consistent with the picture that Cu atoms are deposited unoxidized in the as-deposited films. We modified our anneal as did Mankiewich et al. to include a wet O_2 anneal at 850 °C. Films deposited on $SrTiO_3$ were found to come out after annealing with excellent properties compared to our earlier work, even if the composition was not close to stoichiometrey. Our best results to date (also shown in Fig. 2) on $SrTiO_3$ are T_c of 87-93 K, resistivity ratio of 2.5, and a resistivity of 1 mΩ-cm at 100 K. X-ray measurements show that this film has a mixture of epitaxial regions oriented with the a axis normal to the films, and textured regions with (130)/(101) normal. Films deposited on MgO however, have not been as promising. Using our standard anneal, (Type A, given above) they typically come out with activated resistivity curves and broad or nonexistent transitions. We found that they are

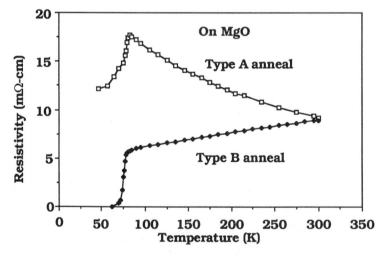

Figure 3: Resistivity traces for YBaCuO films deposited on MgO substrates and using two different anneals, as described in the text.

much more sensitive to the annealing details than the films on SrTiO$_3$. By varying the initial annealing temperature to 950 degrees for 6 minutes, followed by an anneal at 700 °C for 1/2 hour (Type B procedure), we are able to obtain films on MgO with complete transitions by 60 K as shown in Fig. 3. We are currently investigating samples on MgO to understand this problem.

Rutherford Backscattering (RBS) has been used to obtain the film composition and to study substrate/film reactions. 6.2 MeV α particles[3] are used in order to obtain more sensitivity to C and O in the film and clearly separate the signals from the three metal constituent. Figure 4 shows a typical RBS trace on MgO substrate for a 5000 Å thick film with composition $Y_{0.9}Ba_{1.8}Cu_3O_x$. Using RBS on samples deposited on MgO

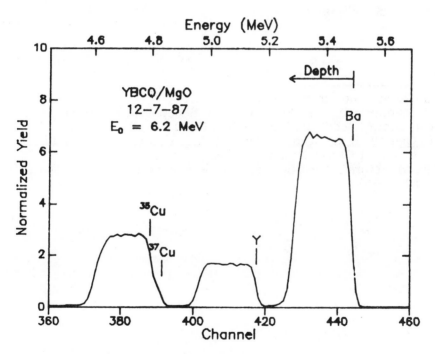

Figure 4: Rutherford Backscattering spectrum obtained on a YBaCuO film using 6.2 MeV α particles. This sample was deposited on MgO using Ba metal.

at this large energy presents a problem in obtaining the composition of oxygen, etc. in the film due to strong Mg resonances, but using a MgO standard it is still possible to obtain this information. RBS has not shown any significant sign of substrate interdiffusion on MgO, while on $SrTiO_3$ the data are not yet complete. Using RBS we have also seen that the level of fluorine in the samples after annealing is below the observable level ($\approx$ 0.1 %). RBS has also shown that using the BaF_2 we have been incorporating a few percent of carbon into the films. We associate this with CO/CO_2 gas dissolved in the BaF_2 source. It is interesting to note that this does not seem to affect the $SrTiO_3$ samples.

CONCLUSIONS

We have prepared superconducting films of YBaCuO on $SrTiO_3$ and MgO and have characterized them through x-ray diffraction, transport measurements, and RBS. Our future plans are to study the films on MgO in order to understand the current difficulties in producing high T_c's on these substrates.

------

[1] John Talvacchio, private communtication.
[2] P. M. Mankiewich, J. H. Scofield, W. J. Skocpol, R. E. Howard, A. H. Dayem, and E. Good, *Appl. Phys. Lett.* **51**, 1753 (1987).
[3] C. R. Gossett, K. S. Grabowski and D. Van Vechten, in "Thin Film Processing and Characterization of High-Temperature Superconductors", AIP Conference Proceedings No. 165, p. 443 (New York, NY 1988).

PARTICIPANTS

Mag.Claudia AMBROSCH
Inst.f.Technische Elektrochemie
Technische Universität Wien
Getreidemarkt 9
A-1060 Wien
Austria

Dr.J.F. BEAUDOIN
E.N.S.P.G.
LMGP
B.P.46
F-38402 St.Martin d'Heres
France

Dr.Peter BLAHA
Inst.f.Techn.Elektrochemie
Technische Universität Wien
Getreidemarkt 9
A-1060 Wien
Österreich

Dr.Merwyn B. BRODSKY
Argonne National Laboratory
Materials Science Division, 223
9700 South Cass Avenue
Argonne, Illinois 60439
U.S.A.

Dr.G.W. CRABTREE
Materials Science Div., 223
Argonne National Laboratory
9700 South Cass Ave.
Argonne, IL 60439
U.S.A.

Prof.Dr.D.K. FINNEMORE
Ames Laboratory
Iowa State University
Ames, Iowa 50011
U.S.A.

Dr.Wolfgang BAUHOFER
Max-Planck Institut für
Festkörperforschung
Heisenbergstraße 1
D-7000 Stuttgart 80
BRD

Prof.Dr.K. BENTE
Mineral.Kristall.Inst.
Universität Göttingen
Goldschmidtstraße 1
D-3400 Göttingen
BRD

Prof.Dr.M. BREITER
Inst.f.Techn.Elektrochemie
Technische Universität Wien
Getreidemarkt 9
A-1060 Wien
Austria

Prof.Dr.J.P. CARBOTTE
Physics Department
McMaster University
Hamilton
Ontario L8S 4M1
Canada

Dr.J. FINK
INFP
Kernforschungszentrum
Postfach 3640
D-7500 Karlsruhe
BRD

Prof.Dr.H.C. FREYHARDT
Institut für Metallphysik
der Universität
Hospitalstraße 3/7
D-3400 Göttingen
BRD

Dipl.Ing.M. FRISCHHERZ
Atominstitut der
Österr.Universitäten
Schüttelstraße 115
A-1020 Wien
Austria

Prof.Dr.Kazuo FUEKI
Dept.of Industrial Chemistry
University of Tokyo
Hongo, Tokyo 113
Japan

Dr.Joze GASPERIC
J.Stefan Institute
Jamova 39
POB 100
Y-61111 Ljubljana
Yugoslavia

Dr.K.E. GRAY
Argonne National Laboratory
Bldg. 223
9700 South Cass Avenue
Argonne, Illinois 60439
U.S.A.

Dr.Hartmut GRUBER
Inst.f.Festkörperphysik
Technische Universität Graz
Petersgasse 16
A-8010 Graz
Österreich

Dr.G. HILSCHER
Inst.f.Experimentalphysik
Technische Universität Wien
Wiedner Hauptstraße 8-10
A-1040 Wien
Austria

Erkki IKONEN
Department of Technical Physics
Helsinki University of Technology

SF-02150 Espoo
Finland

Dr.Luc FRUCHTER
Physique des Solides
Batiment 510

F-91405 Orsay
France

Dr.Jean GALY
LCC-CNRS
205, route de Narbonne
F-31077 Toulouse Cedex
France

A.V. GRABOV
USSR Mission at the
UN Organizations
Erzherzog Karl Straße 182
A-1220 Wien
Austria

Dipl.Ing.P. GREGSHAMMER
Atominstitut der
Österr.Universitäten
Schüttelstraße 115
A-1020 Wien
Austria

Dr.Ulf HELMERSSON
Dept.of Physics
Linköping University
S-58183 Linköping
Schweden

Dr.Osamu HORIGAMI
Toshiba R&D Center
4-1, Ukishima-cho
Kawasaki-Ku
Kawasaki 210
Japan

Dr.A.V. INYUSHKIN
Kurchatov Institute of
Atomic Energy

123182 Moscow
USSR

Maarit KARPPINEN
Helsinki University of
Technology

SF-02150 Espoo
Finland

Dr.M.N. KHLOPKIN
Kurchatov Institute of
Atomic Energy

123182 Moscow
USSR

Dr.Ulf KLEIN
Inst.für Theoretische Physik
Johannes Kepler Universität
A-4040 Linz-Auhof
Austria

Dr.Erich KNY
Metallwerk Plansee

A-6600 Reutte
Austria

Prof.Dr.E. KRAUTZ
Inst.f.Festkörperphysik
Technische Universität Graz
Petersgasse 16
A-8010 Graz
Austria

Dr.Heinz KÜPFER
Kernforschungszentrum
ITP

D-7500 Karlsruhe
BRD

Dr.R.B. LAIBOWITZ
IBM Thomas J.Watson Res.Center
P.O.Box 218
Yorktown Heights, NY 10598
U.S.A.

Dr.Jacek KASPERCZYK
Institute of Physics
Pedagogical University
Al.Zawadzkiego 13/15
PL-42200 Czestochowa
Poland

Prof.Dr.H. KIRCHMAYR
Inst.f.Experimentalphysik
Technische Universität Wien
Wiedner Hauptstraße 8-10
A-1040 Wien

Dr.Peter KNOLL
Inst.f.Experimentalphysik
Universität Graz
Universitätsplatz 5
A-8010 Graz
Austria

Dr.V.D. KOSYNKIN
All Union Research Institute
for Chemical Technology
Moscow
USSR

Dr.Reinhard KREMER
Max-Planck-Institut für
Festkörperforschung
Heisenbergstraße 1
D-7000 Stuttgart 80
BRD

Prof.H. KUZMANY
Inst.f.Festkörperphysik
Universität Wien
Strudlhofgasse 4
A-1090 Wien
Austria

Dr.S. LAVRINENKO
Charkov Physical and
Technological Institute
Charkov
USSR

Dr.Günther LEISING
Inst.f.Festkörperphysik
Technische Universität Graz
Petersgasse 16
A-8010 Graz
Austria

Prof.Dr.G. LIBERTS
Institut für Angewandte Physik
Johannes Kepler Universität

A-4040 Linz
Austria

Dr.D.M.R. LO CASCIO

Natuurkundig Laboratorium
Valckenierstraat 65
NL-1018 XE Amsterdam
The Netherlands

Yasuo MAEDA
The Furukawa Electric
Company Ltd., Suite 410
High Holborn House 52-54
High Holborn WC1V6RP
U.K.

Prof.Dr.M.B. MAPLE
Dept.of Physics, B-019
University of California

La Jolla, CA 92093
U.S.A.

Dr.H. MATTAUSCH
Max-Planck Institut für
Festkörperforschung
Heisenbergstraße 1
D-7000 Stuttgart 80
BRD

Dr.Ulrike MERSITS
CERN

CH-1211 Geneve 23
Switzerland

Dr.Peter MOHN
Inst.f.Techn.Elektrochemie
Technische Universität Wien
Getreidemarkt 9
A-1060 Wien
Austria

Prof.Dr.J.A. MYDOSH
Kamerlingh Onnes Lab.
der Rijksuniversiteit
Postbus 9506
NL-2300 RA Leiden
The Netherlands

Hans NIEDERMAIER
Atominstitut der
Österr. Universitäten
Schüttelstraße 115
A-1020 Wien
Austria

Dr.Anna PAJACZKOWSKA
Institute of Physics of the
Polish Academy of Sciences
Al.Lotnikow 32/46
PL-02110 Warszawa
Poland

Dr.V.I. PANZYRNY
All Union Research Institute
for Nonorganic Materials

Moscow
USSR

Prof.Dr.Martin PETER
Departement de Physique
de la Matiere Condensee
24, quai Ernest-Ansermet
CH-1211 Geneve 4
Switzerland

Dipl.Ing.N. PILLMAYR
Inst.f.Experimentalphysik
Technische Universität Wien
Wiedner Hauptstraße 8-10
A-1040 Wien
Austria

Werner PINT
Inst.f.Theoretische Physik
Technische Universität Graz
Petersgasse 16
A-8010 Graz
Austria

Dipl.Ing.M. PROHAMMER
Inst.f.Theoretische Physik
Technische Universität Graz
Petersgasse 16
A-8010 Graz
Austria

Prof.Dr.Dierk RAINER
Lehrstuhl Theoretische Physik III
Universität Bayreuth
Postfach 3008
D-8580 Bayreuth
BRD

Dr.D. RASSI
Dept.of Physics
Univ.College of Swansea
Swansea SA2 8PP
U.K.

Prof.Dr.Bernard RAVEAU
Universite de Caen
ISM Rayonnement

F-14032 Caen-Cedex
France

Dr.W.L. REITER
Bundesministerium für
Wissenschaft und Forschung
Freyung 1
A-1010 Wien
Austria

Dr.P. ROGL
Inst.f.Physikalische Chemie
Universität Wien
Währinger Straße 42
A-1090 Wien
Austria

Dr.Bernhard RUPP
Inst.f.Festkörperforschung
Kernforschungsanlage Jülich
Postfach 1913
D-5170 Jülich
BRD

Dr.F.M. SAUERZOPF
Atominstitut der
Österr.Universitäten
Schüttelstraße 115
A-1020 Wien
Austria

Dr.E. SCHACHINGER
Inst.f.Theoretische Physik
Technische Universität Graz
Petersgasse 16
A-8010 Graz
Austria

Dr.Christina SCHÖLL
Daimler Benz AG
Abt. FGG
Postfach 600202
D-7000 Stuttgart 60
BRD

Prof.Dr.K. SCHWARZ
Inst.f.Techn.Elektrochemie
Technische Universität Wien
Getreidemarkt 9
A-1060 Wien
Austria

Dr.Erwin SEIDL
Atominstitut der
Österr.Universitäten
Schüttelstraße 115
A-1020 Wien
Austria

Dr.Ernst SEMERAD
Inst.f.Werkstofftechnologie
Österr.Forschungszentrum
A-2444 Seibersdorf
Austria

Dr.Marcin SLASKI
Instytut Fizyki P.K.

Ul.Podchorazych 1
PL-30-084 Krakow
Poland

Dr.R.J. SOULEN
US Naval Research Laboratory
6334

Washington, DC 20375
USA

Gunther SPRINGHOLZ
Institut für Festkörperphysik
Technische Universität Graz
Petersgasse 16
A-8010 Graz
Austria

Reinhard STROH
Inst.f.Chem.Techn.Anorg.Stoffe
Technische Universität
Getreidemarkt 9
A-1060 Wien
Austria

Dr.Karl SVOZIL
Inst.für Theoretische Physik
Technische Universität Wien
Karlsplatz 13
A-1040 Wien
Austria

Dr.Jari VANHATALO
Wihuri Physical Laboratory
University of Turku

SF-20380 Turku
Finland

Dr.C.M. VARMA
AT&T Bell Laboratories
600 Mountain Avenue
Murray Hill, NJ 07974-2070
U.S.A.

Dipl.Ing.Beatrix VLCEK
Atominstitut der
Österr.Universitäten
Schüttelstraße 115
A-1020 Wien
Austria

Michael WACENOVSKY
Atominstitut der
Österr.Universitäten
Schüttelstraße 115
A-1020 Wien
Austria

Dr.Werner WEBER
INFP
Kernforschungszentrum
Postfach 3640
D-7500 Karlsruhe
BRD

Prof.Dr.H.W. WEBER
Atominstitut der
Österr.Universitäten
Schüttelstraße 115
A-1020 Wien
Austria

Prof.Dr.L.E. WENGER
Dept.of Physics and Astronomy
Wayne State University

Detroit, Michigan 48202
U.S.A.

Dr.Lars-David WERNLUND
FOA Research Institute
Box 1165
S-58111 Linköping
Sweden

Dr.J.P.A. WESTERVELD
Natuurkundig Laboratorium
Universiteit van Amsterdam
Valckenierstraat 65
NL-1018 XE Amsterdam
The Netherlands

Heinrich WIESINGER
Atominstitut der
Österr.Universitäten
Schüttelstraße 115
A-1020 Wien
Austria

Dr.Jörg WITTIG
Inst.f.Festkörperforschung
Kernforschungsanlage Jülich
Postfach 1913
D-5170 Jülich 1
BRD

Dr.Sergio ZANNELLA
CISE
P.O.Box 12081
I-20134 Milano
Italy

Dr.R. ZUBEREK
Institute of Physics of the
Polish Academy of Sciences
Al.Lotnikow 32/46
PL-02-668 Warszawa
Poland

Dr.A. WISNIEWSKI
Institute of Physics of the
Polish Academy of Sciences
Al.Latnikow 32/46
PL-02-668 Warszawa
Poland

Prof.Klaus YVON
Lab.de Crist.aux Rayons X
Universite de Geneve
24, quai E.Ansermet
CH-1211 Geneve 4
Schweiz

Prof.Dr.A. ZEILINGER
Atominstitut der
Österr. Universitäten
Schüttelstraße 115
A-1020 Wien
Austria

AUTHOR INDEX

SUBJECT INDEX

Abrikosov-Gor'kov theory, 170-175
Absorption, 67
ac field, 296
  complex susceptibility, 301, 307,
     309, 315
  magnetic susceptibility, 101,
     210, 217, 228, 293, 294,
     299, 301, 311, 313, 316
Ag doping, 84
Alcaline earth ions, 187
Alcylphosphoric acids, 137
Aligned materials, 227-232
Aluminium chlorides, 137
Amorphous materials, 317
Amorphous regions, 61
Anderson transition, 118, 119
Anisotropy
  effect, 163-168, 286, 293, 322
    in films, 333
    of critical current, 247
    of energy gap, 222
    of H_{c1}, 257
    of H_{c2}, 236-238, 293
Anomalous temperature effects,
    188
Antiferromagnetic order, 89, 323
Antiferromagnetism, 14, 16, 30, 37,
    156
Apatite, 135-138
Arrhenius plot, 48
Aslamazov-Larkin theory, 313, 317
Auger electron spectroscopy, 201

Ba, 135-138, 155, 217-220
$BaBiO_3$, 155
Ba-Ca substitution, 124-127
$BaPb_xBi_{1-x}O_3$, 14, 22, 41, 49
Band
  structure, 19-20, 201
  theory, 142, 153, 155
BCS, 13, 121, 141-151, 179, 217
Bean model, 275
Bipolaron, 14, 194

Bi-Sr-Ca-Cu-O, 41
Boson exchange mechanism, 141-151,
    163
Bremsstrahlung-isochromat spectro-
    scopy, 201
Brillouin zone, 121
Bulk, 201, 302, 319, 322
  diffusion, 335

Ca, 136-137
Carbon dioxide, 69
Ce, 137
Cell
  parameters, 72, 75, 100
  volume, 100
Characteristic lengths, 240
      (see also coherence length...)
Charge
  density, 178
  excitations, 153, 158, 195
  fluctuations, 199
Chemistry of copper oxides, 3-12,
    41-52
Clusters, 316
Coexistence, 319
Coherence length, 233, 240, 309,
    317
Correlation effect
  of Cu-d-electrons, 153, 156, 157
Coulomb
  pseudopotential, 141-164
  repulsion, 160
  screening, 178
Covalency, 14
Critical
  currents, 247-250, 267-275, 277,
    281, 285, 289, 293, 301-305,
    320, 333
  flux profile, 296
  magnetic fields, 95, 147-148, 163,
    227-245, 253-257, 285, 287,
    296, 316
  temperature, see T_c